A FIRST COURSE IN BIOPROCESS ENGINEERING

KAUSHIK NATH

Professor of Chemical Engineering
and
Principal-in-Charge
GH Patel College of Engineering and Technology
Vallabh Vidyanagar, Anand
Gujarat

PHI Learning Private Limited

Delhi-110092
2026

In fond memory of ***Shri Asoke K. Ghosh*** *(October 1942 – February 2024), Founder Chairman and Managing Director of PHI Learning, whose vision endlessly inspires.*

The Legacy Continues....

Published by Pushpita Ghosh, PHI Learning Private Limited, Rimjhim House, 111, Patparganj Industrial Estate, Delhi-110092 and Printed by Gopsons Printers Pvt. Ltd., A-14, Sector-60, Noida, Gautambudh Nagar, Noida, Uttar Pradesh-201301.

₹625.00

A FIRST COURSE IN BIOPROCESS ENGINEERING
Kaushik Nath

ISBN-978-81-19364-26-8 (Print Book)
ISBN-978-81-19364-21-3 (e-Book)

I dedicate this book to the fond memories of
my father late Prof. Krishna Kumar Nath
and my mother late Bimalprabha Nath

Contents

Preface .... *xiii*

1. Concept and Importance of Bioprocess Engineering 1–11

1.1 Introduction *1*
1.2 Interdisciplinary Flavor of Bioprocess Engineering *1*
1.3 Fundamental Differences Between Chemical Processes and Bioprocesses *2*
1.4 Historical Perspective of Bioprocess Engineering *2*
1.5 Unit Operations and Unit Processes *3*
1.5.1 Unit Operations in Bioprocess *3*
1.5.2 Unit Processes in Bioprocess *4*
1.6 General Steps of an Integrated Bioprocess *4*
1.7 Major Application Domains of Bioprocess Engineering *7*
Skills Required *9*
Key Points to Remember *9*
Questions for Practice *10*

2. Introduction to Microbial World 12–35

2.1 The Marvels of Microorganisms *12*
2.2 Microbial Diversity *12*
2.2.1 Prokaryotic Cells *14*
2.2.2 Eukaryotic Cells *23*
2.3 Scientific Nomenclature *26*
2.4 Microorganisms and Cycles of Ecosystem *27*
2.4.1 Nitrogen Cycle *27*
2.4.2 Carbon Cycle *29*
2.4.3 Phosphorus Cycle *30*
2.4.4 Sulphur Cycle *30*
2.5 Ecological Interactions at Microbial Level *31*

Key Points to Remember *32*
Questions for Practice *33*

3. Biochemistry of Important Molecules of Life............ 36–54
3.1 Carbohydrates *36*
3.1.1 Monosaccharides *36*
3.1.2 Oligosaccharides *37*
3.1.3 Polysaccharides *38*
3.2 Tests for Carbohydrates *39*
3.3 Proteins *39*
3.3.1 Types of Amino Acids *40*
3.3.2 Types of Proteins *40*
3.3.3 Biological Functions of Proteins *41*
3.3.4 Hierarchy of Protein Structure *41*
3.3.5 Denaturation of Protein *43*
3.3.6 Protein Separation *44*
3.4 Test for Protein *44*
3.5 Lipids *45*
3.5.1 Simple Lipids *45*
3.5.2 Conjugate Lipids *46*
3.5.3 Derived Lipids *47*
3.6 Nucleic Acids *48*
Key Points to Remember *49*
Questions for Practice *50*

4. Metabolic Stoichiometry and Bioenergetics 55–69
4.1 Stoichiometry of Microbial Growth *55*
4.2 Stoichiometric Equations in Bioprocesses *55*
4.3 Respiratory Quotient (RQ) *57*
4.4 Yield Coefficient (Y) *57*
4.5 Concept of Degree of Reductance *58*
4.6 Elemental Material Balance *59*
4.7 Energy Balance in Fermentation Processes *61*
4.7.1 Overall Energy Balance *61*
4.8 Metabolism *62*
4.8.1 Glycolysis *62*
4.8.2 Oxidative Decarboxylation of Pyruvic Acid *64*
4.8.3 Krebs Cycle *64*
4.8.4 Oxidative Phosphorylation or Terminal Oxidation *64*
4.8.5 Fate of Pyruvic Acid *65*
4.9 Metabolism of Nitrogen Compounds *66*
4.10 Metabolism of Hydrocarbons *66*
Key Points to Remember *67*
Questions for Practice *67*

5. Enzymes and Enzyme Catalysis.. 70–104

5.1 Idea of Catalysis by Enzymes *70*
5.2 Chemical Nature of Enzyme *71*
5.3 Naming an Enzyme *71*
5.4 Types of Enzyme *72*
5.4.1 Oxidoreductase *72*
5.4.2 Transferase *73*
5.4.3 Hydrolase *73*
5.4.4 Isomerase *73*
5.4.5 Lyases *73*
5.4.6 Ligases *74*
5.5 Salient Features of an Enzyme Catalyzed Reaction *74*
5.5.1 High Catalytic Activity *74*
5.5.2 Specificity and Selectivity *75*
5.5.3 Versatility *76*
5.5.4 Controlled Expression *76*
5.5.5 Stability *77*
5.6 Factors Affecting Enzyme Activity *77*
5.6.1 Temperature *77*
5.6.2 pH *78*
5.6.3 Mechanical Forces *79*
5.6.4 Chemical Agents *79*
5.6.5 Irradiation (Light, Sound, Ionizing Radiation) *79*
5.7 Various Hypotheses for Enzyme Substrate Reaction *79*
5.7.1 Lock and Key Model *79*
5.7.2 Induced Fit Model *80*
5.7.3 Proximity and Orientation Effects *80*
5.8 Expression of Enzyme Activity *81*
5.9 Enzyme Kinetics *81*
5.9.1 Michaelis-Menten Equation *81*
5.9.2 Evaluation of Kinetic Parameters of M-M Equation *83*
5.10 Inhibition of Enzyme Activity *85*
5.10.1 Competitive Inhibition *85*
5.10.2 Non-competitive Inhibition *86*
5.10.3 Uncompetitive Inhibition *87*
5.11 Industrial Uses of Enzymes *87*
5.11.1 Pharmaceuticals Industry *87*
5.11.2 Food and Beverage Industry *88*
5.11.3 Paper and Pulp Industry *88*
5.11.4 Textile Industries *88*
5.11.5 Leather Industries *89*
5.12 Enzyme Immobilization *91*
5.12.1 Merits and Demerits of Immobilization *91*
5.12.2 Immobilization Techniques *91*

5.13 New and Emerging Support Materials for Enzyme Immobilization *94*
5.14 Enzymes Required for Recombinant DNA Technology *94*
5.15 Performance or Design Equation for Bioreactors for Enzymatic Reactions *95*
5.15.1 Performance Equation by Agitated Batch Reactor *96*
5.15.2 Mixed Flow Reactor (MFR) or Continuous Stirred Tank Reactor (CSTR) *97*
5.15.3 Packed-bed Reactor or Tubular Reactor Considering Ideal Plug Flow Behavior *98*

Solved Numerical Problems *99*
Key Points to Remember *100*
Questions for Practice *101*

6. Biomass Production in Cell Culture and Growth Kinetics................ 105–129

6.1 Culturing of Microbes *105*
6.1.1 Solid Medium *106*
6.1.2 Synthetic and Non-synthetic Medium *107*
6.1.3 Reducing Medium *108*
6.1.4 Selective Medium *108*
6.1.5 Differential Medium *108*
6.1.6 Enrichment Medium *108*
6.2 How to Prepare Culture Media? *108*
6.3 Composition of Some Commonly Used Media *109*
6.4 Inoculation and Subculture *109*
6.5 Phases of Microbial Growth *110*
6.5.1 Lag Phase *110*
6.5.2 Exponential Phase *111*
6.5.3 Deceleration Phase *112*
6.5.4 Stationary Phase *112*
6.5.5 Death Phase *113*
6.6 Measurement of Microbial Growth by Various Methods *113*
6.6.1 Measurement of Cell Number Density *113*
6.6.2 Measurement of Cell Mass Concentration *115*
6.7 Environmental Factors Affecting Microbial Growth *117*
6.7.1 Effect of Temperature *117*
6.7.2 Effect of Medium pH *117*
6.7.3 Effect of DO *117*
6.7.4 Moisture or Humidity *118*
6.7.5 Ionic Strength *118*
6.7.6 Osmotic Pressure *118*
6.7.7 Light *119*
6.8 Microbial Growth Rate *119*
6.8.1 Monod Equation of Growth Kinetics *119*

6.8.2 Kinetic Parameters of Monod Equation *120*
6.8.3 Limitations of Monod Equation *121*

Key Points to Remember *124*
Questions for Practice *125*

7. Transport in Bioprocess Systems.. 130–152

7.1 Introduction *130*
7.2 Mass Transfer *130*
7.2.1 Molecular and Convective Mass Transfer *130*
7.2.2 Fick's Law and other Theories *131*
7.2.3 Correlations for Mass Transfer Coefficients *132*
7.3 Gas Liquid Transport in Cells *133*
7.4 Estimation of Volumetric Mass Transfer Coefficient *133*
7.5 Factors Affecting Oxygen Transfer Rate *134*
7.6 Solid Liquid Mass Transfer *135*
7.6.1 Various Mass Transfer Resistances *135*
7.6.2 Effectiveness Factor, Thiele Modulus and Damkohler's Number *136*
7.7 Rheology of Fermentation Broth *137*
7.8 Heat Generation by Microbial Growth *138*
7.8.1 Evaluation of Metabolic Heat *138*
7.8.2 Evaluation of Heat Exchange with Surrounding Environment *140*
7.9 Thermal Death Rate Kinetics 140
7.10 Sterilization *143*
7.11 Sterilization vs Pasteurization *143*
7.12 Sterilization Methods *144*
7.12.1 Physical Methods *144*
7.12.2 Radiation *144*
7.12.3 Filtration *145*
7.12.4 Chemical Methods *145*
7.13 Methods of Air Sterilization *146*
7.14 Continuous Sterilization *147*

Key Points to Remember *149*
Questions for Practice *150*

8. Bioreactors and Scale-up 153–172

8.1 Classification of Bioreactors: Batch and Continuous Types *153*
8.2 Batch Bioreactor: Basic Features *153*
8.2.1 Culturing in Batch Reactor *153*
8.2.2 An Alternative Method of Representation of Batch Reactor Design Equation *155*
8.2.3 Advantages and Disadvantages of Batch Culture *156*
8.3 Continuous (Mixed Flow) Bioreactor *156*
8.3.1 Culturing in Continuous Reactor *156*

8.3.2 Critical Dilution Rate and Washout Conditions *158*
8.3.3 Merits and Demerits of Continuous Culture *158*
8.4 Continuous Stirred Tank Fermenters in Series *159*
8.5 Continuous Stirred Tank Fermenter with Recycle of Biomass *161*
8.6 Fed-batch Reactor *163*
8.7 Plug Flow Reactor *164*
8.8 Choice of Bioreactor *165*
8.9 Scale-up of Bioreactor *165*
8.10 Scale-down of Bioreactor *167*
Key Points to Remember 168
Questions for Practice 169

9. Fermentation Technology.. 173–196
9.1 Fermentation as a Unit Process *173*
9.2 Basic Functions of a Fermenter *174*
9.3 General Requirements of Fermentation Process *175*
9.3.1 Construction and Mechanical Parts of a Fermenter *175*
9.3.2 Aseptic Operation and Sterilization *178*
9.3.3 Media Design for Fermentation *179*
9.3.4 Inoculum Development *182*
9.3.5 Inoculation Method *182*
9.4 Control of Various Parameters *182*
Key Points to Remember 192
Questions for Practice 192

10. Recovery and Purification of Products.................................. 197–220
10.1 Introduction—Intra and Extra Cellular Products *197*
10.2 Separation of Insoluble Products from the Broth *197*
10.2.1 Filtration *198*
10.2.2 Centrifugation *200*
10.2.3 Sedimentation *201*
10.3 Separation of Soluble Products *202*
10.3.1 Extraction *202*
10.3.2 Distillation *204*
10.3.3 Precipitation *204*
10.3.4 Adsorption *205*
10.3.5 Membrane Processes *205*
10.4 Separation of Intracellular Products by Cell Disruption *208*
10.4.1 Mechanical Methods *208*
10.4.2 Non-mechanical Methods *210*
10.5 Finishing Steps for Purification *210*
10.5.1 Crystallization *210*
10.5.2 Drying *212*

10.6 Chromatorgraphy 213
10.7 Electrophoresis 215
Key Points to Remember 216
Questions for Practice 217

11. Important Industrial Bioprocesses.. 221–240

11.1 Anaerobic Bioprocesses 221
11.1.1 Ethanol 221
11.1.2 Lactic Acid 223
11.1.3 Acetone and Butanol 224
11.2 Aerobic Bioprocesses 225
11.2.1 Citric Acid 225
11.2.2 Penicillin 227
11.2.3 Single Cell Proteins 230
11.2.4 Bakers' Yeast 231
11.3 Flavor and Aroma Production 231
11.4 Important Isolates from Plants or Animals 232
11.5 Nutritive Sweeteners (Glucose/Fructose Syrups) 233
11.6 Biosafety in Industrial Operations 233
11.7 Good Manufacturing Practices (GMP) 234
11.8 Quality Control of Biological Products 235
Key Points to Remember 235
Questions for Practice 236

12. Biological Wastewater Treatment.. 241–252

12.1 Introduction 241
12.2 Overview of General Parameters of Wastewater 242
12.3 General Treatment Strategy of Wastewater 243
12.4 Aerobic Processes 244
12.4.1 Activated Sludge Process 244
12.4.2 Rotating Biological Contactor 245
12.4.3 Trickling Biofilter 246
12.4.4 Oxidation Pond 247
12.5 Anaerobic Process 247
12.6 Biological Nitrification and Denitrification 249
Key Points to Remember 249
Questions for Practice 250

***Suggested Readings*.. 253–254**

***Index*.. 255–265**

Preface

Bioprocess engineering is a specialization of chemical engineering exploiting the combined knowledge-base of biology and engineering to deliver value added products and processes for end users or serves as components for other industries. It embraces the aspects of biochemistry, cell and molecular biology, industrial microbiology but has at its core the discipline of chemical engineering. A bioprocess is one that uses living cells to deliver a product, while engineering is the science of designing and building complex machines or processes. To that end, bioprocess engineering is the design and development of equipment and processes for large scale commercial manufacture of a variety of biological products, such as biopharmaceuticals, industrial enzymes, biofuels, biopolymers, and other materials. The practice of biological processes has a long history including early applications as the fermentation of grapes for wine, brewing of beer, cheese manufacture, effluent treatment and so on.

With the increasing emphasis on life sciences in engineering curricula, many students get inclined to credit a few courses of interdisciplinary flavor. The biology, biotechnology, chemistry, pharmacy and chemical engineering students at various universities and engineering institutions are required to take the biochemical or bioprocess engineering course either as an elective or compulsory subject. This book is written keeping in mind the need for a text book on afore subject for students from both engineering and biology backgrounds. Bioprocess engineering and its practitioners are essential in every industry that relies on biological products. This course is built on core concepts such as thermodynamics, stoichiometry, reactor design, transport phenomena, and process control, and it helps students familiarize the state-of-the-art knowledge in topics such as metabolic engineering, enzyme kinetics, biomass growth and propagation, fermentation and other industrial bioprocesses.

This book is a humble endeavor to provide the fundamental overview of bioprocess engineering in a simple and straightforward manner with emphasis on their scope and applications. It provides students with the exposure that will enable them to contribute in various professional fields, including bioprocess development, modeling and simulation, and environmental engineering. Each chapter begins with a fundamental explanation for general readers and ends with key points to remember as a ready-reckoner for the students. Chapter wise objective questions, higher order thinking problems and short numericals are also appended at the end for the readers to ponder on and to verify his or her level of understanding of the subject.

Applying biology for technology development is a multidisciplinary challenge requiring an appreciation of the engineering aspects of process analysis, design, and scale-up. Consistent with this overall aim, a little bit of exposure of microbiology has been provided in the initial chapters. Despite presenting the major aspects of bioprocess engineering quantitatively, and using several theoretical derivations and simplifications to demonstrate how empirical kinetic models are applicable to complicated bioprocess systems, priority has been given to minimizing the use of complex mathematics that may be beyond the comfort zone of non-engineering readers. Accordingly, the material has a descriptive focus without a heavy reliance on mathematical details.

I am profoundly grateful to many publishers and authors of books and journals and equipment manufacturers who have kindly given their consents to use their copyrighted figures and images. I also record my sense of appreciation to my students whose inquisitive questions during my lectures and laboratory sessions provided valuable input into this effort. The moral and mental support received from my wife and daughter was immeasurable. I dedicate this book to the fond memories of my father late Professor. Krishna Kumar Nath and my mother late Bimalprabha Nath. Finally, I am thankful to PHI learning Pvt. Ltd and its entire team for timely production of the book. The book has taxed me several years of hard work and dedication. My efforts will be successful only to that extent to which it receives patronage from the wider section of the students and teachers alike. Constructive comments and suggestions are always welcome from users of this book.

KAUSHIK NATH

1 Concept and Importance of Bioprocess Engineering

1.1 INTRODUCTION

Mother Nature has beautifully constructed biological systems on the earth. They are unique yet ubiquitous and are not only governed by the rules of physical sciences but can also be analyzed by the engineering principles. Living cells can do wonder if their potentials can be exploited on a commercial scale for the benefit of mankind. Microorganisms have been exploited in the production of industrial chemicals for a long time, through microbial fermentation. The versatility of microorganisms is attributed to their capacity to synthesize a vast array of chemical compounds, which range from simple amino acids, fatty acid and monosaccharides to the very complex alkaloids, polysaccharides, nucleic acids and proteins.

1.2 INTERDISCIPLINARY FLAVOR OF BIOPROCESS ENGINEERING

Bioprocess engineering deals with the design, development and interpretation of biological processes and systems using engineering principles. It is an interdisciplinary subject interfacing the knowledge of chemistry, physics, biology and engineering for the development of products and processes. Biotechnology is a life-science driven industry sector that harnesses living organism and their derivatives to produce value-added products and process for the benefit of society. Biochemical engineering is the application of chemical engineering principles in the biological systems on industrial scale. Although biologist and engineers differ in their approaches to understand and interpret natural processes, their skills are complementary. A synergic confluence and interaction of knowledge bases of biological sciences, basic sciences and engineering principles can go a long way in the advancement of modern civilization. Large-scale applications are not possible without understanding the engineering principles. The promises of molecular biology and genetics can be converted into newer processes to make usable products and that essentially requires integration of these skills. A successful bioprocess engineer needs to be familiarized with the fundamentals of microbiology, biochemistry, physics and chemical engineering.

1.3 FUNDAMENTAL DIFFERENCES BETWEEN CHEMICAL PROCESSES AND BIOPROCESSES

Although both bioprocess and chemical process are destined to produce useful products, they are fundamentally different on several counts. These are stated below:

(i) A bioprocess always takes place in presence of a living system (mostly microorganism) while chemical processes do not.

(ii) Majority of bioprocesses take place at mild physico-chemical conditions (ambient temperature and atmospheric pressure) while chemical processes mostly occurs at extremities of conditions (high temperature and pressure).

(iii) As the bioprocesses take place at mild physico-chemical conditions compared to chemical processes, these are less energy intensive than their chemical counterparts.

(iv) Given exact reaction conditions and reactant amount, chemical reactions can yield products instantaneously, while bioprocesses or biochemical reactions need longer time (several hours or days) to yield products. This is because of the fact that microorganisms take long time to grow, activate and initiate the bio-reaction.

(v) To carry out chemical reactions and processes high purity chemical reagents (reactants) are required, whereas less pure carbonaceous materials can be used as substrates for biochemical reactions or bioprocesses.

1.4 HISTORICAL PERSPECTIVE OF BIOPROCESS ENGINEERING

The art of bioprocesses, in its rudimentary forms, were in practice in the primitive ages of human civilization. Ancient scripts and literatures cite numerous examples of processes and products which were primarily the outcome of microbial fermentation. Caveman observed that meat became tastier if it were consumed after a day, instead of freshly hunted meat. Grains and fruits could be used for making intoxicating drinks. Ancient tribes of central Asia, Balkan regions and China found pleasure in sour camel's milk, soy sauce and cheese. Mesopotamian and Assyrian people used to make wine and beer from fruits, barley, malts and many other naturally available cereals and grains. Loaves of bread were traced with the mummies inside the pyramid of Egyptian civilization built 6000 years ago. Moldy soybean curd was used for healing of wounds and curing of infections in the time of ancient Chinese civilization. All these traditional knowledge and customs were in vogue from time immemorial, without the understanding of their underlying reasons or mechanisms.

Ancient man was unaware of the fact that while getting products via fermentation he was dealing with a variety of living entity, revealed and recognized much later as microorganisms–our invisible friends. In 1857, Pasteur proved that the living cells of yeast bring about alcoholic fermentation. His research also shed light on curing of diseases by applying microorganisms selectively. In 1923, the first commercial plant of citric acid fermentation was established by Pfizer, using a mold *Aspergillus niger*. Then came the epoch-making discovery of penicillin by Alexander Fleming in 1928. This was followed by further development of large-scale production facilities of other antibiotics such as streptomycin, tetracycline, erythromycin, kanamycin and many more. Fermentation industry continued to grow to ensure successful production of vitamin B_2, B_{12}, steroid molecules and derivatives, amino acids, plant growth regulators, etc.

Biotechnology industry received a major boost with the emergence of recombinant DNA technology involving deliberate alteration of genetic materials of an organism for many beneficial purposes. Any biotechnological venture based on biological catalysis such as bacteria, yeast and fungi, is vulnerable to alteration of its genetic structure. Whereas traditional genetic methods of strain (species of any microorganism) improvement depend only on the genetic potentials of the species involved but during past 30 years recombinant DNA technology which included gene cloning techniques has revolutionized the basic concept of microbe-based biotechnological industries. Microbial cells can now be engineered genetically into a desired variant. Thus, the recombinant DNA technology together with biochemical engineering and downstream processing methods provide excellent tools to produce large amount of biotechnological products.

1.5 UNIT OPERATIONS AND UNIT PROCESSES

A chemical or biochemical process can be subdivided into a number of steps of unit operations and unit processes. A coordinated series of these operations and processes forms the premise of an integrated bioprocess.

1.5.1 Unit Operations in Bioprocess

Unit operation is a traditional yet significant term in chemical engineering. Unit operation implies an operation which involves physical changes of materials (alteration of physical properties) such as distillation, crystallization, drying, absorption, adsorption, leaching, etc. Unit operation is a collective name of individual treatment encompassing heat and mass transfer, size reduction and other physical separation processes.

Majority of bioprocesses involve fermentation at the core. Upstream fermentation includes production or synthesis of products via conversion of substrates while downstream fermentation process includes separation, purification, recovery and packaging of products. In both upstream and downstream processes unit operations play a major role. Commonly used unit operations in bioprocesses are listed in Table 1.1.

Table 1.1 Commonly used unit operations

	Name of unit operation	Specific examples
Upstream processes	Mixing and agitation	Mixing of oxygen in fermentation broth
	Heat transfer (heating and cooling)	Sterilization, Pasteurization, process heating and cooling during fermentation.
	Gas-liquid mass transfer	Oxygen transfer to cells
Downstream processes	Filtration	Filtration of fermentation broth
	Distillation	Separation of ethanol water mixture obtained in fermentation
	Liquid-liquid extraction	Separation of antibiotic from fermentation broth

(Contd.)

Name of unit operation	Specific examples
Drying	Freeze drying of cells, drying of cells for determining concentration Spray drying of aminobutyric acid during fermentation
Crystallization	Separation of active pharmaceutical ingredients (API)
Adsorption	Chromatographic separation

Even non-traditional separation processes such as ultrafiltration, microfiltration, nanofiltration membrane processes, and chromatography are increasingly being used for product recovery and purification. These can also be considered as unit operations.

1.5.2 Unit Processes in Bioprocess

Unit processes are those where chemical transformation takes place with the formation of newer products or derivatives. Popular examples of unit processes are nitration, sulfonation, hydrolysis, esterification, hydrogenation, polymerization, etc. In a broad sense, even fermentation can be considered to be a unit bioprocess. However, during fermentation a number of other unit processes like enzymatic hydrolysis, polymerization, isomerization and complex biosynthesis of materials are encountered.

1.6 GENERAL STEPS OF AN INTEGRATED BIOPROCESS

An integrated bioprocess to produce a product consists of a series of several steps. The major steps are as follows:

(i) Selection and isolation of the strain of microorganism
(ii) Preservation of strain
(iii) Growing of inoculum
(iv) Pre-fermentation growth
(v) Media preparation and sterilization
(vi) Fermentation or product formation
(vii) Product recovery and purification
(viii) Wastewater treatment
(ix) Quality control

The above steps are discussed in the following section and are schematically represented in Figure 1.1.

(i) Selection and isolation of the strain of microorganisms: Selection of suitable strain of microorganism is the first step of any industrial bioprocess. The process of selection is based on literature review, earlier performance history such as substrate utilization, yield, cost of substrate, conditions, etc. The steps of detection, isolation, and separation of desired microorganism from natural environment (like soil, rocks, water, air, milk, compost, etc.) are collectively known as screening. There are two types of screening, primary and secondary. In primary screening,

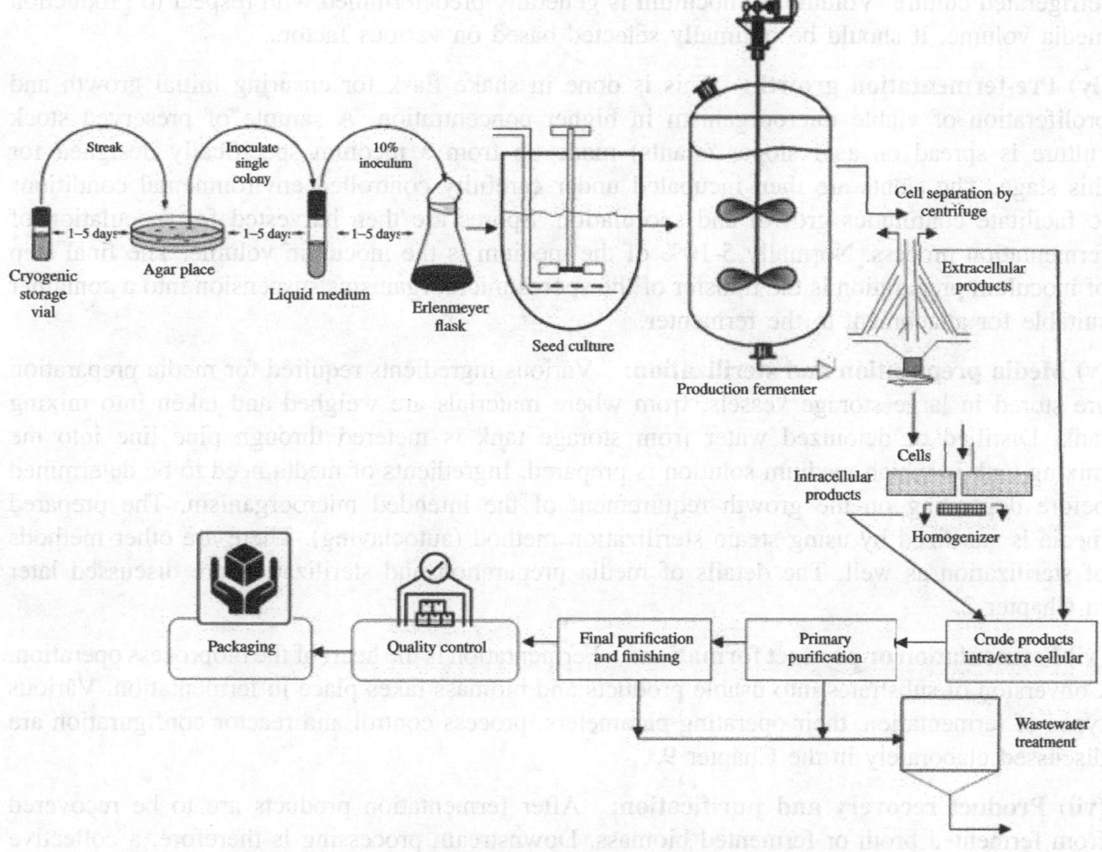

Figure 1.1 General steps of an integrated bioprocess.

preliminary separation of a few potentially important microorganisms is done, while discarding the less useful ones. Some dyes are used for screening of acid producing microbes.

Secondary screening is a systematic procedure to isolate distinctively useful microorganism. It is done by some advanced techniques such as paper or thin layer chromatography. Economic feasibility of the process using the isolated strain needs to be carried out at pilot scale to gather more information. Various culture repositories are available from where one can procure suitable microorganism. These depositories can supply good quality authentic strain of microbes for ready use in the laboratory. Examples of such depositories in India are MTCC (Microbial Type Culture Collection and Gene Bank), Chandigarh, NCMR (National Centre for Microbial Resources), Pune, Indian Type Culture Collection (ITCC), New Delhi and few others.

(ii) Preservation of culture (strain): After isolation the strain needs to be preserved in the laboratory for further uses. For short-term storage slant, stabs and petri dishes are sufficient. They are stored at 4°C in refrigerator. For longer storage, lyophilization (freeze drying) can be done, and these can be stored at sub-zero temperature (cryopreservation) at 18 to −80°C.

(iii) Growing of inoculum: Prior to culturing in large-scale bioreactor, inoculum is to be prepared for pre-fermentation media. Lyophilized cultures take longer time than frozen or refrigerated culture. Volume of inoculum is generally predetermined with respect to production media volume. It should be optimally selected based on various factors.

(iv) Pre-fermentation growth: This is done in shake flask for ensuring initial growth and proliferation of viable microorganism in higher concentration. A sample of preserved stock culture is spread on agar slopes (slants) made up from a medium specifically designed for this stage. The slants are then incubated under carefully controlled environmental conditions to facilitate continuous growth and sporulation. Spores are then harvested for inoculation of fermentation process. Normally 5-10% of the medium is the inoculum volume. The final step of inoculum preparation is the transfer of the spores/microorganisms suspension into a container suitable for attachment to the fermenter.

(v) Media preparation and sterilization: Various ingredients required for media preparation are stored in large storage vessels, from where materials are weighed and taken into mixing tank. Distilled or deionized water from storage tank is metered through pipe line into the mixing tank in which medium solution is prepared. Ingredients of media need to be determined before depending on the growth requirement of the intended microorganism. The prepared media is sterilized by using steam sterilization method (autoclaving). There are other methods of sterilization as well. The details of media preparation and sterilization are discussed later in Chapter 2.

(vi) Fermentation or product formation: Fermentation is the heart of the bioprocess operation. Conversion of substrates into usable products and biomass takes place in fermentation. Various types of fermentation, their operating parameters, process control and reactor configuration are discussed elaborately in the Chapter 9.

(vii) Product recovery and purification: After fermentation products are to be recovered from fermented broth or fermented biomass. Downstream processing is therefore, a collective term of all these steps that are involved in harvesting and purification of desired product. Separation and purification of a desired product from the fermentation broth or cell culture supernatant is a crucial challenge in commercial biotechnology. The high cost of product recovery stems from many factors, e.g. presence of very small amount of desired product in a complex fermentation broth, recovery of products of desired purity and requirement of several discrete steps. Thus the ultimate objective to devise an overall strategy that meets the set standard of purity and minimize the products cost. For example, recombinant DNA-derived protein products (erythropoietin, human growth hormone, etc.) for therapeutic application the purity requirement is 99.99%, on the other hand, the enzymes can be used as industrial catalyst in relatively crude form. Once the purification criteria have been established, the specific purification procedure can be selected.

For the products derived from living cells, the recovery procedure will be influenced by the producing organism (bacteria, yeast, fungi mammalian cells) and the location of the protein (intracellular, inclusion bodies, extracellular) after completion of the fermentation is very important. The general sequence of protein recovery operations are as follows:

(a) Separation of insoluble (whole cells, cell debris, pellets of aggregated proteins, undissolved nutrients, etc.)
Operations involved: sedimentation, centrifugation, and filtration

(b) Isolation and concentration of the desired products
Operation involved: extraction, ultrafiltration, precipitation

(c) Primary purification
Operation involved: fractional precipitation, chromatography, electrophoresis

(d) Final purification and polishing
Operation involved: drying, crystallization, packing

These processes are further discussed in details in Chapter 10.

(viii) Wastewater treatment: Biological processes generate a copious quantity of wastewater containing various components such as suspended cell mass fragments, other non-cellular products and unconverted nutrients. Microorganisms can be used for reclamation and degradation of such wastewater. Examples of various processes include activated sludge process, trickling bed filter, rotating biological contactors, oxidation ponds, etc.

(ix) Quality control (QC): Quality control deals with detecting defects and deviations and conforming to the stipulated quality standards of various governing agencies. It involves creating a product and testing the product itself in such a way that all parameters are meticulously maintained. In fermentation and pharmaceutical industries various references such as US Pharmacopeia or European Pharmacopoeia or any other guidelines of the respective countries are followed. Depending on previous manufacturing history or other literature data, some additional testing is also carried out. Materials such as yeast extracts or plant-derived hydrolysates may exhibit variability of their quality parameters from one batch to another. To address this, there must be a biological performance assay implemented on incoming lots of the raw material. QC testing of liquid cell culture media involves checking of pH and osmolality. Tests are also carried out for any endotoxins or other contaminants. Further qualitative or quantitative testing may be required depending upon the product requirement or to meet the standard criteria.

1.7 MAJOR APPLICATION DOMAINS OF BIOPROCESS ENGINEERING

Bioprocess engineering dwells upon the fundamental premise of a careful understanding of the optimum physico-chemical conditions and its engineering applications for scaled-up production of value added chemicals. Advances in genetic engineering have, over the past two decennia, results in the genesis of a plethora of novel molecules and their interactions with microbes, and other living systems, to solve many issues pertaining to agriculture, environment and pharmaceuticals. Table 1.2 summarizes some major products of bioprocessing industries.

Table 1.2 Some major products of bioprocessing

Product types	Examples	Microorganisms used
Organics	Alcohol	Yeast
	Acetone/Butanol	*Clostridium acetobutylicum*
Organic acids	Citric acid	*Aspergillus niger*
	Gluconic acid	*Aspergillus niger*
	Lactic acid	*Lactobacillus* sp.

(*Contd.*)

Product types	Examples	Microorganisms used
Amino acids	L-glutamic acid	*Corneybacterium glutamicum*
	L Arginine	*Bravibacterium flavum*
Antibiotics	Penicillin	*Penicillium chrysogenum*
	Tetracyclines	*Streptomyces aureofaciens*
Enzymes	Protease	*Bacillus* sp.
	Alpha amylase	*Bacillus streothermophilus*
	pectinase	*Aspergillus niger*
Vitamins	B-12	*Pasudomonas dentrificans*
Therapeutic proteins	Insulin	Recombinant *E. coli* cells
	Growth hormone	Recombinant *E. coli* cells
	Interferon alpha 2	Recombinant *E. coli* cells
	Erythropoietin	Recombinant mammalian cell culture

Pharmaceuticals

Biochemical engineers contribute to the research and development of safe trial of medicines, vaccines and bulk drug manufacturing. They also assist in laboratory and pilot plant studies of various compounds to develop vaccines, clinical trials and even veterinary medicines. Bioprocess engineers in the pharmaceutical industry need to work with medical professionals and research scientists to devise effective diagnostic methods.

Biomedicine

Bioprocess engineers are engaged in development of medical devices for various therapeutic applications. For example, development of synthetic material 3D print organs and its biocompatibility study is carried out by bioprocess engineering professional. Bioprocess engineers may help developing medical imaging equipment, life support systems and many other devices for precise administering of drugs.

Agriculture and food science

Agricultural applications are also dependent on bioprocess engineering in the areas like production of foodstuffs and commercial farming. These include maintaining and supplementation of nutrients to foods, designing appropriate fermentation system for safe food production and many more. In other areas, such as development of drought and disease-resistant crops, maximizing nutrient content and creating more efficient pest management the role of bioprocess engineering assumes paramount importance. Many agricultural applications are improved with emergence of alternative food sources and sustainable planting, and harvesting methods. Bioprocess engineers ensure development of abundant yet nutritious food while maintaining the environmental sustainability.

Conservation of environment

Bioprocess engineers can also contribute immensely in improving conservation of environment and resources by devising and applying eco-friendly control methods to improve environmental sustainability. Enrichment of fertility soil by cultivating bio-fertilizers, bioremediation of wastewater and various out bio-mediated treatment processes are developed by bioprocess engineers.

SKILLS REQUIRED

Some of the essential and desirable skills a bioprocess engineer should have includes the following:

- Ability to collaborate and network across the disciplines
- Effective knowledge of biology
- Effective knowledge of chemistry
- Understanding of engineering principles particularly chemical engineering
- Organization
- Time management
- Leadership
- Project management
- Written and verbal communication

KEY POINTS TO REMEMBER

- ➢ Bioprocess engineering deals with the design, development and application of biological processes with engineering principles.
- ➢ Bioprocess engineering is an interdisciplinary study interfacing with biology, chemistry, and engineering.
- ➢ A bioprocess or a biochemical reaction takes place in presence of living microorganism, at ambient temperature and pressure, however consuming more time.
- ➢ Historically, fermentation finds its application in the days of ancient civilization for the production of food and drinks.
- ➢ Unit operations (such as distillation, absorption, crystallization, drying, etc.) are those where physical changes take place while in unit processes (such as hydrolysis, nitration, esterification, etc.) chemical changes occur.
- ➢ An integrated bioprocess consists of several steps starting from the selection and screening of microorganisms, its cultivation in laboratory using appropriate media, fermentation in a bioreactor (fermenter) followed by separation and purification of products.
- ➢ Bioprocess engineering is widely used in pharmaceutical industries, food and beverage industries, agriculture and medicinal purposes and production of high-value biochemicals.

QUESTIONS FOR PRACTICE

Multiple Choice Question

1. Identify the correct sequence during the industrial production of substances
   (a) Inoculation, screening, fermentation, downstream processing, removal of waste
   (b) Screening, inoculation, fermentation, downstream processing, removal of waste
   (c) Fermentation, screening, inoculation, removal of waste, downstream processing
   (d) Fermentation, inoculation, inoculation, removal of waste, downstream processing

2. In general biological processes take place at ambient temperature and pressure. This is because of
   (a) Chances of deactivation of enzymes are more at high temperatures.
   (b) Changes in temperature impact the stability, flexibility, and structure of molecules involved with biological processes.
   (c) Denaturation of protein may occur at high temperature.
   (d) All of the above.

3. Which of the following is not considered to be naturally occurring biomass?
   (a) Rice hulls (b) Hemp
   (c) Coconut coir (d) Rayon

4. In which of the following process, chemical changes occur?
   (a) Filtration (b) Distillation
   (c) Oxidation (d) Extraction

5. Identify the 'odd man out' from the lists given below: (State reason)
   (a) Filtration (b) Sedimentation
   (c) Centrifugation (d) Distillation

6. In order to become a/an ____________, an atom of chlorine must ____________ an electron.
   (a) Ion; gain (b) Molecule; lose
   (c) Ion; lose (d) Molecule; gain

7. Salts and sugars work to preserve foods by creating
   (a) Depletion of nutrient (b) hypertonic environment
   (c) hypotonic environment (d) lower pH

8. Freeze drying for preservation of microbial starins is also know as
   (a) Lyophilization (b) Cryopreservation
   (c) Sublimation drying (d) All of the above

9. Disease that affects many people at different countries is termed as
   (a) Sporadic (b) Pandemic
   (c) Epidemic (d) Endemic

10. First Pasteur conducted fermentation experiments in
   (a) Milk (b) Food material
   (c) Fruit juices (d) Both A and C

11. Which of the forestry materials listed below can be utilized as biomass?

(a) Logging residues (b) Manure
(c) Tallow (d) Fish oil

12. ____________ is made from the aerobic decomposition of sewage.

(a) Biomass (b) Synthetic fuels
(c) Metal articles (d) Bio fuels

Answer Key

1. (b) 2. (d) 3. (d) 4. (c) 5. (d) (Distillation, as the other methods are used for solid liquid separation)

6. (a) 7. (b) 8. (d) 9. (b) 10. (c) 11. (a) 12. (d)

Descriptive Questions

1. Name three industrially important biochemical reactions and mention the name(s) of the microorganism involved therein.
2. State and briefly explain various steps, either sequential or concurrent, of an integrated bioprocess in general.
3. Name various unit operations used in upstream and downstream processing in fermentation
4. Identify the following phenomena either as physical or chemical changes and justify assigning reasons (Also write down the conformational change(s) or product(s) formation, if any/or name the process, if any).
   (i) Boiling of rice
   (ii) Curdling of milk
   (iii) Boiling an egg
   (iv) Frying an egg
   (v) Toasting a piece of bread
   (vi) Ripening of fruit

2 Introduction to Microbial World

2.1 THE MARVELS OF MICROORGANISMS

Louis Pasteur, widely acknowledged as the pioneer of microbiology once said 'The role of the infinitely small in nature is infinitely great'. Microorganisms are those tiny living entities which can be viewed and characterized under microscope. Their sizes are in the range of several micrometers. Probably, microbes are the most diverse, ancient, and ubiquitous members of the living world. Microbial activity is the key to the basic functioning of the earth—from plants to humans we all survive due to microbial activity. These microbes closely influence each of nutrient cycles on earth, be it carbon, nitrogen, oxygen, sulphur and so on. Photosynthetic microorganisms produce oxygen in water bodies, and provide necessary nutrients like sulphur and nitrogen for living beings. Microbial activities in soil increase the soil fertility and helps in production of food grains. Many microorganisms are used in various food processing industries, dairies, confectionaries and even in chemical processing units. The art of fermentation was recognized over several thousand years ago. Our ancient scripts cite production of wines, health drinks, sanitized beverages, curd, yoghurt, cheese and so on through fermentation of naturally available ingredients.

Microorganisms like bacteria, fungi, algae, etc. constitute nearly 17% of the living world while the human population counts for only 0.01%. As a matter of fact, we are surrounded by an invisible cloud of microorganisms. Our skins, stomach and gut tracts are home to countless microbes that are in continuous interaction with the host bodies. Surprisingly among one trillion species of microorganisms on earth, only a few thousands are pathogenic. The remaining all are beneficial either one way or the other. Industrial microbiology and biotechnology aims to exploit these huge numbers of beneficial microorganisms for various useful purposes.

2.2 MICROBIAL DIVERSITY

It is mind-boggling to observe that microorganisms can survive and sustain though geologic ages. Starting from fossilized ancient archaea to the relatively new eubacteria can sustain and adapt to diverse, extreme and harsh physico-chemical environmental conditions. Lot of new information is coming up in various studies on microbial life traced in the ancient rocks dating billions of years back. Very recently, in 2022 a huge bacterium named *Thiomargarita magnifica*, which is almost 5000 times bigger than the ordinary microbes, has been isolated by scientists in the Caribbean mangrove forest.

Microorganisms belonging to the protist kingdom are categorized in various groups and subgroups (Figure 2.1).

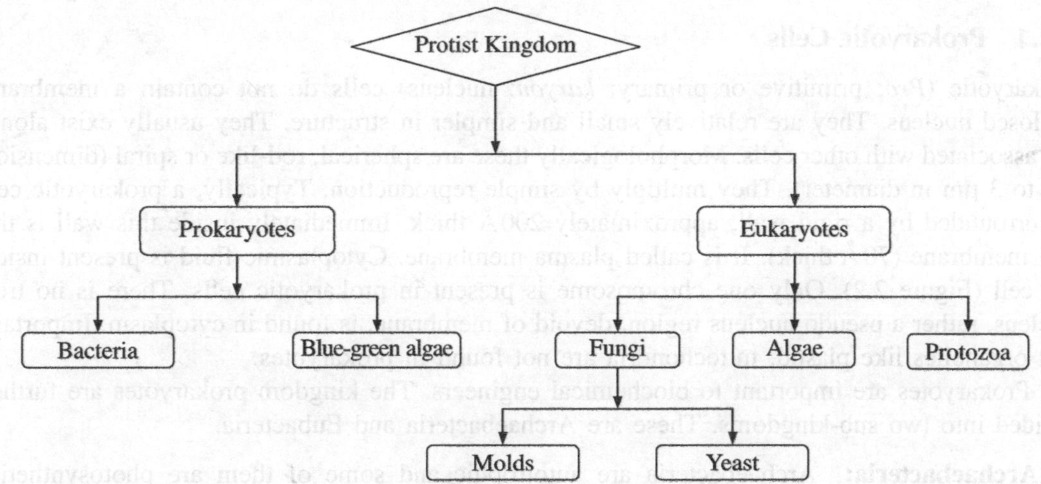

Figure 2.1 Important members of the protist kingdom.

According to Figure 2.1 Protist kingdom is broadly divided into two types, prokaryotes or prokaryotic cells and eukaryotes or eukaryotic cells. Other than protist there are entities in the non-protist world as well, for example virus. Virus is considered as a tiny, infectious particle often at the borderline of the living and the non-living. The notable differences between the prokaryotes and the eukaryotes are outlined in the Table 2.1.

Table 2.1 Comparison of prokaryotic and eukaryotic cells

Features	Prokaryotic cells	Eukaryotic cells
Number of cells	Mostly unicellular	Mostly multicellular
Size	1–10 μm in length	5–100 μm in length
Cell wall	Peptidoglycan (murein or muco-peptide) as component	Absence of peptidoglycan.
Membrane enclosed nucleus	Not present	Present
Internal organelles	No membrane separated compart-ments (Mitochondria, Golgi bodies, ER, etc.) are present	Membrane separated compartments are distinctly present.
Genetic level	Single DNA molecule	Multiple DNA molecules
Reproduction	Asexual recombination	Sexual recombination
Cell division	Only fission and budding instead of mitosis.	Predominantly miotic cell division

(Contd.)

Features	Prokaryotic cells	Eukaryotic cells
Ribosomes	70S ribosomes are present.	80S ribosomes are present in cytoplasm and 70S ribosomes are present in mitochondria and plastid
Metabolic mechanisms	Largely anaerobic and exothermic reactions	Follows glycolysis route for energy generation

2.2.1 Prokaryotic Cells

Prokaryotic (*Pro*: primitive or primary; *karyon*: nucleus) cells do not contain a membrane enclosed nucleus. They are relatively small and simpler in structure. They usually exist alone, not associated with other cells. Morphologically these are spherical, rod-like or spiral (dimension 0.5 to 3 µm in diameter). They multiply by simple reproduction. Typically, a prokaryotic cell is surrounded by a rigid wall, approximately 200 Å thick. Immediately inside this wall is the cell membrane (70 Å thick). It is called plasma membrane. Cytoplasmic fluid is present inside the cell (Figure 2.2). Only one chromosome is present in prokaryotic cells. There is no true nucleus, rather a pseudo nucleus region, devoid of membrane, is found in cytoplasm. Important cell organelles like plastid, mitochondria are not found in prokaryotes.

Prokaryotes are important to biochemical engineers. The kingdom prokaryotes are further divided into two sub-kingdoms. These are Archaebacteria and Eubacteria.

(i) Archaebacteria: Archaebacteria are autotrophic and some of them are photosynthetic. These bacteria generate energy from the oxidation of reduced sulphides, hydrides, hydrocarbons like H_2S, NH_3, CH_4, etc.

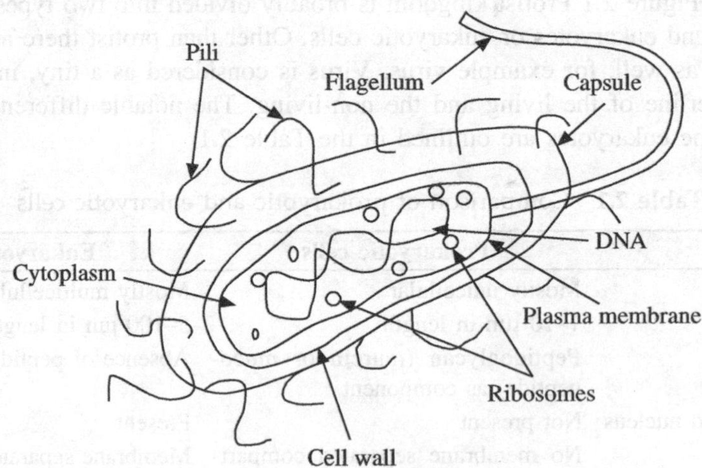

Figure 2.2 Schematic diagram of a prokaryotic cell.

They are of three major types:

(i) *Methanogens*: As their name implies, methanogens can synthesize methane by their metabolic activities. These are anaerobic organisms and are found in swampy and marshy lands. Examples: *Methanospirillum*, *Methanococcus*, etc.

(ii) *Thermoacidophiles*: These bacteria are adapted to high temperature and low pH conditions. Many thermoacidophiles use hydrogen sulphides as their energy sources. Examples: *Thermoproteus*, *Thermoplasma*, etc.

(iii) *Halophiles*: These bacteria grow in water bodies having high concentration of halide salts. There are several salt water lakes around the world like the Caspian Sea, Chilika Lake, etc. They are adapted to high pH (strongly alkaline) environment. Examples: *Halobacte halobium.*

Archaebacteria are definite types of prokaryotes that have the ability to sustain populations in extreme and typically harsh environments. Those surviving in the most hostile environment are known as *extremophile archaea.*

(ii) Eubacteria: They are also commonly termed as bacteria—an important prokaryote of immense interest to the industrial microbiologists. These are discussed in the next section.

Bacteria

Bacteria (singular: bacterium) are relatively small prokaryotic cells, enclosed by rigid walls that protect them from osmotic rupture in an aqueous environment. The outermost sticky and glutinous layer of the bacterial cell wall is known as capsule. Nucleus and other important cell organelles like chloroplast, mitochondria are conspicuously absent in bacteria. They have a coiled DNA in nucleoid which acts as a single chromosome. Energy-yielding metabolic activities of bacteria are carried out in the inner side of plasma membrane.

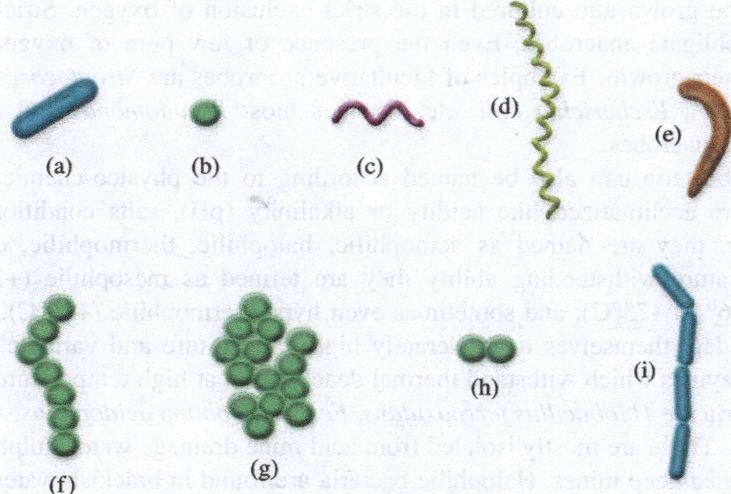

Figure 2.3 Typical morphologies of bacteria (a) *Bacillus* (b) *Coccus* (c) *Spirillum* (d) *Spirochaete* (cockscrew) (e) *Vibrios* (comma) (f) chain of *Cocci* (g) cluster of *Cocci* (h) pair of *Cocci* (i) chain of *Bacilli*.

Bacteria have typically a single cell, but they may exist in different patterns such as: sphere shaped (*Cocci*), rod shaped (*Bacilli*), spiral shaped (*Spirilla*), comma shaped (*Vibrios*) or corkscrew shaped (*Spirochaetes*) (Figure 2.3). Bacteria are found as singular cells or in pairs. Other formations like clusters, chains are also found. In addition, star shaped, square shaped are less frequently available. The cell walls of bacteria are composed of a special type of cross-linked polymer, called 'peptidoglycan' with high molecular weight. Most of them cannot utilize light energy, as they lack the presence of photosynthetic pigments. They are, however, capable of motion (motility[#]) with the help of flagella. Flagella are composed of flagellin—a type of protein and are responsible for the locomotion of bacteria. Some bacteria possess finely dispersed hair-like frills on the cell. These are called 'pili', which perform functions like locomotion, attaching pathogens and helping in reproduction. They are reproduced by division in to two daughter cells, the process of which is known as binary fission.

\# **Motility** *is a term which refers to the ability of any living entity to move spontaneously. Although, The terms Brownian motion and motility are apparently similar in meaning but these are not exactly identical. Brownian motion refers to unidirectional zigzag movement, while motility is directional movement under a gradient. There are various types of gradient such along a chemical gradient (chemotaxis), along a light gradient (phototaxis), along an electric field (galvanotaxis), along a rigidity gradient (durotaxis) and so on.*

Depending on the oxygen (air) utilization feature bacteria can be grouped into two types namely aerobic (or oxic), which can utilize oxygen and anaerobic (or anoxic) being unable to utilize oxygen. The anaerobes are further classified as facultative and strict anaerobes. Facultative anaerobes can tolerate small amount (a few ppm) of oxygen, however strict anaerobes must be grown and cultured in the strict exclusion of oxygen. Strict anaerobes are also known as obligate anaerobes. Even the presence of few ppm of oxygen proves to be detrimental for their growth. Examples of facultative anaerobes are *Streptococcus pneumoniae*, *Staphylococcus spp.*, *Escherichia coli*, etc. whereas most *Pseudomonas* and *Mycobacterium* species are strict anaerobes.

Sometimes, bacteria can also be named according to the physico-chemical environment to which they are acclimatized like acidity or alkalinity (pH), salts condition, temperature, etc. Accordingly, they are named as acidophilic, halophilic, thermophilic, etc. Depending upon the temperature withstanding ability they are termed as mesophilic (+11° to +45°C), thermophilic (+46° to +75°C); and sometimes even hyperthermophilic (+76°C). These bacteria can effectively adapt themselves to moderately high temperature and variable pH conditions. They produce enzymes which withstand thermal deactivation at high temperatures. Examples of acidophilic bacteria are *Thiobacillus ferrooxidans, Ferromicrobium acidophilus, Acidimicrobium ferrooxidans*, etc. These are mostly isolated from acid mine drainage water, sulphur mining sites and from abandoned deep mines. Halophilic bacteria are found in brackish water particularly in marine ecosystems and also in food products, preserved with salts. Some anaerobic *halophiles* can be sulphate-reducing, acetogenic, and even fermentative species.

Gram Reaction

Gram reaction refers to the action of dye staining on the bacterial cells. It was first introduced by a Danish Bacteriologist Hans Christian Gram while identifying Pneumonia causing organisms. Gram-staining process differentiates two types of bacteria as gram-positive and gram-negative.

Cells are first coloured with the crystal violet dye, and subsequently mixed with an iodine solution. It is then washed in alcohol and tap water and counterstained with safranin. Bacteria retaining the blue crystal-violet colour after this process are called gram-positive and loss of colours indicates gram-negative. The cell membrane of Gram-negative bacteria is made of lipid bilayers, backed by another layer of peptidoglycan. The cytoplasm is protected by another lipid bilayer called plasma membrane (similar to cytoplasmic membrane) or inner membrane. There is periplasmic space to separate it. However, there is no exterior membrane in gram-positive bacteria excepting only a thick layer of peptidoglycan. An example of a typical gram-positive bacterium is *Bacillus subtilis* and a gram-negative bacterium is *Escherichia coli.*

Bacteria in industrial uses

A large number of bacterial species have been identified and are commercially used in a variety of industrial production process. Industrial uses of some selected bacteria are presented in Table 2.2.

Table 2.2 Useful bacteria for commercial applications

Type of industry	Product	*Species of bacteria*
Food and Dairy	Cheese, yoghurt	*Lactobacillus casei*
	Butter milk	*Lactobacillus lactis*
	Vinegar	*Acetobater aceti; Gluconobacter suboxidans*
	Soy sauce	*Enterococcus faecium*
	Chocolate	*Gluconobacter oxydans*
	Enhance shelf-life of breads	*Bacillius stearothermophilus, Bacillus subtillis*
Beverage	Beer	*Acetobacter cerevisiae*
	Coffee	*Aceobacter fabarum*
	Wine, rum	*Lactobacillus mali*
	Starch hydrolysis	*Bacillus, Streptomyces, Rhizopus,*
	Starch saccharification	*Bacillus* sp., *Klebsiella* sp.
Enzymes	Amylase	*Thermophilic bacillus* sp.
	Cellulase	*Clostridium thermocellus*
	Protease	*Thermus aquaticus/Bacillus* sp.
	Streptokinase, urokinase	*Streptococci* sp., *Bacillus subtilis*
Organic chemicals	Acetic acid	*Acetobacter aceti*
	Butyric acid	*Clostridium butylicum*
	Lactic acid	*Lactobacillus* sp.
	d-glucaric acid	Recombinant *E. coli*
	Butanol	*Clostridium butyricum*
	Dye manufacturing, beauty products, pigments, artificial flavours, psticides	*Trametes versicolour, Bacillus subtilis*
	Synthesis of acrylamide, butyramide, nicotinamide	*Rhodococcus rhodochrous PA-34, Bacillus* sp. *APB-6*

(Contd.)

Type of industry	Product	Species of bacteria
Pharmaceuticals	Teteracyclin, streptomycin	*Streptomyces* sp.
Textile	Non-chlorine Bleaching, fabric dyeing	*Bacillus subtilis*
	Wool finishing	*Clostridium histolyticum*
	Desizing	*Bacillus* sp., *B. licheniformis*
	Cotton scouring, synthetic fibre modification	*Pseudomonas mendocina, Fusarium solanipisi, Thermomonospora fusca*
Waste management	Cyanide waste treatment	*Rhodococcus erythropolis*
	Reclamation of keratin waste	
	Bioremediation of keratinic wastes	*Chrysosporium keratinophilum*
	Degradation of halogenated contaminants	*Pseudomonas* sp., *Rhodococcus sp.*
	Degradation of phenolic compounds	*Phanerochaete chrysosporium, Coprinus cinereus, Pseudomonas putida*
Minerals processing	Microbial leaching of iron ores/lead or zinc flotation	*Acidithiobacillus ferrooxidans*
	Microbial leaching of suphide minerals	*Acidithiobacillus thiooxidans*
	Mineral flotation of calcite, apatite, magnesite, etc.	*Rhodococcus opacus*
	Flotation and flocculation of iron ore minerals	*Paenibacillu spolymyxa*
	Sulphate reduction	*Desulfovibrio* sp.
Energy and fuel	Biological methane production	*Methanobacterium formicicum, Methanospirillum hungati*
	Biological hydrogen production	*Clostridium, Enterobacter, Klebsiella, Citrobacter, and Bacillus*

BGA (Blue-green algae)

The other name of blue-green algae is cyanobacteria. They contain photosynthetic pigment such as chlorophyll-*a*, β-carotene; xanthophyll, zeaxanthin, etc. and they take part in nitrogen cycle. Cyanobacteria are common in fresh waters, marine waters, in the soil and on moist surfaces. They form large visible colonies. They are unicellular, and sometimes filamentous. They are thought to be one of the oldest forms of fossilized microbes with age more than 3×10^9 years. Cyanobacteria were present in the evolution of early life on the earth during proteozoic and archaean era. These were responsible for providing oxygen atmosphere. Blue-green algae blooms are visible in ponds and lakes. They contain only one form of chlorophyll *a*, a green pigment. In addition, the red pigment phycoerythrin. The characteristic blue-green colour is due to the presence of two pigments phycobilin and chlorophyll together. Even other pigments are also found in blue-green algae. Thus many of such species are brown, red or even yellow in colour.

Biotechnological applications of cyanobacteria have made them much popular in recent times. These are increasingly used in cultivation of aquatic plants for food, wastewater remediation, and production of organic fertilizers. Cyanobacteria can also generate secondary metabolites like toxins, vitamins, exopolysaccharides and a few pharmaceutical ingredients. A lot of work is currently underway on the photo-production of hydrogen by cyanobacteria.

These organisms can be adapted to produce hydrogen photosynthetically by growing them in a medium free from both oxygen and nitrogen such as one argon and carbon dioxide.

Algae (singular alga)

Algae are usually unicellular green organisms, having chlorophyll to carry out photosynthesis (Figure 2.4). Some algae contain pigments of red, brown and purple colours. The size of a typical unicellular alga is 10 to 30 μm. They are enclosed by cell wall made of cellulose layers. Mechanical and conducting tissues are absent. Majority of algal growth is found in aquatic environment.

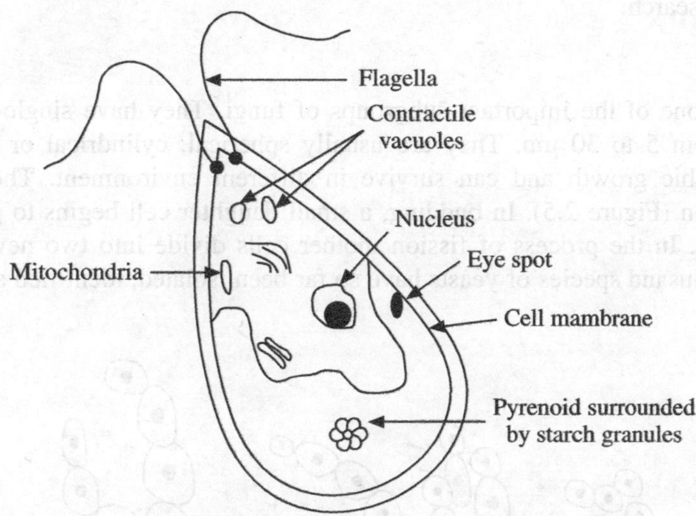

Figure 2.4 Diagram of a typical unicellular alga.

A few of them grows on land near water bodies. Many diatoms (a kind of sea algae) have exterior skeletons made of silicates. These are used as adsorbents (known as diatomaceous earth); filter aids in various industrial operations particularly in wastewater treatment. Some algae are utilized as foodstuffs and food supplements. A few algae (like *Chlorella*, *Spirulina*, etc.) take part in in the reclamation of effluent water with concurrent production of single-cell-protein which is widely used as food or feed.

Unicellular green alga *Chlorella* has been commercialized as an important source for production of healthy food. Currently several plants for *Chlorella* and *Spirulina* (blue-green bacterium) production are operating in Japan, Taiwan, Mexico, Thailand and USA. The production of food from microalgae holds much promise for future applications. Some selected algal products useful in food industry are presented in Table 2.3.

Table 2.3 Some useful products from algae

Algal genus	Products	Uses
Dunaliella	β-carotene	Food supplement and food colour
Porphyridium	Polysaccharides	Viscosifers gum

(*Contd.*)

Algal genus	Products	Uses
Chrysophytes	Bivalva feeds	Cultivation of seeds, aquaculture
Chlamydomonas	Soil inoculum	Conditioner
Chlorella	Amino acids	Proline, arginine, aspartic acid.

Algae can also trap solar energy and produce hydrogen gas as a potential future energy source. A large number of algae are being exploited for hydrogen production. Some of the important algae include *Chlorella*, *Chlamydomonas*, *Scenedesmus*, *Ankistrodesmus*, *Dunaliella*, *Corralina* and *Porphyridium*. Green algal hydrogen production is one of the emerging areas of biohydrogen research.

Yeasts

Yeasts constitute one of the important subgroups of fungi. They have single small cells with length ranging from 5 to 30 μm. They are usually spherical, cylindrical or oval. Yeasts are capable of anaerobic growth and can survive in different environment. They reproduce by budding and fission (Figure 2.5). In budding, a small daughter cell begins to grow on the side of the mother cell. In the process of fission mother cells divide into two new daughter cells. Approximately thousand species of yeasts have so far been isolated, identified and characterized scientifically.

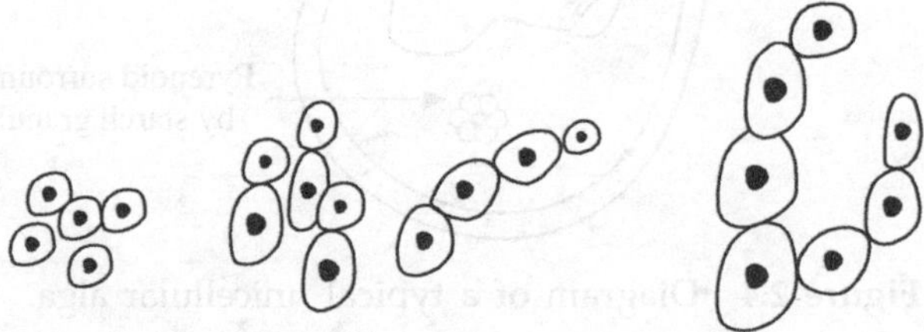

Figure 2.5 Budding of yeast cells.

Yeasts are credited with numerous uses largely in food, beverage, distillery and baking industries. *Saccharomyces cerevisiae* is well-known as 'Baker's yeast' used in distillery industries under anoxic conditions [production of wine, beer and whiskey (distilled after fermentation)]. The particular role of yeast in baking is to soften and leaven the batter or dough. Carbon dioxide effervescence in fermentation increases the porosity and fluffiness of bread and bread products. It is also used in biomass production in the form of single-cell-proteins (SCP). Recent application of yeasts includes sustainable production of biofuels such as bioethanol from carbohydrates and heterologous compounds (foreign proteins). Some other types of yeasts include *Saccharomyces italicus, Candida pulcherrima, Endomycopsis lindneri,* etc. most of which are commonly found on grapes, wines and musts.

Molds

Mold is a multicellular filamentous fungus generally grown in damp and musty locations. Fine thread-like structures or filaments grown on its body are called hyphae (Figure 2.6). They multiply by asexual reproduction of spores. Different types of spores include conidiospore, sporangiospore and arthrospore. Bread molds are specific type of fungus that grows on moist bread. Usually bread molds appear gauzy and finely threaded with greyish-black colour. These are mostly the species of *Aspergillus, Penicillium, Fusarium, Mucor, and Rhizopus.*

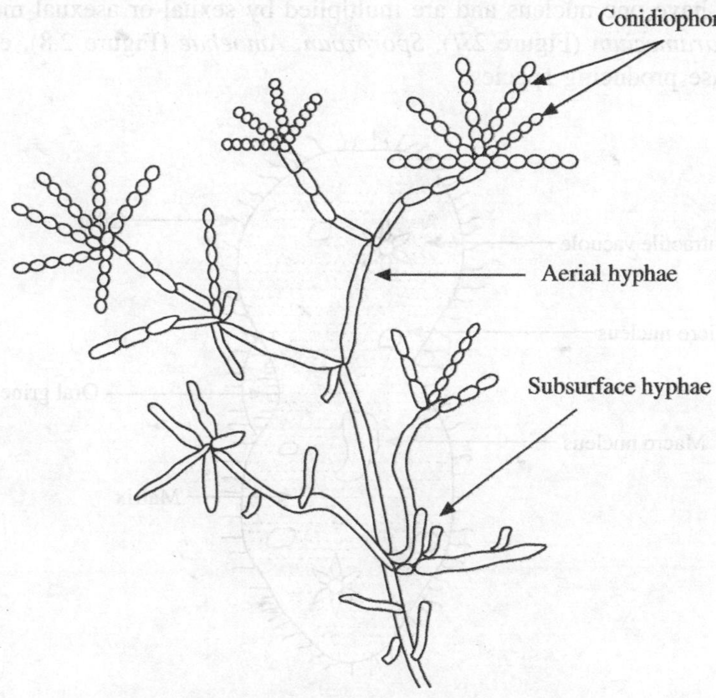

Figure 2.6 Mycelial structure of molds.

Molds cause infections such as tinea or ringworm, histoplasmosis, etc. Molds are used in the production of both food and feed. They help improvement properties of processed food products (e.g. *Roquefort, Camembert*). Molds are used to produce enzymes like amylase which is used in bakery and beverage industries. These are called amylolytic molds. One species of *Bothrytis cinerea*, finds its application in the decomposition of grapes for wine production (Table 2.4).

Table 2.4 Some useful products from molds

Type of mold	Product	Uses
Penicillium notatum	Penicillin	Antibiotic
Aspergillus niger	Citric acid	Food products, medicinal citrates, in blood transfusion
Rhizopus nigricus	Fumaric acid	Manufacture of alkyd resin, wetting agents
Aspergillus niger	Gluconic acid	Pharmaceutical products, textile, leather, photography
Fusarium moniliforme	Gibberellic acid	Setting of fruit, food production
Rhizopus oryzae	Lactic acid	Food and pharmaceuticals
Aspergillus oryzae,	Beverage/juice	Depectinization
Aspergillus flavus	Medicinal product	Gout
Candida tropicalis	Lipase enzyme	Degradation of crude oil hydrocarbons

Protozoa

Protozoa are unicellular or multicellular, motile and relatively large (1 mm to 50 mm) eukaryotic cells without having any cell walls. Ingestion of other smaller organisms as food is a feature of protozoa. They have one nucleus and are multiplied by sexual or asexual means. Examples of protozoa are *Paramecium* (Figure 2.7), *Sporozoan*, *Amoebae* (Figure 2.8), etc. Most of the protozoa are disease producing species.

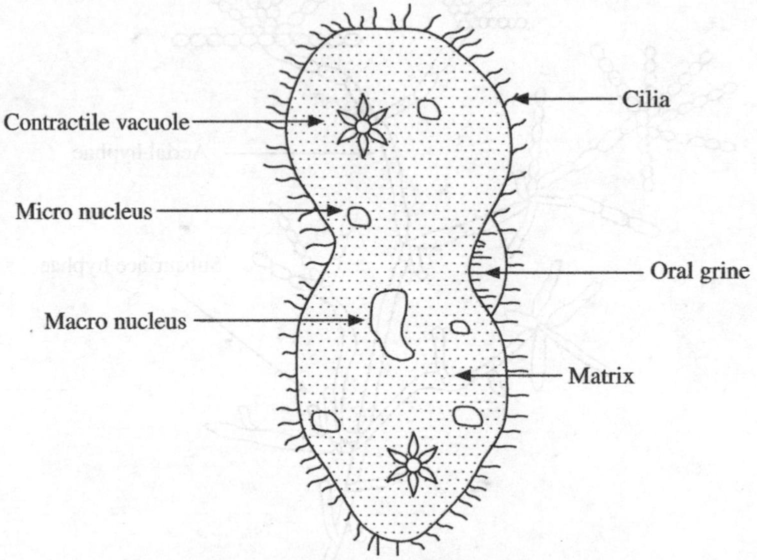

Figure 2.7 A typical structure of *Paramecium*.

However, it can assist in the activated sludge, trickling or percolating filter, and rotating biological disc reactor where they colonize the biofilm. Amongst all microorganisms, bacteria in particular, play the major role in degradation, bioremediation and mineralization of carbonaceous waste. But the presence of protozoa in wastewater augments the remediation process.

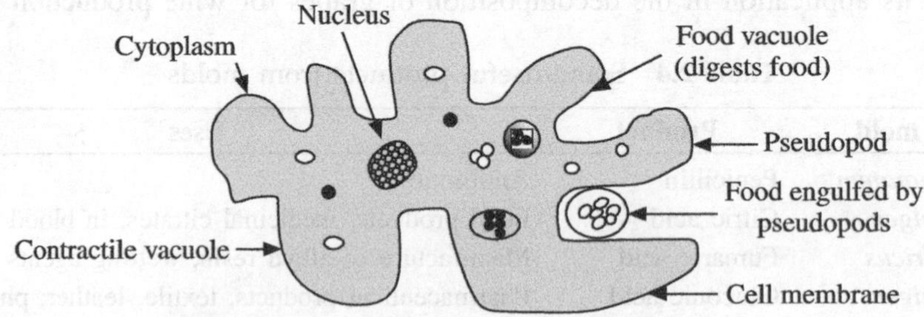

Figure 2.8 A typical structure of *Amoeba*.

Protozoa excretes micronutrients and other compounds which stimulates bacterial growth for biodegradation. Protozoa is known as an important bacteriovorous. It consumes smaller size bacterial population. This is known as grazing which plays an important role in controlling bacterial population in aquatic ecosystem. In essence, protozoa despite being pathogenic assist in the secondary treatment of wastewater.

2.2.2 Eukaryotic Cells

Eukaryotic (*Eu*: true or developed; *karyon*: nucleus) cells contain nucleus which is enclosed by membrane. They are 1000 to 10000 times larger than prokaryotes. The internal cellular structure of eukaryotes is more complex than prokaryotic cells. They contain cell organelles within cytoplasm. These are mitochondria, ribosomes, chloroplast, etc. The nucleus here is surrounded by a porous membrane. The function of nucleus is to control the catalytic activity at the ribosomes. Eukaryotic cells include both plant and animal cells. Cell membranes are composed of 55% proteins, 40% lipids and 5% carbohydrates. Schematic diagram of a typical eukaryotic cell is presented in Figure 2.9.

Cytoplasm: Cytoplasm is composed of fluid called cytoplasmic sol or cytosol and various organelles dispersed or embedded into it. Cytoplasmic sol is aqueous in nature with nearly same density and viscosity as water. It contains several metal ions and biomolecules such as proteins, lipids, carbohydrates, etc. A number of metabolic functions are carried out by eukaryotic cells. Various organelles of a typical eukaryotic cell can be grouped into three types:

(i) Organelles with two membrane layers (mitochondria and plastids)
(ii) Organelles with one membrane layer (Endoplasmic reticulum (ER), Golgi bodies, Lysosomes, vacuoles)
(iii) Organelles with zero membrane layer (Ribosome and centrosome)

These are briefly discussed in the following section.

Mitochondria

These are tiny bodies of tubular shapes enclosed by a double membrane envelope and are distributed in the cytoplasm. Process of respiration takes place in mitochondria. During respiration, oxidation of food releases energy in the form of ATP (adenosine triphosphate). It is known as the 'energy-currency' of the cell. The stored energy as ATP is utilized for carrying out various metabolic and other functions of cell. In the true sense it is called powerhouse of the cell.

Plastids

Plastids are present in plant cells and in some protists. Like mitochondria, plastids are also double membraned cell organelles. There are three types of plastids based on their pigments:

(i) *Chromoplast*: These are coloured plastids, other than green. They occur in flowers, fruits, and roots e.g., carrot. They contain coloured pigments called carotenoids including carotene and xanthophyll which impart yellow, orange, red and other colours to the fruits, vegetables and flowers. Colourful flowers attract insects and help pollination.
(ii) *Leucoplasts*: These are non-pigmental plastids. They are of three types depending upon the storage product—amyloplasts (store starch), aleuroplast/proteinoplast (store protein) and elaioplast (store oils).

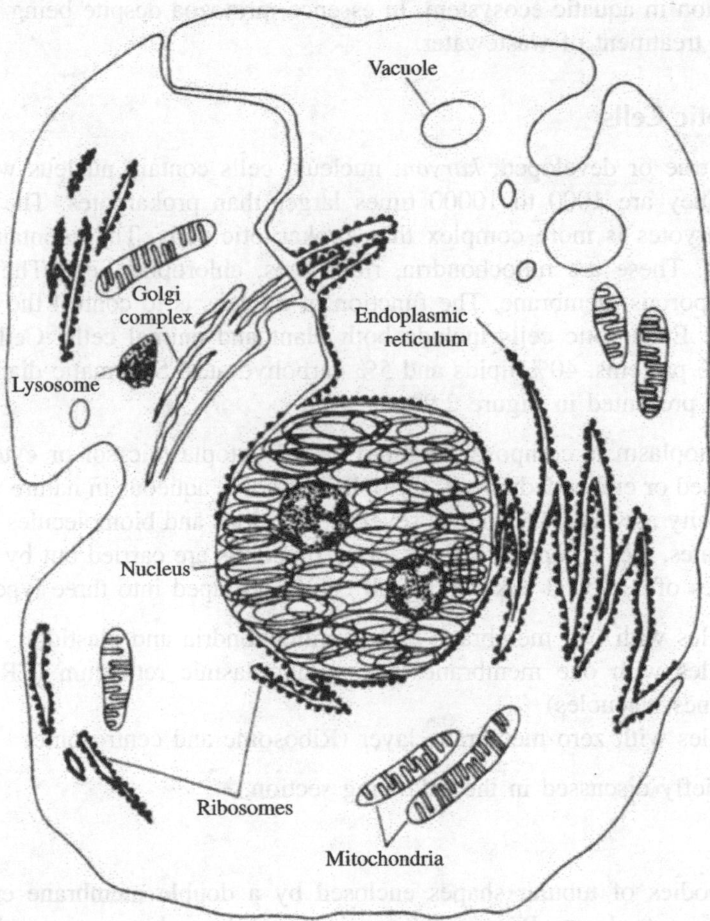

Figure 2.9 Schematic diagram of a typical eukaryotic cell (Adapted from Biochemical Engineering Fundamentals, J.E. Bailey and D.F. Oillis, 2nd Edition, 1986, McGraw Hill Book Company).

(iii) *Chloroplasts*: These are green coloured oval shaped plastids. They contain photosynthetic pigments called chlorophyll. It provides green colour to the plants and helps in photosynthesis. The internal organization of chloroplast consists of numerous internal spaces filled with dense fluid known as stroma which surrounds the grana, a sac-like staking arrangement. These are crucial structural features of chloroplast. Aged chloroplasts are known as gerontoplast.

Chromoplasts impart colours to fruits and flowers thereby facilitating pollination and fruit dispersal by attracting insects. Leucoplasts can store variety of biomolecules such as proteins, fats, etc. Chloroplasts capture sunlight to synthesize food through the process of photosynthesis. Hence, chroroplast are also known as the 'kitchen of the cell'.

Endoplasmic reticulum (ER)

ER is a large interconnected system of single membrane-enclosed tubules and sheets present in the cytoplasm. It is the largest organelles of cell. The ER membrane has close structural similarity with plasma membrane. ER acts as channels for movement of nutrients and stored biomolecules across different locations within cytoplasm. It provides surface for various biochemical synthesis, transportation, storage, and metabolism. It is a major site for protein synthesis.

Golgi apparatus

The Golgi apparatus or Golgi bodies were first described by Camillo Golgi. It consists of a set of membrane-bound, fluid-filled vesicles, tubules and flattened cisternae, which are ordered along a stack with parallel fashion. Golgi body's membranes are interconnected with ER membrane. They are specialized to modify, package and secrete complex biomolecules. The functions of Golgi apparatus helps store modify and pack various cellular products in vesicles from ER.

Lysosome (lysis–digestive, soma–body) (digestive bags)

Lysosomes are simple, tiny (diameter: 0.2 to 0.8 μm), spherical, single-membrane bound sacs filled with digestive enzymes (called lysozymes) for intracellular digestion and waste disposal. These enzymes (hydrolytic enzymes) facilitate degradation of carbonaceous materials. Lysosomes have almost fifty types of enzymes such as protease, lipase, nuclease, glycosidase, lysozyme and so on. It digests foreign materials and clean up cells by consuming debris of dead organelles. Thus, they are truly called 'Cellular scavengers'. Important functions of various cell organelles are presented in Table 2.5.

Table 2.5 Summary of important functions of various cell organelles

Organelles	Important functions
Nucleus	Controlling of major cellular activities and cell division; synthesis of RNA, sending genetic instruction
Mitochondria	Energy production, ATP synthesis, detoxification, calcium storage, etc.
Plastid	
ER	Synthesis of proteins (by rough ER) and lipids and steroids (by smooth ER), help in metabolism
Golgi bodies	Modifying, packaging and secreting complex biomolecules.
Ribosome	Protein synthesis
Lysozome	Degradation of worn-out organelles, and own cells
Centrosome	Ensures movement of chromosomes during cell division
Peroxisomes	Breakdown of excess fatty acids
Vacuoles	Osmoregulation and excretion, storage of salts, sugars, amino acids, etc.

Vacuoles

Vacuoles are membrane-bound sacs for storage of solid or liquid contents. Plant cell vacuoles are larger than those in animal cells. In some plant cells, 70–75% of cell volume is occupied by vacuole due to which nucleus and other cell organelles are pushed to sides. Vacuoles help in osmoregulation and excretion and also in storage of a number of substances like salts, sugars, amino acids, etc. They function as dump house of waste products in plant cells and provide turgidity (water swelling potency) and rigidity (toughness) to the plant cells.

Ribosomes

Ribosomes are membrane-less, thick, globular and grain-like particles which are present in both prokaryotic and eukaryotic organisms. These are the smallest organelles which occur either freely in the cytosol or stay combined to ER and the outer nuclear membrane. Protein synthesis in the cell takes place in ribosomes. Therefore, they are also called protein factories. They are absent in mammalian RBCs.

Centrosome

Centrosome is generally observed in animal cells and in some of the lower plant cells. It is formed by centrioles and centrosphere and is not bound by any membrane. Centrioles appear as granule like structures. Their shape is empty cylindrical-like composed of various tubules of micron sizes. It helps in the regulation of cell cycle.

2.3 SCIENTIFIC NOMENCLATURE

In view of the growing diversity of the types of living microorganisms, a formal system of naming them became essential for their systematic identification and appropriate applications in microbiological and biotechnological research. Historically, the pioneering effort to classify microbes was put forward by Carl Linnaeus, a botanist from Sweden. He put forward this naming protocol in the middle of seventeenth century. All known plants, animals, and minerals were named using Latin and Greek names. Microorganisms are scientifically recognized using a binomial nomenclature using two words the prefix being 'genus' and the suffix is the 'species'. The names assigned to microorganisms use Latin grammatical forms, although some of them are derived from other languages as well. Such a name, called binomial nomenclature is printed in italics by convention or underlined. The first alphabet of the genus name is written in capital letters, while the species name is written with small letters.

For example, in the name bacteria *Lactobacillus casei*, 'Lactobacillus' is the genus while 'casei' is the species. In *Escherichia coli, Escherichia* is the genus while *coli* is the species name. In one genus there could be hundreds of species. The genus *Lactobacillus* currently contains over 170 species covering a wide variety of organisms (such as *Lactobacillus casei*, *Lactobacillus bombi*, *Lactobacillus bulgaricus*, *Lactobacillus lactis*, *Lactobacillus mobilis*, *Lactobacillus pontis*, *Lactobacillus florum* and so on). Nomenclature also is indicative of certain typical characteristics of the microorganisms. Examples of salient characteristics as a part of microorganism name are presented in Table 2.6.

Table 2.6 Examples of salient characteristics as a part of microorganism name

Microorganism	Salient or Indicative characteristics
Micrococcus	Size
Staphylococcus, Bacillus	Shape
Staphylococcus, Streptococcus	Arrangement

(Contd.)

Microorganism	Salient or Indicative characteristics
Micrococcus roseus	Colour of colony
Escherichia coli, Staphylococcus epidermidis	Habitat; (The word 'coli' is obtained from colon, which is the home to these bacteria.)
Haemoplilus aegypticus	Nutritional requirements (found living in human epithelial cell linings)
Neisseria meningitidis, V. cholerae	Disease
Escherichia sp., *Neisseria* sp.	Researcher; (named after Theodor Escherich a German bacteriologist, who, isolated it in 1885.) The genus 'Neisseria' was named after Albert Ludwig Sigesmund Neisser, a German physician who discovered *Neisseria gonorrhoeae* in 1879.

2.4 MICROORGANISMS AND CYCLES OF ECOSYSTEM

Cycles of ecosystems refer to biogeochemical cycles, whereby an element or compound transforms into various chemical states in lithosphere, hydrosphere and atmosphere under the influence of certain environmental factors.

2.4.1 Nitrogen Cycle

Nitrogen is an indispensable constituent of several cellular substances such as amino acids, proteins, enzymes, nucleic acids, alkaloids, vitamins and so on. Nitrogen is essential of all living organisms to grow and live. Although air contains almost 79% nitrogen, most of this becomes unavailable for use by organisms. Although present in large quantities in atmosphere, it is unfortunately unavailable to living systems easily. Molecular nitrogen is a highly stable and inert gas, by virtue of the strong triple bond between the N atoms (N≡N). Nitrogen is the only diatomic gas having triple bond with one sigma and two Pi bonds.

The nitrogen bond has the shortest bond length of 1.095 Å, the highest ionization potential (15.58 eV) and the highest stretching frequency. Energy requirement for the conversion of molecular nitrogen to atomic nitrogen is about 160 kcal. Thus, barring a few natural phenomena involving very high temperature such as lightning, volcanic eruption, forest fires, etc. dissociation of nitrogen becomes extremely difficult. Actually, plants and animals can only use nitrogen, N_2 if it is chemically or biochemically available to plants in some usable and soluble forms. Five major steps of nitrogen cycle include fixation, uptake, decay, nitrification and denitrification. Let us briefly discuss all these processes in the following section.

Nitrogen fixation

Nitrogen fixation is the process wherein atmospheric nitrogen is transformed to ammonium (NH_4^+) or nitrate (NO_3^-) by naturally occurring nitrogen-fixing organisms. Fixing of atmospheric nitrogen in the soil by bacteria is one of the wonderful biological phenomena. Bacteria such as *Clostridium, Azotobacter*, etc. stay in the nodules of leguminous plants (peanuts, soybeans, lentils, alfalfa, beans, clovers, etc.). They get soil nutrients to live and in exchange they supply nitrogen to the plants. This is called symbiosis, where both the participating species gets mutual benefits. Certain blue-green algae, fungi and mosses can assimilate free nitrogen, and a bacterium, *Azotobacter chroococcum*, present in the soil can fix free nitrogen directly in presence of calcium carbonate.

Approximately about 200 million metric tons of nitrogen is added to the earth's surface every year by biological method. Man produces only about 28 million metric tons in the form of fertilizers. Microorganisms such as *Azotobacter* and *Clostridium* can add approximately 6 to 7 kg of nitrogen per hectare of land per annum. Rainwater also brings down about 71 kg of nitrogen per hectare per year. Additionally leguminous plants can fix about 100 to 200 kg of nitrogen in one hectare land in one year.

Biotechnology of nitrogen fixation: *Besides interest in the fundamental aspects of the chemistry and biology behind the process of nitrogen fixation, recent developments in the field of plant biotechnology focuses in the thrust areas such as transfer of nitrogen fixing gene to other plants, improvement of existing nitrogen fixing plants and microbes and also production, release and utilization of bio-ammonia by nitrogen fixing microbes.*

Nitrification

Microbial conversion of ammonium ions into nitrate ions is known as nitrification. Nitrosomonas bacteria (a genus of Gram-negative rod-shaped chemoautotrophic bacteria, for example, *Nitrosococcus, Nitrobacter and Nitrococcus* sp.) convert nitrogen into nitrite [NO_2^- and subsequently to nitrate (NO_3^-)]. Even some of the ammonium produced by decomposition of organic biomass material is converted to nitrate through nitrification. Nitrification is catalyzed by oxygen-rich surface of flowing water bodies or soil layer. Clay particles in soils are negatively charged. Hence these particles can bind positively charged ions and radicals like NH_4^+ but cannot bind positively charged ions and radicals like NO_3^-, NO_2^-, etc. As a result nitrate ions can be easily removed from the soil by leaching. Soil gets deprived of nitrates but water bodies get enriched with these nutrients.

Nitrogen uptake or assimilation

In the process of assimilation, nitrate and ammonia formed during first two steps of the cycle are incorporated into the plant. Plants receive these nutrients through soil via root transport system and slowly transform these into complex nitrogenous compounds, primarily proteins and nucleic acids. Herbivores are dependent on green plants for this nitrogen. On the other hand, carnivores, which feed on herbivores, can also get the nitrogen by this way. As a result, atmospheric nitrogen gets transferred to plants and animals.

Ammonification or nitrogen mineralization

Microbial conversion of organic nitrogen from plant or animal bodies into ammonia through the process of decay and decomposition is known as ammonification. Decomposer bacteria convert nitrogen nutrients from decaying animal and plant bodies back into ammonium salts and ammonia. *Bacillus, Clostridium, Proteus, Pseudomonas*, etc are some of the examples of ammonifying bacteria which helps in the ammonification process of nitrogen. Schematic representation of nitrogen cycle is given in Figure 2.10.

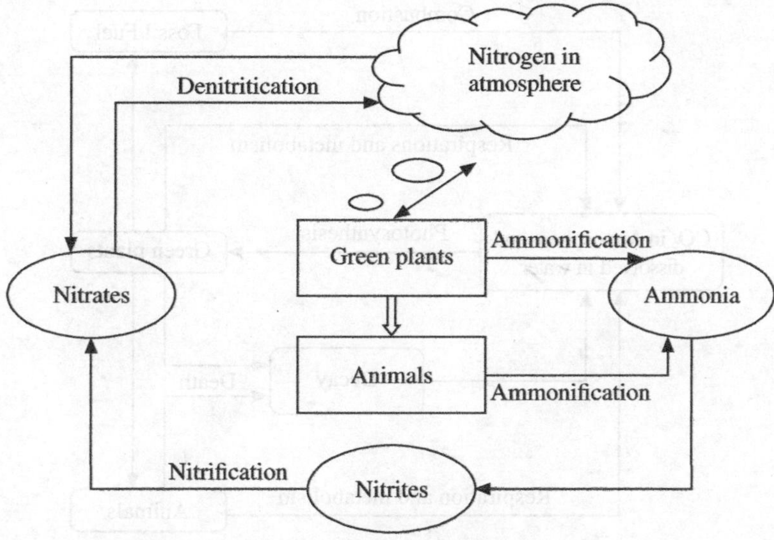

Figure 2.10 Schematic diagram of nitrogen cycle.

Denitrification

This is predominantly a process of conversion of nitrates and nitrites into nitrogen and partly to gaseous nitrous oxide. Also denitrifying bacteria from soil help conversion of nitrates into nitrogen gas according to the following steps:

$$NO_3 \rightarrow NO_2 \rightarrow NO \rightarrow N_2O \rightarrow N_2.$$

The process is responsible for maintaining proportion of nitrogen in atmosphere.

2.4.2 Carbon Cycle

Right from the fossil fuels to the complex biomolecules of DNA and RNA, carbon is the omnipresent chemical constituent of most organic matter on earth. Almost all living organisms obtain carbohydrate (glucose) produced by photosynthesis. During respiration metabolic breakdown of glucose generates energy required to carry out life processes. Plants consume CO_2 during photosynthesis and release it during respiration. During respiration breakdown of glucose takes place to produce CO_2, water and energy while in photosynthesis glucose is produced along with evolution of oxygen.

Chlorophyll containing green plants converts atmospheric carbon dioxide into carbohydrates (sugars) by photosynthesis in presence of sunlight. Breakdown of this carbohydrates takes place in a reverse process, known as respiration, where plants and animals consume these carbohydrates (and other products derived from them) for energy generation. Respiration leads to oxidation and breakdown of carbohydrates in presence of oxygen and releases carbon dioxide. The amount of carbon that is consumed in the photosynthesis process and is released back during respiration in the form of carbon dioxide is much higher than carbon moving through geological cycle on yearly basis. Schematic diagram of carbon cycle is given in Figure 2.11.

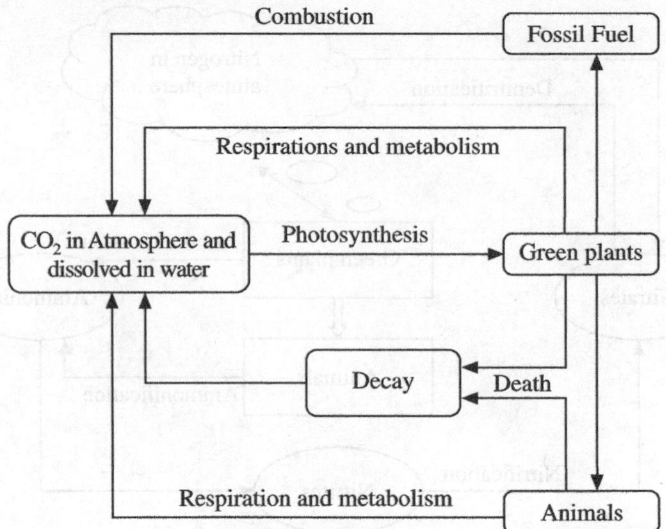

Figure 2.11 Schematic representation of carbon cycle.

Phytoplanktons in the marine ecosystem have the ability to consume carbon to form their exterior shell. These shells are chemically composed of chalk or calcium carbonate. When phytoplanktons die, this cell debris get buried and deposited over time. Carbonaceous organic matter deposited and buried over long period of time leads to formation of fossil fuels like coal and petroleum.

2.4.3 Phosphorus Cycle

Phosphorus cycle is important because phosphorus is a limiting nutrient in the ecological system. Unlike C and N, phosphorus does not have any stable gaseous form. So, phosphorous cycle is endogenic involving various subsurface rocks in lithosphere and hydrosphere. Phosphorus cycle is independent of atmosphere. The key sources of phosphorus are rocks (mainly phosphates such as apatite, monazite, etc.), natural phosphate deposits (such as phosphorites, bioclastic and pebble bed phosphorites and marine sediments).

Phosphorus is extracted from its source rocks and minerals either by natural process of erosion-corrosion, weathering or by man-made metallurgical processes like leaching, lixiviation, etc. While plants can absorb phosphorus from soil, it enters into the herbivores via food chain. Finally decomposition of plant and animal bodies after their death enriches the soil with phosphorus. It gets deposited in the rock strata or sediments may be released again as a result of weathering activities (Figure 2.12).

2.4.4 Sulphur Cycle

Sulphur is one of the essential elements for all organisms. Much of the sulphur is bound to underground rocks and is primarily released through geological activity. Subsequently it is used by plants and then by animals (Figure 2.13).

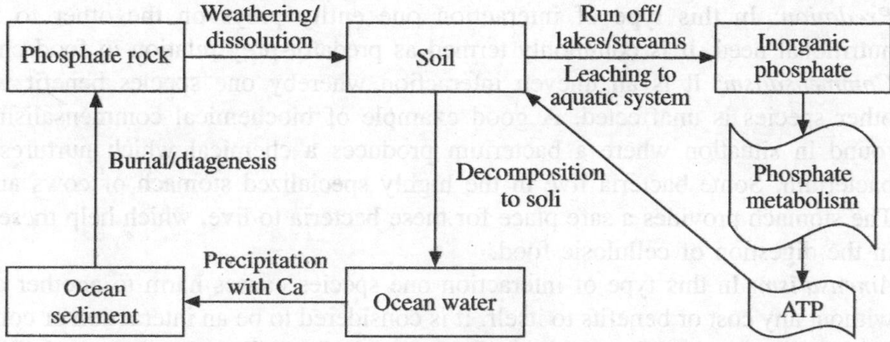

Figure 2.12 Schematic representation of phosphorus cycle.

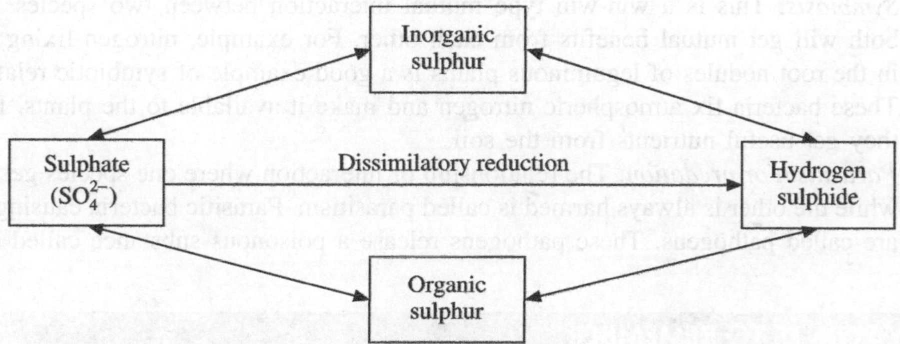

Figure 2.13 Schematic representation of sulphur cycle.

Sulphur is mostly taken up by plants as sulphates and is one of the structural ingredients of amino acids—the basic unit of protein. Sulphur of amino acids is transformed into hydrogen sulphides by soil microorganism. Eventually it is redeposited in the ground. Sulphate is reduced to sulphides by dissimilatory reduction by anaerobic process and is biologically oxidized to elemental sulphur or sulphate. Such inter-conversion of sulphate, sulphide and elemental sulphur is facilitated by a number of chemical reactions such as disproportionation, desulphurylation, oxidation and reduction. Various microorganisms such as *Proteus*, *Pseudomonas*, *Salmonella*, purple sulphur bacteria, etc. play an important role in sulphur cycle.

2.5 ECOLOGICAL INTERACTIONS AT MICROBIAL LEVEL

Nature presents different types of biological interactions among various living entities of the earth. These interactions are closely associated not only with the survival of microorganisms but also for the effective functioning of ecosystem as a whole. Important possible mutual interactions of various species are summarized below:

(i) *Neutralism*: It is most commonly encountered inter-species interaction, where neither of the species gets harmed or even benefitted. For example, a few species of *Lactobacillus* and *Streptococcus* can co-survive without any interference either positive or negative.

(ii) *Predation*: In this type of interaction one entity preys on the other to fulfil its nutritional need. It is commonly termed as predator-prey relation in food chain.

(iii) *Commensalism*: It is an uneven interaction whereby one species benefit while the other species is unaffected. A good example of biochemical commensalism can be found in situation where a bacterium produces a chemical which nurtures another bacterium. Some bacteria live in the highly specialized stomach of cows and goats. The stomach provides a safe place for these bacteria to live, which help these animals in the digestion of cellulosic food.

(iv) *Amensalism*: In this type of interaction one species causes harm to another organism without any cost or benefits to itself. It is considered to be an interaction or competitive behaviour among other organisms. For example *Penicillium* secretes penicillin, which kills various bacteria.

(v) *Symbiosis*: This is a win-win type mutual interaction between two species whereby both will get mutual benefits from each other. For example, nitrogen-fixing bacteria in the root nodules of leguminous plants is a good example of symbiotic relationship. These bacteria fix atmospheric nitrogen and make it available to the plants. In return they get useful nutrients from the soil.

(vi) *Parasitism or predation*: The relationship or interaction where one species get benefits while the other is always harmed is called parasitism. Parasitic bacteria causing disease are called pathogens. These pathogens release a poisonous substance called toxin.

KEY POINTS TO REMEMBER

- Protist kingdom is divided into two types, prokaryotes or prokaryotic cells and eukaryotes or eukaryotic cells.
- Prokaryotic cells are unicellular, smaller in size, devoid of membrane enclosed nucleus and other organelles while eukaryotic cells are multicellular, bigger in size and possess nucleus and other organelles.
- Bacteria and blue-green algae belong to prokaryotes while fungi, algae and protozoa belong to eukaryotes.
- Morphologically bacteria are of four major types-spherical (*Cocci*), rod shaped (*Bacilli*), spiral shaped (*Spirilla*), and comma shaped (*Vibrio*),
- Based on their response to Gram staining bacteria are of two types, gram-positive and gram-negative.
- Based on oxygen utilization ability bacteria can be aerobic or anaerobic.
- BGA contains photosynthetic pigment such as chlorophyll-*a*, and takes part in nitrogen cycle.
- Algae contain chlorophyll to carry out photosynthesis. They are enclosed by cell wall made of cellulose layers. They trap solar energy and can produce hydrogen gas.
- Yeasts are largely used in food, beverage, distillery and baking industries.
- Mold is a multicellular filamentous fungus generally grown in damp and musty locations and is used in the production of antibiotics.

- Protozoa help in the activated sludge, trickling or percolating filter, and rotating biological disc reactor for treatment of wastewater.
- A number of metabolic functions are carried out by eukaryotic cells.
- Major organelles present in eukaryotic cells are mitochondria, plastids, ER, Golgi bodies, lysosome, ribosome and vacuoles.
- Process of respiration takes place in mitochondria, it is known as the powerhouse of cell.
- Chloroplasts capture sunlight to synthesize food through the process of photosynthesis and are known as the 'kitchen of the cell'.
- Protein synthesis in the cell takes place in ribosomes.
- Lysosomes carry out degradation of worn-out organelles, and own cells
- Microorganisms are scientifically recognized using a binomial nomenclature using two words the prefix being 'genus' and the suffix is the 'species.
- Five major steps of nitrogen cycle include fixation, uptake, decay, nitrification and denitrification.
- During respiration breakdown of glucose takes place to produce CO_2, water and energy while in photosynthesis glucose is produced along with evolution of oxygen.
- Nature presents different types of biological interactions among various living entities of the earth. These are neutralism, predation, commensalism, amensalism, and symbiosis.

? QUESTIONS FOR PRACTICE

Multiple Choice Questions

1. A bacterium does not have a

(a) Cytoplasm (b) True nucleus
(c) Protoplasm (d) DNA

2. Spherical-shaped bacteria are referred as

(a) *Cocci* (b) *Bacilli*
(c) *Spirilla* (d) None of these

3. Bacteria do not always swim aimlessly but are attracted by such nutrients such as sugar and amino acids, and are repelled by harmful substances and bacterial waste products. Movement toward chemical attractants and away from repellents is called

(a) Gliding motility (b) Tumbling.
(c) Chemotaxis (d) None of these

4. Prokaryotic cells that can withstand very high salt concentrations are called

(a) Capnophiles (b) Barophiles
(c) Halophiles (d) Microaerophiles

5. Which is the correct list of elements found in 98% of all living cells?

(a) Ca, N, O, Na, P, S (b) C, H, O, N, P, S
(c) Cu, O, S, K, I, P (d) Ca, H, O, Na, P, S

6. Salts and sugars work to preserve foods by creating

(a) Depletion of nutrients (b) Hypertonic environment
(c) Hypotonic environment (d) Lower pH.

7. In the name *Esherichia coli*, *Escherichia* is the
   (a) Specific epithet (b) Genus
   (c) Kingdom (d) Family
8. Which of the following is important for the effectiveness of an antimicrobial treatment?
   (a) Initial numbers of microbes (b) Time of exposure
   (c) Microbial characteristics (d) All of the above
9. An organism has an optimal growth rate when the hydrogen ion concentration is very high. This organism is
   (a) Osmotolerant (b) Acidophile
   (c) Halophile (d) Aerotolerant anaerobe
10. Cell wall is
    (a) Thick in gram-positive than gram-negative
    (b) Thick in gram-negative than gram-positive
    (c) Equal in both
    (d) In gram-negative cell wall is absent
11. The solidifying agent used most successfully in bacterial nutrition media is
    (a) Gelatin (b) Peptone
    (c) Agar (d) Starch
12. Which term refers to a decomposition reaction in which a complete water molecule is added back into a compound?
    (a) Condensation (b) Dehydration
    (c) Hydrolysis (d) Denaturization
13. Lyophilization means
    (a) Sterilization (b) Freeze drying
    (c) Burning to ashes (d) Inoculation
14. Cork-screw shaped forms of bacteria are
    (a) *Bacilli* (b) Stalked bacteria
    (c) *Spirochaetes* (d) *Actinomycetes*
15. Which of the following is called as the 'brewer's yeast?
    (a) *Saccharomyces ludwigi* (b) *Saccharomyces cerevisiae*
    (c) *Saccharomyces boulardii* (d) *Saccharomyces pastorianus*
16. In which of the following form nitrogen is absorbed by the plant?
    (a) Nitrate (b) Nitrite
    (c) Ammonium (d) All of the above
17. Nitrogen fixation is the conversion of
    (a) N_2 to N (b) N_2 to NH_3
    (c) N_2 to NO_3^- (d) N_2 to urea
18. Conversion of nitrates to nitrogen is known as
    (a) Ammonification (b) Nitrification
    (c) Nitrogen fixation (d) Denitrification

19. Which of the following statements about the nitrogen cycle is false?

(a) Nitrogen fixing bacteria exist on the root nodules of legumes and in the soil.

(b) Denitrifying bacteria convert nitrates (NO_3^-) into nitrogen gas (N_2).

(c) Ammonification is the process by which ammonium ion (NH_4^+) is released from decomposing organic compounds.

(d) Nitrification is the process by which nitrites (NO_2^-) are converted to ammonium ion (NH_4^+).

Answer Key

1. (a)	2. (a)	3. (c)	4. (c)	5. (b)	6. (b)	7. (b)	8. (d)	9. (b)	10. (a)
11. (c)	12. (c)	13. (b)	14. (c)	15. (b)	16. (d)	17. (c)	18. (d)	19. (d)	

Descriptive Questions

1. What are the dyes that are used for gram staining test?
2. Classify bacteria based on the morphology.
3. What is the difference between sterilization and pasteurization?
4. Write down the conditions for stem sterilization.
5. What is the difference between motility and Brownian motion?
6. State the types of anaerobes with examples.
7. A new species of a prokaryote has been isolated by a microbiologist. It is found to be of *Bacillus* morphology with a lipid bilayer and cell wall that stains positive for peptidoglycan. It contains circular chromosome which replicates from a single origin of replication. How could you predict whether the specimen is an *Archaea*, a gram-positive bacterium, or a gram-negative bacterium?
8. Crop rotation is an old agricultural practice to increase the land fertility. A farmer decided to plant soybeans in a field in a particular season, after which the field would be used to grow carrots. Do you think that it would help maintain nitrogen in the soil? Explain.
9. Imagine a region of soil became contaminated, killing bacteria that decompose dead plants and animals. How would this affect the carbon cycle in the area? Be specific in stating where carbon would accumulate in the cycle.
10. One of your lab mates says that prokaryotes are always detrimental and pathogenic. How would you explain to him that he is wrong?
11. Many people use antimicrobial soap to kill bacteria on their hands. However, overuse may actually increase the risk of infection. How could this occur?

3 Biochemistry of Important Molecules of Life

Aaron Ciechanover, the Israeli biologist and the Nobel laureate said 'Biochemistry is the science of life. All our life processes—walking, talking, moving, feeding are essentially chemical reactions.' Ciechanover won the Nobel Prize in Chemistry in 2004 for characterizing the method of degradation and recycling of proteins using ubiquitin.

Biomolecules, also known as biological molecules, are those substances that are produced by cells and living organisms. Biomolecules are of varied sizes and structures and perform a number of functions in any living system. There are four important biomolecules namely carbohydrates, proteins, lipids and nucleic acids. Apart from these vitamins can also be considered as biomolecules. Without any of these four molecules, a cell and organism would cease to exist. In this chapter we briefly study the salient structural and functional features of important biological molecules.

3.1 CARBOHYDRATES

Carbohydrates are considered as the storage and structural compound in cell. For example cellulose and peptidoglycan are the important structural carbohydrates, the former being the component of plant cell wall while the latter that of bacterial cell wall. They store energy, transport stored energy and serve as carbon skeleton. They also help in recognizing certain signals in cells like biosensors. These are synthesized in a natural process called photosynthesis. Chemically these are represented as $C_n(H_2O)_n$ (hydrates of carbon) where *n* is greater than or equal to 3. Chemically they are polyhydroxy aldehydes or ketones. Carbohydrates are of three types—monosaccharides, oligosaccharides, and polysaccharides.

3.1.1 Monosaccharides

They are simple sugars, denoted by the general formula $C_nH_{2n}O_n$, where *n* varies from 3 to 7. All the simple monosaccharides are white crystalline solids, sweet in taste and are freely soluble in water. The carbon skeleton of common monosaccharides is unbranched, and each carbon atom except one, contains a hydroxyl group (Figure 3.1). The carbon atom has a carbonyl oxygen which is combined in an aldehyde or ketose linkage. Examples of monosaccharides are glyceraldehyde, glucose, fructose, ribose, etc. Pentose and hexose are most common in nature. Glucose is also called blood sugar, corn sugar, grape sugar or dextrose (as it is dextrorotatory). Mannose and galactose are the isomers of glucose. Fructose is found in all fruits except glucose. Some monosaccharides are known as derived monosaccharides. For example, deoxysugar e.g., deoxyribose, sugar acid e.g., Glucuronic acid, sugar alcohol like mannitol.

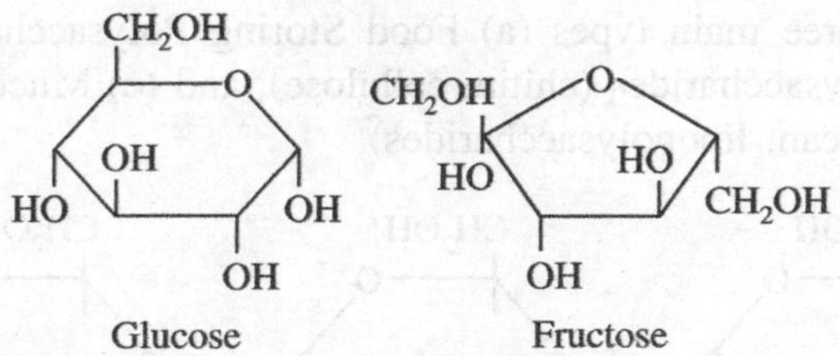

Glucose Fructose

Figure 3.1 Structure of glucose and fructose.

Monosaccharides undergo enolization reaction with dilute alkali at low temperature leading to the formation of enediol. However, in presence of concentrated alkali solution and at high temperature monosaccharides undergo oxidative degradation and polymerization. Glucose is considered to be a reducing sugar as it contains free aldehyde groups. It readily gets oxidized by oxidizing agents or by enzymes to gluconic acid. Monosaccharides can also be reduced in presence of $NaBH_4$ or enzymatically in to sugar alcohol. For example glucose is reduced to sorbitol.

3.1.2 Oligosaccharides

Oligosaccharides (*oligo* means a few) are the condensation products of 2 to 6 monosaccharides joined by glycosidic linkage. Examples of disaccharides: sucrose (disaccharide of α-D glucose and β-fructose) also known as cane sugar; lactose (disaccharides of β-D glucose and β-D galactose), found in milk and whey; maltose (disaccharides of two molecules of glucose) (Figure 3.2). Sucrose is widely used as sweetening agent in food industry.

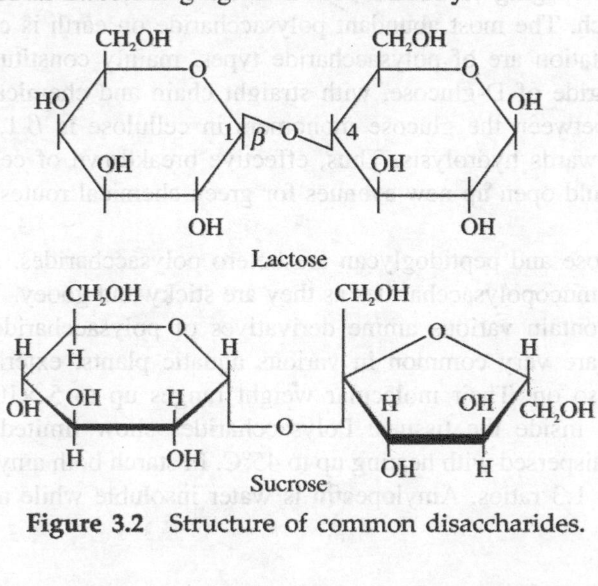

Lactose

Sucrose

Figure 3.2 Structure of common disaccharides.

3.1.3 Polysaccharides

More than two monosaccharides undergo condensation to form polysaccharides. These monosaccharides are joined by covalent glycosidic linkage where H atom from one monosaccharide reacts with OH group of the other monosaccharide to form a bond (Figure 3.3). They are insoluble in water, not sweet in taste and do not disturb osmolarity in cell due to their relatively large molecular weight. They are involved in bulk reduction and storage. Important polysaccharides are of three main types (a) Food Storing Polysaccharides (starch, glycogen, inulin), (b) Structural Polysaccharides (chitin, cellulose), and (c) Mucopolysaccharides (pectin, hemicellulose, peptidoglycan, lipopolysaccharides).

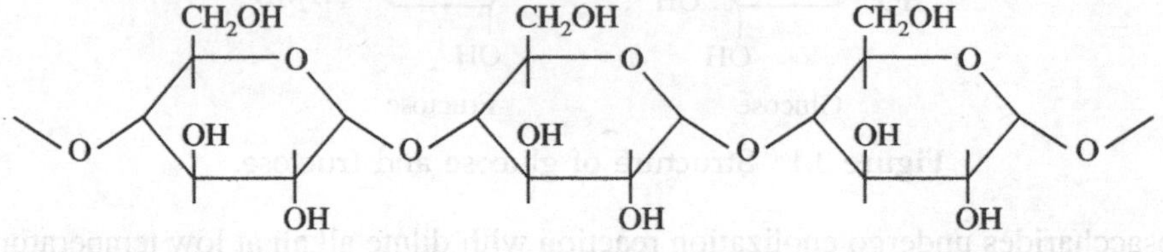

CH_2OH CH_2OH CH_2OH
O O O
O OH O OH O OH O
OH OH OH

Starch

CH_2OH OH CH_2OH
O O O O O
O OH OH OH
O
OH CH_2OH OH

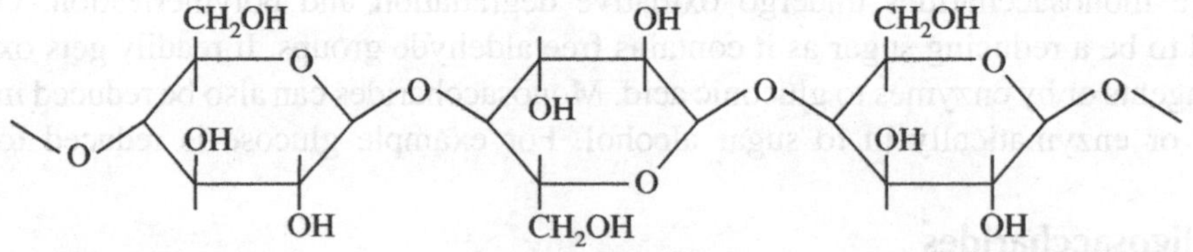

Cellulose

Figure 3.3 Structure of common polysaccharides: starch and cellulose.

Starch is the principal storage compound of plants. Plant starch is known as amylose with molecular weight ranging between a few thousands and 5 lakh daltons. Amylopectin is branched form of starch. The most abundant polysaccharide on earth is cellulose. Most of the carbohydrates in vegetation are of polysaccharide types, mainly constituting plant cell walls. It is homo polysaccharide of D-glucose, with straight chain and chemically more stable than starch. The bonding between the glucose monomers in cellulose is β-1, 4 glycosidic bonds. This bond is stable towards hydrolysis. Thus, effective breakdown of cellulose by hydrolysis into simpler sugars could open up new avenues for green chemical routes of synthesis of fuels or useful chemicals.

Pectin, hemicellulose and peptidoglycan are hetero polysaccharides. Agar (extracted from algae) and isabgul are mucopolysaccharides as they are sticky and gooey. These are also known as mucilages which contain various amine derivatives of polysaccharides, uronic acids, etc. Mucopolysaccharides are very common in various aquatic plants, exterior of bacterial cells, blue-green algae and so on. Their molecular weight ranges up to 5×10^6 Da. They produce cementing body fluid inside the tissues. Polysaccharides show limited solubility in water, however these can be dispersed with heating up to 45°C. In starch both amylose and amylopectin are present in roughly 1:3 ratios. Amylopectin is water insoluble while amylose is soluble.

3.2 TESTS FOR CARBOHYDRATES

Sugars can be classified into two types: reducing and non-reducing sugars. The reducing sugars are those, which have free aldehyde or ketone group and, which have the capacity to reduce certain compounds like potassium ferrocyanide. Examples of reducing sugars are glucose, fructose and maltose. Non-reducing sugars can also be estimated by this method by first hydrolyzing them in to reducing sugar.

Glucose can be estimated using standardized Fehling's solution by a titrimetric method. Grape juice/fruit juice contains glucose and fructose. Their presence can be tested by Fehling's test. Take 5 mL of fruit juice in a test tube. Add an equal quantity of Fehling's solution I and II and boil. A brick red precipitate of cuprous oxide indicates the presence of glucose or fructose in fruit juice. ($Cu^{2+}{}_{(blue)} \rightarrow Cu^{+}{}_{(red)}$). However, sucrose gives negative with Fehling's test. It is first hydrolyzed by con. H_2SO_4. Boil 5 mL of sucrose solution with a few drops of con. H_2SO_4. It is then neutralized with NaOH. Boil again and then add Fehling's solution I and II drop wise. A brick red precipitate is formed due to hydrolysis of sucrose into glucose and fructose.

Glucose in very small concentration can be determined using DNS (dinitrosalicylic acid) reagent spectrophotometrically at a wavelength of 540 nm. Glucose can also be determined by enzymatic method using glucose oxidase enzyme, extracted from the growth medium of *Aspergillus niger*. Glucose is enzymatically oxidized in presence of oxygen into gluconolactone with oxygen being reduced to hydrogen peroxide. Slow hydrolysis of gluconolactone produces gluconic acid. While enzymatic dissociation of hydrogen peroxide in presence of peroxidase forms oxygen and water. This oxygen reacts with orthotoluidine to produce a coloured compound, which can be estimated spectrophotometrically. Fructose gives a red colour with Seliwanoff's reagent (resorcinol + conc. HCl) while glucose gives no colouration.

3.3 PROTEINS

Proteins are considered as chemicals of life. Typically, protein constitutes 20–60% of the microbial cells. Carbon, hydrogen, oxygen and nitrogen are the most prevalent elements present in proteins. Besides, it may also contain di-sulphide bonds in its structure. The building block of protein is amino acid. Simplest amino acid is formed by the substitution of two hydrogens of methane with amino and carboxylic acid functional groups. Alpha carbon of amino acid is asymmetrical. Most prevalent form of amino acid is L amino acid while D-amino acid is rare in nature. Peptide linkage, formed by linear long chain of many amino acids, is the backbone of protein.

Amino acids can exist as acidic, basic and even neutral on the basis of the pH of the solution. The carboxylic acid group (–COOH) is neutral at low pH while at high pH it ionizes as $-COO^-$. There is a certain pH value at which amino acids do not show any net charge. This is called isoelectric point which is a function of the nature of alkyl group present in amino acid. Amino acids are considered as zwitter ions, i.e. they carry both positively and negatively charged terminals. They carry both acidic (COOH) and basic (NH_2) groups.

3.3.1 Types of Amino Acids

There are twenty coded amino acids. These are the alphabets of protein language. However, there are several other amino acids but they are never a part of proteins because their codes are not available with the DNA. They exist only as monomers.

Amino acids can be classified based on *R*-groups. These are:

(i) Amino acids with aliphatic and non-polar *R* groups. They are neutral molecules: Examples: Glycine, Alanine
(ii) Amino acids with polar but uncharged *R* groups. Examples: Serine, Cysteine
(iii) Amino acids with charged *R* groups (this can be again acidic and basic amino acids). Examples: Glutamic acid, Histidine
(iv) Aromatic amino acids. Example: Tyrosine, Tryptophan

In addition, there are essential amino acids and non-coded amino acids. Functions of some amino acids are given in Table 3.1.

Table 3.1 Functions of some important amino acids

Amino acids	Major Functions
Glycine	Synthesis of heme.
Tysosine	Synthesis of thyroxine, adrenaline, nor-adrenaline, melanin
Tryptophan	Precursor for nicotinamide, melatonin synthesis
Histidine	Precursor for histamine synthesis
Alanine	Precursor for pantothenic acid and CoA synthesis.

3.3.2 Types of Proteins

According to chemical composition proteins are of two types, simple protein and conjugate proteins. Simple proteins consist only of amino acids. Albumin, globulin, histone are the examples of simple protein. Conjugate proteins contain other organic or inorganic non-protein parts, called prosthetic group besides amino acids. Conjugate proteins are of following types:

(i) Nucleoproteins (prosthetic group–nucleic acid), e.g. Protamines
(ii) Metalloproteins (prosthetic group–metals), e.g. Ferritin
(iii) Chromoproteins (prosthetic group–pigment), e.g. Cytochromes
(iv) Prosphoproteins (prosthetic group–phosphoric acid), e.g. Casein of milk.
(v) Lipoproteins (prosthetic group-lipids), e.g. Chylomicron
(vi) Glycoprotein (prosthetic group-carbohydrates), e.g. Mucin

Structurally proteins are also typed as globular proteins and fibrous proteins. In globular proteins polypeptide chain is firmly puckered into compact spheroidal or globular shapes. They are usually soluble in aqueous medium. Enzymes, antibodies are the examples of such type of proteins. In fibrous proteins, the polypeptide chains are extended along one axis to make a rope-like structure. These are insoluble in water. They serve mostly a structural or protective role. For example, α-keratin in hair, fibroin in silk, etc. On the basis of chemical nature, proteins can be acidic or basic. Acidic proteins are rich in aspartic and glutamic acid. Blood protein is an example of acidic protein. Basic proteins are rich in lysine and arginine. For example, histone is basic in nature. On the basis of nutrition proteins are classified as Class I and Class II proteins. Class-I proteins supply all essential amino acids when heated, whereas class-II proteins do not provide all essential amino acids. Generally animal proteins are class I while plant proteins are class II proteins. Plant proteins do not have lysine and tryptophan and cereals do not have methionine and cystenine. The most abundant protein in biospehere is RuBisCo.

3.3.3 Biological Functions of Proteins

Proteins are responsible for carrying out several important functions in living systems. Some of the important biological functions of proteins are stated below:

(i) *Enzymes*: They are sophisticated biocatalysts which accelerate the rate of biochemical reactions in living system. There are close to 1200 enzymes of which many are industrially important. Examples: Hexokinase, dehydrogenase, glucose oxidase, urease, etc. We shall study enzymes in details later.

(ii) *Storage proteins*: It is stored in cells. Example: casein (milk), ovalbumin (egg), ferritin (spleen).

(iii) *Transport protein*: It takes part in transport phenomena across the cells. Example: Hemoglobin, Myoglobin, serum albumin, GLUT-4 (enables glucose transport into the cells).

(iv) *Contractile protein*: It is located in muscles and cilia thus helping in contraction and expansion of muscular tissues. Example: Myosin, actin, dynein.

(v) *Defense protein*: It protects body from toxicity. Example: antibody, fibrinogen, thrombin.

(vi) *Structural protein*: These form the structural part of the living system. Example: ligament, keratin, fibrin.

(vii) *Hormones*: Hormones are also made of proteins. Example: Insulin, growth hormone, etc.

3.3.4 Hierarchy of Protein Structure

Protein is made up of monomers of amino acid. These amino acids are linked by each other forming a bond between carbons of carbonyl group (>C=O) of one amino acid and nitrogen of amino group of another amino acid, after removal of one molecule of water via condensation. This is called peptide bond. These bonds are covalent, rigid and planar. These bonds bring stability to the protein structure. Sometimes besides amino acid several other compounds such as metal ions, oligosaccharides (chains of maltose and other sugars–these proteins are said to be glycosylated), small functional molecules (e.g. heme group–hemoglobin).

Structurally proteins can be categorized as primary (1°), secondary (2°), tertiary (3°) and quaternary (4°)

(i) *Primary structure*: In primary structure proteins have covalent backbone with linear sequence of amino acid residues. A long chain of polypeptide backbone joins thousands of amino acids monomer units via peptide link. These bonds are planar and trans.

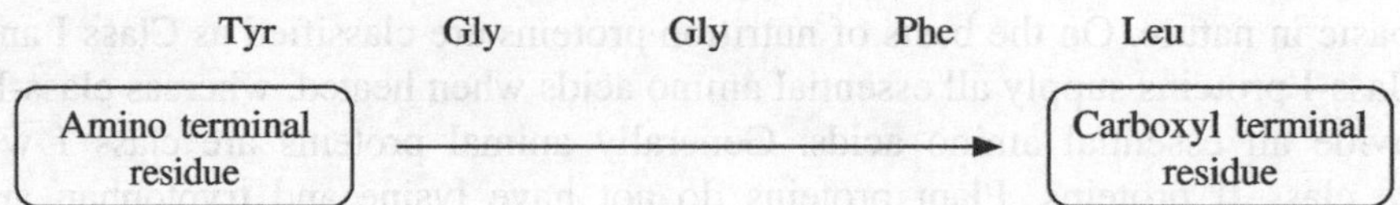

(ii) *Secondary structure*: In the secondary structure polypeptide chains are extended further. This is due to the presence of hydrogen bonds between the residues in close proximity. Two important kinds of secondary structures are helixes and sheets or pleated. Helical structure can be either α-helix or triple-helix (Figure 3.4). In α-helix structure polypeptide backbone is firmly coiled around the longitudinal axis of molecular structure.

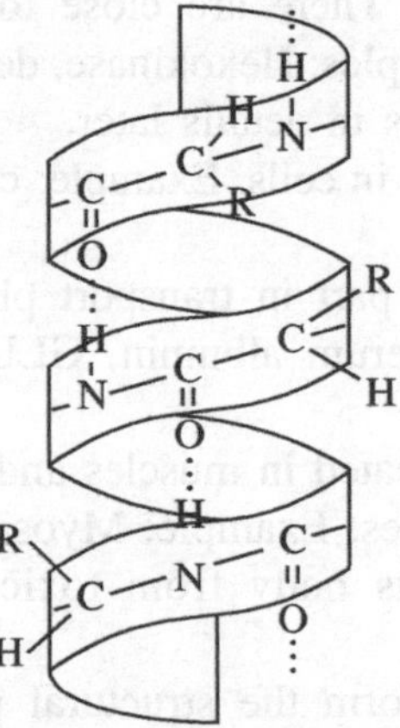

Figure 3.4 Alpha helix structure of protein.

It is a stable and favored conformation of polypeptide chain. This is because it permits a hydrogen bond to form between every H atom attached to the electronegative N atom of the peptide linkage and the electronegative oxygen atom of the carbonyl group of the fourth amino acid behind it in the helix. X-ray studies have supported the fact. Each successive coil of α-helix is held to the adjacent coils by several hydrogen bonds giving the entire structure a considerable stability. In β-sheet structure the backbone of the polypeptide chain is expanded into a zig-zag instead of helical structure. The zig-zag polypeptide is arranged alongside to form β-sheets.

Examples of α-helix: α keratin (present in hair, nails); β-sheet: Fibroin (natural silk), β-keratin (found in scales of reptiles); Triple helix: proline, collagen.

(iii) *Tertiary structure*: In tertiary structure polypeptide chains of globular proteins are tightly folded with little or no interior space for water molecules. Nearly all the hydrophobic R groups are inside the molecules, shielded from water and most of the ionic *R* groups are outside. There are different forces such as hydrogen bonding; ionic interaction between oppositely charged *R* groups, ionic covalent resonance stabilization, hydrophobic interaction stabilizes the tertiary structure. Sometimes disulphide bonds are formed between side chains of amino acids containing sulphur.

Disulphide bonds are considered to act like 'safety pins' to tightly fix polypeptide bonds. Examples of tertiary structure of proteins: cytochrome, myoglobin (Figure 3.5).

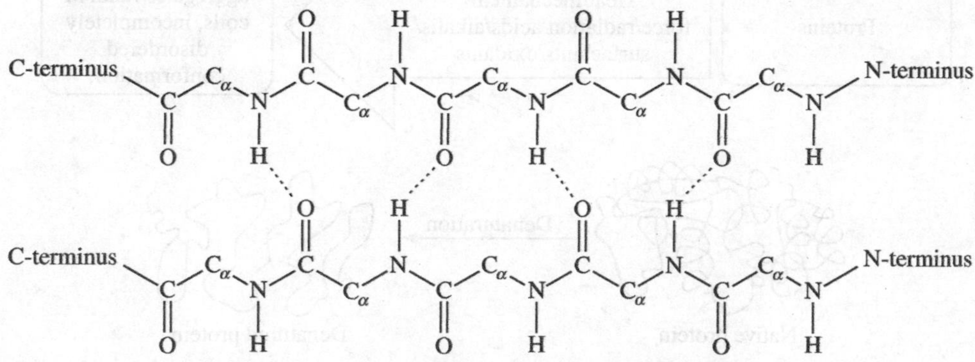

Figure 3.5 Parallel beta pleated sheet structure.

(iv) *Quaternary structure*: Quaternary structures are complex forms of protein configuration. It is the aggregation of more than one subunit of polypeptide chains. These subunits are strongly associated by hydrogen bonds, van der Waals force, hydrophobic interactions or even salt bridges between oppositely charged peptide chains. Example of quaternary structure: Hemoglobin. Primary, secondary tertiary and quaternary structures are given in Figure 3.6.

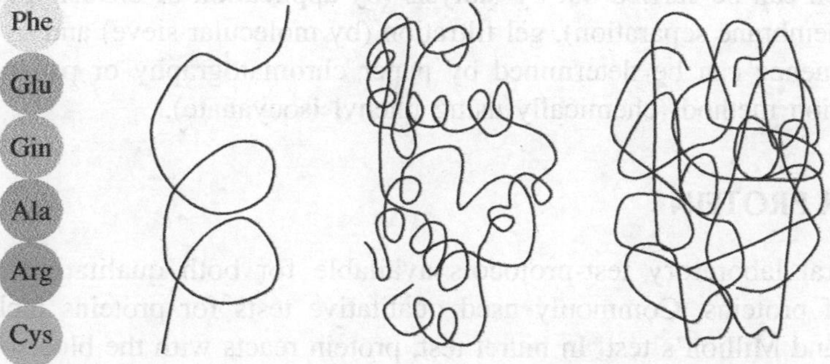

Figure 3.6 Primary (1°), secondary (2°), tertiary (3°) and quaternary (4°) structure of proteins.

3.3.5 Denaturation of Protein

Globular proteins on application of heat or exposure to irradiation or in presence of certain reagents like urea, detergents, strong alkalis, solvents become insoluble and lose their biological activity without damage to the polypeptide backbone. This process called denaturation is due to unfolding of polypeptide chains. Disordering of protein molecules take place leading to malfunctioning or even non-functioning of proteins (Figure 3.7).

Important driving forces of protein denaturation are high temperature, induced thermal diffusion, irreversible covalent modification (disulphide interchange), decreased structure forming and functional interaction, and changes in solvation and void volume shearing.

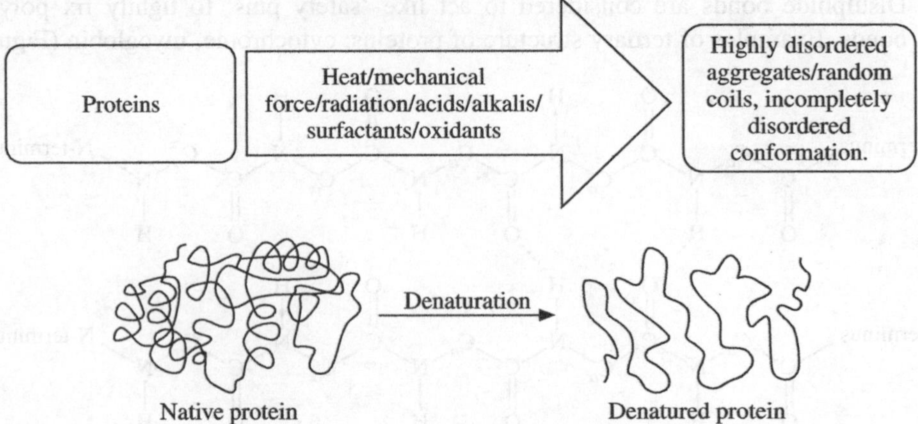

Figure 3.7 Denaturation of proteins.

In some of the cases protein denaturation could be a reversible process. Once the denaturing agent is removed, it regains its original structure. But under extreme physico-chemical conditions denaturation becomes irreversible.

3.3.6 Protein Separation

Protein separation can be carried out by dialysis (by application of diffusion as driving force), ultrafiltration (membrane separation), gel filtration (by molecular sieve) and by electrophoresis. Amino acid sequence can be determined by paper chromatography or paper electrophoresis; Edman degradation method (chemically using phenyl isocyanate).

3.4 TEST FOR PROTEIN

There are several laboratory test-protocols available for both qualitative and quantitative determination of proteins. Commonly used qualitative tests for proteins include Biuret test, Ninhydrin test, and Million's test. In biuret test, protein reacts with the blue solution of $CuSO_4$ (1%) and turns it to deep purple colour. If the protein is not present, no colour change is observed. In this test protein undergoes a complexation reaction with Cu^{2+}.

In ninhydrin test 2-2 dihydroxy indane 1, 3 dione or nynhydrin (a mixture of 0.35 g ninhydrin in 100 mL ethyl alcohol) is added to the sample solution in a test tube. A deep blue colour formation is indicative of the presence of proteins or secondary amino acids or even NH_3. This test is used for fingerprint detection in forensic investigations. In Million's test the presence of tyrosine is detected. Compounds containing phenolic OH groups react with Million's reagent (acidified $HgSO_4$) to form yellow precipitates of mercury-amino complex.

Protein can be estimated quantitatively by spectrophotometric method proposed by Lowry using Folin-Ciocalteu reagent. This reagent is a mixture of phosphotungstic acid and phosphomolybdic acid.

3.5 LIPIDS

Lipids are hydrophobic macrobiomolecule that are insoluble in water but soluble in non-aqueous solvents like n-hexane, alcohol, chloroform, benzene, etc. They are waxy or oily in nature and contain large proportion of C–H bonds. They release a lot of energy on break down and hence are considered as the energy storage media. Lipids undergo saponification and produce glycerol and fatty acid. They are synthesized by the cells from the sugars.

The functions of lipids are many. They serve as biological fuel molecule, act as structural components of cell membranes, serve as cellular metabolic regulator and also protect internal organs by serving as insulator. Some typical examples of lipids are waxes, fats, oils, a few vitamins (like vitamins A, D E and K) which are fat soluble. A part of cell membrane is also made of lipid. Lipids are of three principal types, namely, simple lipids, conjugate lipids and derived lipids. Lipids can be divided into three types, namely simple lipids, conjugate lipids and derived lipids (Figure 3.8).

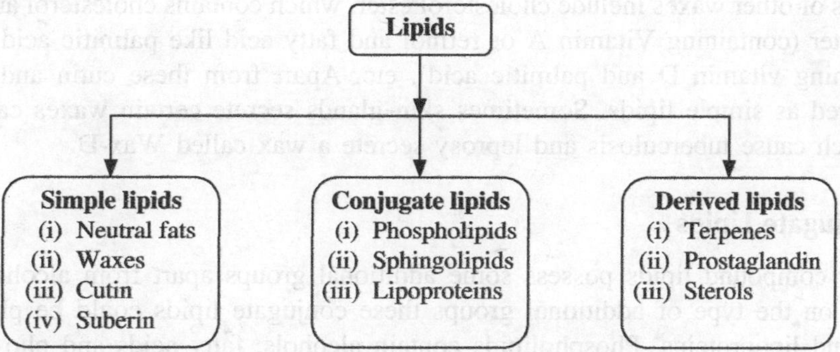

Figure 3.8 General classification of lipids.

3.5.1 Simple Lipids

The name itself indicates that they are having very simple composition. They contain alcohol and fatty acids. Depending on the type of alcohols present in simple lipids they are again sub-classified into fats or oils, and waxes. Fat differs from oil in the fact that the former is solid at room temperature while the latter is liquid. Major components of most simple lipids are fatty acids with general formula CH_3–$(CH_2)_n$–COOH, where the values of 'n' varies between 12 and 20 (even numbers) in biological systems. One common example of fat is triacyl glycerol. Three fatty acids are linked to glycerol by ester linkage. Examples of oils which contain triacyl glycerol are vegetable oils such as groundnut oils, mustard oils, sunflower oils, etc. Palmitic, stearic and lauric acids are the examples of saturated fatty acids while oleic acid is the example of unsaturated fatty acid. There are three essential fatty acids. These are linoleic acid, linolenic acid and arachidonic acid. The deficiency of these fatty acids in diet leads to phrynoderma. Fats are esters formed by the condensation of fatty acids with glycerol (Figure 3.9).

Waxes (for example, bee wax) are esters formed from one fatty acid and one long chain monohydric alcohol. The most common components are straight chain alkanes and 25 to 35

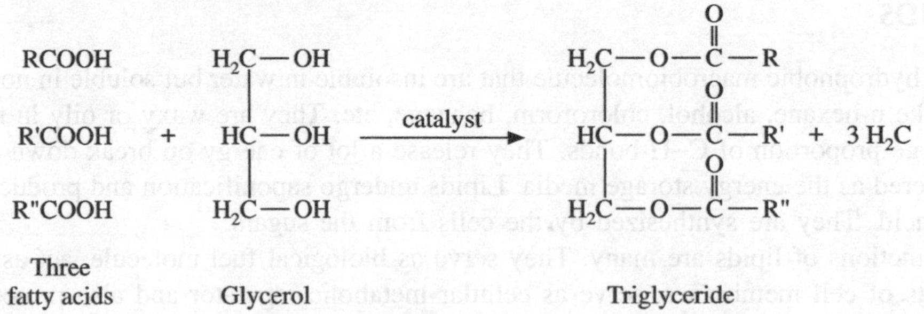

$$
\begin{array}{lclcl}
RCOOH & & H_2C-OH & & H_2C-O-\overset{\displaystyle O}{\overset{\|}{C}}-R \\
R'COOH & + & HC-OH & \xrightarrow{\text{catalyst}} & HC-O-\overset{\displaystyle O}{\overset{\|}{C}}-R' \;+\; 3\,H_2C \\
R''COOH & & H_2C-OH & & H_2C-O-\overset{\displaystyle O}{\overset{\|}{C}}-R'' \\
\text{Three fatty acids} & & \text{Glycerol} & & \text{Triglyceride}
\end{array}
$$

Figure 3.9 Formation of triglycerides by the reaction between fatty acids and glycerol.

carbon of alcohol. Waxes are divided into two groups—true waxes and other waxes. True waxes contain monohydric long chain alcohols with high molecular weight and not glycerols. Examples of true waxes are bees wax, spermaceti oil, etc.

Examples of other waxes include cholesterol ester, which contains cholesterol and fatty acid, vitamin A ester (containing Vitamin A or retinol and fatty acid like palmitic acid), vitamin D ester (containing vitamin D and palmitic acid), etc. Apart from these cutin and suberin are also considered as simple lipids. Sometimes skin glands secrete certain waxes called lanolin. Bacteria which cause tuberculosis and leprosy secrete a wax called Wax-D.

3.5.2 Conjugate Lipids

Conjugate or compound lipids possess some additional groups apart from alcohols and fatty acids. Based on the type of additional groups these conjugate lipids could be phospholipids, glycolipids and lipoproteins. Phospholipids contain alcohols, fatty acids and phosphoric acid. These are conjugate lipids consisting of glycerol connected to a phosphate group. Bilayer structure of phospholipid is schematically presented in Figure 3.11 They are generally white crystalline solids and characterized by presence of nitrogen base such as choline and ethanol amine. Based on the type of alcohol they are further grouped into glycerophospholipids and sphingophospholipids. If alcohol present in phospholipid is glycerol it is called glycerophospholipid and if alcohol is sphingosin, it is called sphingophospholipid. Sphingocin is an amino alcohol and the combination of sphingocin with any fatty acids is called ceramin. Glycolipids containing sphingosine as the alcohol, fatty acids and one or more monosaccharide are also known as glycol-sphingolipids. Micelles formation in lipid molecules is schematically represented in Figure 3.10.

Steroids are also considered as a type of lipids. Naturally occurring steroids are hormones that are important regulators of animal development. Cholesterol is a common steroid present in membrane of animal tissues. It provides stability to cell membrane.

Sphingolipids are similar to glycerophospholipids and they contain sphingosine (an 18-carbon amino alcohol with an unsaturated hydrocarbon chain) rather than glycerol, a fatty acid, phosphate and an amino alcohol.

Lipoproteins are conjugate lipids consisting of lipids and proteins. Lipoproteins are the basic framework of all types of membranes in cell. Many enzymes, structural proteins, transporters, adhesins, toxins are composed of lipoproteins. Examples of lipoproteins are high density

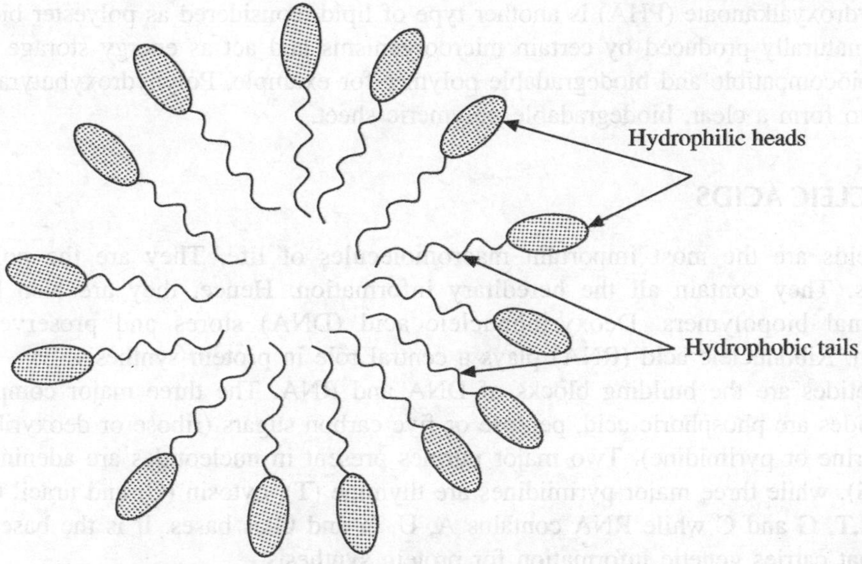

Figure 3.10 Micelles formation of lipid molecules.

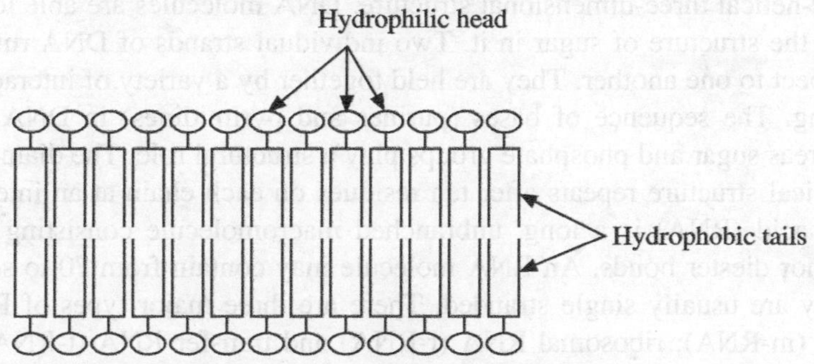

Figure 3.11 Phospholipid bilayers.

lipoproteins (HDL), low density lipoproteins (LDL) and very low-density lipoproteins (VLDL). HDL is considered as good cholesterol while LDL is considered as bad cholesterol.

3.5.3 Derived Lipids

Derived lipids are formed by fused hydrocarbon rings and a long hydrocarbon side chain. Important derived lipids include terpenes, prostaglandins and sterols. Terpenes are strongly odorous natural products obtained from either plants or some insects. It is constituted from isoprene. Prostaglandin is synthesized enzymatically from a reaction between arachidonic acid and a 20-carbon polyunsaturated fatty acid. It is found in almost every tissue in humans and other animals. Sterols, also known as steroid alcohols, are an important type of naturally lipids in plants and animals. For example, cholesterol is animal sterol (also called zoosterol) which provides stability to cell membranes. Plant sterols are called phytosterols, for example, diosgenin (similar to oestrogen, used in birth control pills) and digitalis (obtained from *Digitalis purpurea*, and is used as a cardio tonic).

Polyhydroxyalkanoate (PHA) is another type of lipid, considered as polyester biopolymer. These are naturally produced by certain microorganisms and act as energy storage materials. They are biocompatible and biodegradable polymer for example, Polyhydroxybutyrate (PHB). It is used to form a clear, biodegradable polymeric sheet.

3.6 NUCLEIC ACIDS

Nucleic acids are the most important macromolecules of life. They are the polymers of nucleotides. They contain all the hereditary information. Hence, they are also known as informational biopolymers. Deoxyribonucleic acid (DNA) stores and preserves genetic information. Ribonucleic acid (RNA) plays a central role in protein synthesis.

Nucleotides are the building blocks of DNA and RNA. The three major components in all nucleotides are phosphoric acid, pentose or five carbon sugars (ribose or deoxyribose), and a base (purine or pyrimidine). Two major purines present in nucleotides are adenine (A) and guanine (G), while three major pyrimidines are thymine (T), cytosin (C) and uracil (U). DNA contains A,T, G and C while RNA contains A, U, G and C as bases. It is the base sequence in DNA that carries genetic information for protein synthesis.

DNA is a very large thread-like macromolecule (Molecular weight: 2×10^9 D in *E. coli*) and has a double-helical three-dimensional structure. DNA molecules are able to form a double helix because of the structure of sugar in it. Two individual strands of DNA run in antiparallel fashion with respect to one another. They are held together by a variety of interactions including hydrogen bonding. The sequence of bases (purines and pyrimidines) in DNA carries genetic information whereas sugar and phosphate groups play a structural role. The diameter of the helix is 2 nm. The helical structure repeats after ten residues on each chain at an interval of 3.4 nm.

Ribonucleic acid (RNA) is a long, unbranched macromolecule consisting of nucleotides, joined by phosphor diester bonds. An RNA molecule may contain from 70 to several thousand nucleotides. They are usually single stranded. There are three major types of RNA. These are messenger RNA (m-RNA); ribosomal RNA (r-RNA) and transfer RNA (t-RNA). In RNA and DNA phosphate groups are the same, but they have slightly different sugar. In RNA the sugar is ribose and in DNA the sugar is deoxyribose with one less oxygen, hence the names are derived. Differences between DNA and RNA are given in Table 3.2.

Table 3.2 Differences between DNA and RNA

DNA	RNA
Located inside the nucleus	Mostly located outside the nucleus
DNA is double stranded nucleic acid	RNA is single stranded nucleic acid
DNA stores genetic information	RNA transfer genetic information from one generation to the next.
Contains A, G, C T bases	Contains A, G C U bases
Antiparallel helix structure	Hairpins and loops structure
Long life time	Short life time
Contains deoxyribose sugar	Contains Ribose sugar

KEY POINTS TO REMEMBER

- Important biomolecules include carbohydrate, protein, fats or lipids, nucleic acids and vitamins.
- Carbohydrates (general formula $C_nH_{2n}O_n$) are the storage and structural compound in cell.
- Carbohydrates are of three types namely monosaccharides (examples glucose, fructose, etc.), disaccharides (examples sucrose, lactose, etc.) and polysaccharides (examples starch, cellulose, etc.)
- The building block of protein is amino acid. Simplest amino acid is formed by the substitution of two hydrogens of methane with amino and carboxylic acid functional groups.
- Proteins perform a number of biological functions in a living system such as transport, contraction, defense, catalysis, role as hormone and so on.
- Amino acids in protein are linked by peptide bond formed between carbons of carbonyl group (>C=O) of one amino acid and nitrogen of amino group of another amino acid, after removal of one molecule of water via condensation.
- Structurally proteins can be categorized as primary (1°), secondary (2°), tertiary (3°) and quaternary (4°).
- In primary structure proteins have covalent backbone with linear sequence of amino acid residues.
- In the secondary structure polypeptide chains are extended further due hydrogen bonds. Two important kinds of secondary structures are helixes and sheets or pleated.
- In tertiary structure polypeptide chains of globular proteins are tightly folded with little or no interior space for water molecules.
- Protein denaturation occurs due to unfolding of polypeptide chains in presence of heat, irradiation or certain chemicals.
- Lipids are hydrophobic macromolecules that are insoluble in water but soluble in non-aqueous solvents.
- Lipids are of three types, namely simple lipids, conjugate lipids and derived lipids.
- Simple lipids are again sub-classified into fats or oils, and waxes.
- Major components of most simple lipids are fatty acids with general formula $CH_3-(CH_2)_n-COOH$
- Waxes (for example, bee wax) are esters formed from one fatty acid and one long chain monohydric alcohol.
- Conjugate or compound lipids possess some additional groups apart from alcohols and fatty acids. These could be phospholipids, glycolipids and lipoproteins.
- Derived lipids are formed by fused hydrocarbon rings and a long hydrocarbon side chain. Examples are terpenes, prostaglandins and sterols.
- Deoxyribonucleic acid (DNA) stores and preserves genetic information. Ribonucleic acid (RNA) plays a central role in protein synthesis.
- The three major components in all nucleotides are phosphoric acid, pentose or five carbon sugars (ribose or deoxyribose), and a base (purine or pyrimidine).

➢ Two major purines present in nucleotides are adenine (A) and guanine (G), while three major pyrimidines are thymine (T), cytosin (C) and uracil (U).

? QUESTIONS FOR PRACTICE

Multiple Choice Questions

1. Which of the following monosaccharide is a pentose sugar?
(a) Glucose (b) Galactose
(c) Xylose (d) Mannose
2. What is the number of isomers glucose has?
(a) 2 (b) 4
(c) 8 (d) 16
3. Which of the following is a triose sugar?
(a) Ribose (b) Erythrose
(c) Fructose (d) Glycerose
4. Identify the sugar-alcohol among the following compounds?
(a) Mannitol (b) Trehalose
(c) Xylulose (d) Arabinose
5. Identify the reducing sugar from the following list?
(a) Glucose (b) Sucrose
(c) Maltose (d) Lactose
6. Which of the following is a disaccharide of two units of glucose?
(a) Maltose (b) Galactose
(c) Sucrose (d) Trehalose
7. Which is most abundantly available in nature?
(a) Starch (b) Glycogen
(c) Cellulose (d) Chitin
8. In carbohydrate the ratio of hydrogen and oxygen is
(a) 2:1 (b) 3:1
(c) 2:2 (d) 1:1
9. The sugar found in RNA is
(a) Ribose (b) Deoxyribose
(c) Ribulose (d) Erythrose
10. Amylose is a constituent of
(a) Starch (b) Cellulose
(c) Glycogen (d) None of these
11. Iodine solution produces no colour with
(a) Cellulose (b) Starch
(c) Dextrin (d) Glycogen

12. Which of the following statements about amino acid is correct?
(a) All amino acids contain both positive and negative charges
(b) All amino acids contain positively charged side chains
(c) Some amino acids contain only positive charge
(d) All amino acids contain negatively charged side chains

13. An aromatic amino acid is
(a) Lysine (b) Tyrosine
(c) Taurine (d) Arginine

14. The protein present in hair is
(a) Elastin (b) Prolamine
(c) Keratin (d) Gliadin

15. Identify the 'odd one out' from the lists given below (stating reasons)
(i) Enzyme, Coenzyme, Hormone, Hemoglobin
(ii) Slant, Broth, Petri dish, Stab
(iii) Bacteria, Virus, Fungi, Algae

16. A process by which glucose is synthesized from fats is called
(a) Krebs Cycle (b) Glycolysis
(c) Saponification (d) Gluconeogenesis

17. Name the process in which fats are hydrolyzed into fatty acids and glycerol.
(a) Coagualtion (b) Esterification
(c) Saponification (d) Alkylation

18. Identify the phospholipid
(a) Sterol (b) Steroid
(c) Lecithin (d) Cholesterol

19. Which of the following has the highest content of triglycerides?
(a) LDL (b) VLDL
(c) HDL (d) Chylomicrons

20. Identify the protein which is not fibrous.
(a) Keratin (b) Collagen
(c) Carbonic anhydrase (d) Fibrinogen

21. Select the most suitable set of characteristics for amino acid Glycine.
(a) Hydrophilic, basic and charged
(b) Hydrophilic, acidic and charge
(c) Optically active
(d) Hydrophobic

22. The product of reduction of glucose in presence of $NaBH_4$ is
(a) Xylitol (b) Sorbitol
(c) Mannitol (d) Dulcitol

23. Which of the following is not a hybrid biochemical?
(a) Peptidoglycan (b) Lipopolysaccharide
(c) Isozyme (d) Glycolipid

24. Which of the following is not considered to be an environmental precursor in the construction of cellular organization?
(a) O_2 (b) CO_2
(c) H_2O (d) N_2

25. Which of the following molecules or substances contain, or are derived from, fatty acids?
(a) Glycerolphospholipids (b) Beeswax
(c) Triacylglycerols (d) Sphingolipids
(e) All of the above are derived from fatty acids.

26. When a bacterium such as *E. coli* is shifted from a warm growth temperature to a cooler growth temperature, it compensates by:
(a) Putting longer-chain saturated fatty acids into its membranes.
(b) Putting more unsaturated fatty acids into its membranes.
(c) Increasing its metabolic rate to generate more heat.
(d) Synthesizing thicker membranes to insulate the cell.

27. When the linear form of glucose cyclizes, the product is a (*n*):
(a) Glycoside (b) Hemiacetal
(c) Anhydride (d) Lactone
(e) Oligosaccharide.

28. Starch and glycogen are both polymers of:
(a) α-D-glucose (b) β-D-glucose
(c) Glucose-1-phosphate (d) Sucrose
(e) Fructose

Answer Key

1. (c) 2. (d) 3. (d) 4. (a) 5. (a) 6. (a) 7. (c) 8. (a) 9. (a) 10. (a)
11. (a) 12. (a) 13. (b) 14. (c) 15. [(i) Coenzyme (as the others have a protein part); (ii) Broth (as the others are solid media); (iii) Virus (as the others are members of protist kingdom)]
16. (d) 17. (c) 18. (c) 19. (d) 20. (c) 21. (d) 22. (b) 23. (c) 24. (a) 25. (e)
26. (a) Putting longer-chain saturated fatty acids into its membranes. 27. (b) 28. (a)

Match the Following:

1. Match the columns:

Column I	Column II
1. Ovalbumin	(a) Structural protein
2. Miocin	(b) Transport protein
3. Antibody	(c) Contractile protein
4. Keratin	(d) Protective protein
5. Mioglobin	(e) Storage protein

2. Match these molecules to their biological roles.

1. Carbohydrate storage in animal liver	(a) Proteoglycan
2. Extracellular matrix of animal tissues	(b) Starch
3. Blood clotting factor	(c) Chitin
4. Carbohydrate storage in plants	(d) Cellulose
5. Polyanionic acid-sugar that forms core of proteoglycans	(e) Glycogen
6. Structural component of plant cell walls	(f) Hyaluronic acid
7. Disaccharide	(g) Sucrose
8. Exoskeleton of insects	(h) Heparin

Answer Key

(i) 1–e; 2–c; 3–d; 4–a; 5–b.
(ii) 1–e; 2–a; 3–h; 4–b; 5–f; 6–d; 7–g; 8–c.

Descriptive Questions

1. Explain why cellulose, even though abundantly available is difficult to use as raw material for synthesis of useful chemicals.
2. Name the most abundant protein in biosphere.
3. Adduce reasons for the following:
   (i) Starch has limited solubility in water:
   (ii) Glycine is a Zwitter ion
   (iii) Woollen clothes shrink in hot water, while silk clothes do not.
   (iv) Lipids are mostly present in non-aqueous biological phases.
   (v) Phophoglycerides and fats are structurally similar but not in function.
4. Name various forces that stabilize tertiary structure of protein.
5. Is protein denaturation a reversible process? Explain.
6. Name a biodegradable polymer used in packaging.
7. Give two examples of disaccharides mentioning their constituent monosaccharides.
8. Give examples (one each) of transport protein and protective protein.
9. Define residue of a protein molecule.
10. Are proteins and polypeptides same? Explain with reason.
11. Name the most abundant protein in higher animals.
12. Draw the differences between globular and fibrous proteins.
13. What is hybrid biochemical? Cite three examples,
14. A mixture consisting predominantly of glycine shows a pH near neutral, if pH is lowered how it affects the ionization. Write down probable reactions.
15. A membrane of a plant cell contains 45% proteins and the rest is phophoglycerides (both on weight basis). The average molecular weight of lipid is 800 daltons, while that of protein is 50,000 dalton. Estimate the molar ratio of protein to lipid in the membrane.
16. What is denaturation of protein? Name 3 factors responsible for denaturation of protein.
17. What are the two different types of secondary structure of proteins?

18. Write the appropriate answers:

(i) Write the Wave length used for the spectrophotometric determination of glucose by DNS method.

(ii) What is the Indicator used for the iodine value test of oil?

(iii) What is the Indicator used for the determination of CO_2 in water sample?

(iv) Name the experiment where $MnSO_4$ and alkaline iodide-azide solution are used.

(v) Write the name of the cabinet where inoculation of culture is done in laboratory.

(vi) Write the name of the instrument where liquid medium is kept for growth.

(vii) Name the solid media (03) which are used in the laboratory.

4 Metabolic Stoichiometry and Bioenergetics

4.1 STOICHIOMETRY OF MICROBIAL GROWTH

Microorganisms under suitable physico-chemical conditions multiply by cell divisions and yield various useful metabolic products. They can utilize substrates such as carbon, nitrogen, etc. and different other minerals to form cellular materials and other intracellular or extracellular products. There are four major types of useful products from biochemical reaction involving fermentation. These are microbial cells or biomass, large molecules such as enzymes and polysaccharides, products of primary and secondary metabolism, processes which are not directly linked with growth.

4.2 STOICHIOMETRIC EQUATIONS IN BIOPROCESSES

Different metabolic end products are either released into the fermented broth or accumulated inside the cell body. The relevant balance equations can be classified under the following four categories:

1. The main product results as a function of primary energy metabolism: say for example, ethanol production during anaerobic growth of yeast. Here, the process of substrate utilization, cell mass synthesis and product formation are linked in a simple, single chemical reaction.
2. The main product appears indirectly for energy metabolism; for example, citric acid formation during aerobic mold cultivation. Here product formation is not necessarily proportional to substrate utilization or increase in cell mass. For product formation, an independent reaction is required.
3. The product is a secondary metabolite. Example is penicillin production in aerated mold culture. This depends on particular substrates and product is independent of cell growth.
4. *Biotransformation*: The product is obtained from substrate through one or more reactions catalyzed by enzymes in the cell body. For example, steroid hydroxylation

The principle of mass conservation holds good in the complex processes of cellular growth, substrate consumption and product formation. Therefore, the relation between the utilization of substrate (carbon source, nitrogen source, oxygen, minerals, etc.) and production of biomass, biochemicals and carbon dioxide is obtained by performing elemental material balances. C, H, O, N and other elements of biomolecule substrates undergo rearrangement during the course of biochemical reaction but elemental balance between feed or substrate molecules and the products is exactly maintained. A schematic diagram of substrate input and products output in aerobic cellular system is shown in Figure 4.1. Moreover, the quantity of any other metabolic products produced during the reaction becomes proportional to the substrate consumption.

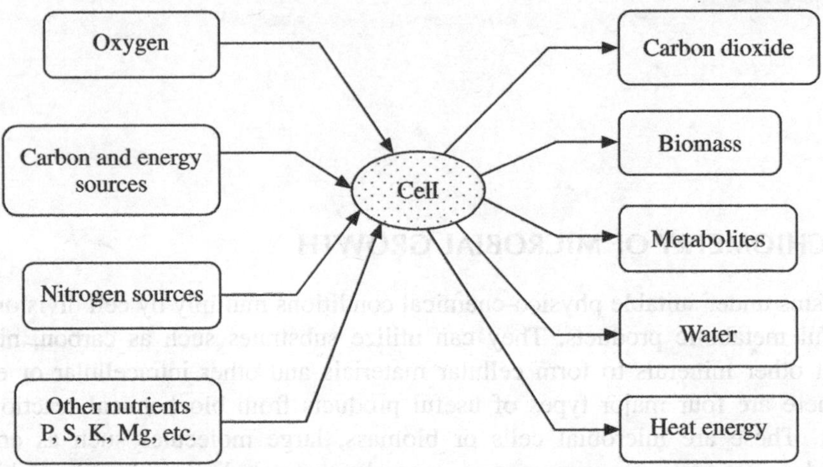

Figure 4.1 A schematic diagram of substrate input and products output in aerobic cellular system.

These material and energy balance considerations have wide implications in biochemical engineering ranging from medium formulation to computer control and cooling requirements in fermenter. Moreover, the knowledge of material and energy balance relationships is quite important in formulating bioreactor material balances and systematic use of reaction kinetics.

Quantitative information on the growth of microorganisms is an important consideration in fermentation, biological waste treatment or in biogeological processes. The growth of microorganisms fundamentally involves the process of catabolism and anabolism, collectively known as metabolism. During the process available C, and N sources, H^+ ions, bicarbonate, water and other ingredients also have important roles.

In general microorganisms have similarity in their compositions with generalized empirical formula of $CH_{1.8}O_{0.5}N_{0.2}$. However, it does not give an exact stoichiometric composition.

Complete knowledge of growth stoichiometry, biomass yield based on substrate consumption, O_2 requirement, CO_2 production and heat production for designing an optimal process for fermentation or any other biochemical reaction are important. The knowledge of material and energy balance relationships is quite important in formulating bioreactor material balances and systematic use of reaction kinetics.

We know that the major elemental demands of a growing cell are carbon, oxygen, hydrogen and nitrogen. Let us assume a general biochemical reaction of the following type where cells are grown on ammonia as the only source of nitrogen and a carbohydrate as sole carbon and energy source:

$$aCH_2O + bNH_4^+ + cO_2 \rightarrow C_\alpha H_\beta O_\gamma N_\delta + dCO_2 + eH_2O$$
(Biomass)

In this equation stoichiometric coefficients a, b, c, d, e, α, β, γ, and δ can be determined by constructing set of algebraic equation and solving:

$$\begin{aligned} &\text{O-balance:} && a + 2c = \gamma + 2d + e \\ &\text{C-balance:} && a = \alpha + d \\ &\text{H-balance:} && 2a + 4b = \beta + 2e \\ &\text{N-balance:} && b = \delta \end{aligned}$$

There are 9 variables and 4 equations. If cell composition is given, i.e. the values of α, β, γ, and δ are known, number of unknown variables would be 5. In that case we need to construct 5 equation or relation to get unique solutions. Here are two such additional empirically-based relationships that can be used for finding out the values of the stoichiometric coefficients.

4.3 RESPIRATORY QUOTIENT (RQ)

In aerobic growth, respiratory quotient (RQ) is defined as ratio of the volume of carbon dioxide evolved to that of oxygen consumed. Following the same, RQ for the above empirical equation will be: $RQ = \frac{e}{c}$

The respiratory quotient is a function of substrates used. Generally for carbohydrates RQ is considered to be 1 due to its complete oxidation according to the following reaction:

$$C_6H_{12}O_6 + 6O_2 \rightarrow 6\ CO_2 + 6\ H_2O + \text{Energy}$$

$$RQ = \frac{6\ CO_2}{6\ O_2} = 1.0$$

When fats are used in respiration, the RQ is less than 1. Calculation for a fatty acid, tripalmitin, if used as substrate is shown:

$$2\ (C_{51}H_{98}O_6) + 145O_2 \rightarrow 102\ CO_2 + 98H_2O + \text{Energy}$$

Tripalmitin

$$RQ = \frac{102\ CO_2}{145\ O_2} = 0.7$$

RQ using protein substrates is approximately 0.9.

4.4 YIELD COEFFICIENT (Y)

Overall growth yield coefficient (or simply yield coefficient) is defined as the ratio of maximum yield of biomass or cell mass to mass of substrate consumed. It is calculated considering complete conversion of substrate without maintenance energy requirement. Yield coefficient can be expressed based on substrate consumption, oxygen consumption or any other specific nutrients. Accordingly the expression for the yield coefficient, based on substrate utilization will be:

$$Y_{X/S} = \frac{\text{M.W of biomass}}{\text{M.W of substrate}}\left(\frac{\text{gdw cells}}{\text{g substrate}}\right)$$

(MW molecular weight; gdw: gram dry weight)

Yield coefficient (on the basis of oxygen consumption)

$$Y_{X/O_2} = \frac{\text{MW of biomass}}{\text{MW of Oxygen}}\left(\frac{\text{gdw cells}}{\text{g } O_2}\right)$$

Yield coefficient (on the basis of ammonia consumption)

$$Y_{X/NH_3} = \frac{\text{MW of biomass}}{\text{MW of ammonia}}\left(\frac{\text{gdw cells}}{\text{g } NH_3}\right)$$

4.5 CONCEPT OF DEGREE OF REDUCTANCE

The degree of reductance of a molecule represents the amount of electrons becoming available from that molecule upon oxidation to the reference compounds such as carbon dioxide, water and ammonia. For organic molecules, the number of available electrons is usually normalized per C atom for inorganic molecules it is per mole. The degree of reductance for organic compounds is denoted as γ and it is defined as the number of equivalent of the available electrons per gram atom carbon. The degree of reductance of any element in a compound is equal to the valence of this element, for instance, the valence of carbon in CH_4 is +4 and the valence of nitrogen in NH_3 is −3.

Let us calculate degree of reductance for glucose, ethyl alcohol and methane.

For glucose ($C_6H_{12}O_6$): 6(4) +12(1) + 6(−2) = 24; $\gamma_{Glucose}$ = 24/6 (since there are 6 carbon atoms) = 4

For ethyl alcohol (C_2H_5OH): 2(4) + 6(1) + (1)(−2) = 12; $\gamma_{\text{ethyl alcohol}}$ = 12/2 (since there are 2 carbon atoms) = 6

Similarly, for methane (CH_4), $\gamma_{methane} = 8$

When degree of reductance is high, there will be a low degree of oxidation, i.e. $\gamma_{CH_4} > \gamma_{C_2H_5OH} > \gamma_{Glucose}$

Typical values of degree of reductance for some common organic compounds are listed in Table 4.1.

Table 4.1 Degree of reductance of some common organic compounds

Compounds	Molecular formula	Degree of reductance (γ)
Glycerol	$C_3H_8O_3$	4.67
Mannitol	$C_6H_{14}O_6$	4.33
Acetic acid	CH_3COOH	4.0
Lactic acid	$C_3H_6O_3$	4.0
Citric acid	$C_6H_8O_7$	3.0
Oxalic acid	$C_2H_2O_4$	1.0
Gluconic acid	$C_6H_{12}O_7$	3.67

4.6 ELEMENTAL MATERIAL BALANCE

EXAMPLE 4.1 Aerobic degradation of 3-hydroxy propionic acid obtained from the effluent of a resin unit is to be carried out in presence of consortia of organisms in wastewater. The generalized reaction can be represented as follows:

$$C_3H_6O_3 + aO_2 + bNH_3 \rightarrow \underset{\text{(biomass)}}{cC_5H_7NO_2} + dH_2O + eCO_2$$

(i) Starting from the construction of elemental material balance, determine *a*, *b*, *c*, *d*, and *e*, if $Y_{X/S} = 0.4$ gX/g S.

(ii) Determine the yield coefficients based on oxygen and ammonia consumption.

(iii) Determine the degree of reductance for the substrate and biomass.

Solution:

(i) By making elemental balance for C, H, O and N we can write:

C balance: $3 = 5c + e$

H balance: $6 + 3b = 7c + 2d$

O balance: $3 + 2a = 2c + d + 2e$

N balance: $b = c$

$Y_{X/S} = 0.4$

$$= \frac{c(5 \times 12 + 7 + 14 + 32)}{90} = 1.25c$$

Therefore, $c = 0.318$

By solving the set of algebraic equations we get:

$$a = 1.4;\ b = c = 0.318;\ d = 2.36,\ e = 1.4$$

(ii) Yield coefficient based on oxygen consumption $Y_{X/O2}$

$$= \frac{c(\text{MW of biomass})}{(\text{MW of substrate } O_2)} = \frac{(113 \times 0.318)}{32} = 1.12 \frac{\text{g cells}}{\text{g } O_2}$$

Yield coefficient based on ammonia consumption Y_{X/NH_3}

$$= \frac{c(\text{MW of biomass})}{(\text{MW of substrate } NH_3)} = \frac{(113 \times 0.318)}{17} = 2.11 \frac{\text{g cells}}{\text{g } NH_3}$$

(iii) Degree of reductance of substrate:

$$C_3H_6O: 3(4) + 6(1) + (-2)\ (1)/3 = 4$$

Degree of reductance of biomass:

$$C_5H_7NO_2: 5(4) + 7(1) + 1(-3)+2\ (-2)/5 = 4$$

EXAMPLE 4.2 A yeast ($CH_{1.66}N_{0.13}O_{0.40}$) is growing aerobically on arabinose ($C_5H_{10}O_5$) and ammonium hydroxide (NH_4OH) with a respiratory quotient of 1.4. Estimate the stoichiometric coefficient of the equation:

$$aC_5H_{10}O_5 + bO_2 + cNH_4OH \rightarrow CH_{1.66}N_{0.13}O_{0.40} + dCO_2 + eH_2O$$

Hints:

C balance: $5a = 1 + d$ (i)

H balance: $10a + 5c = 1.66 + 2e$ (ii)

O balance: $5a + 2b + c = 0.4 + 2d + e$ (iii)

N balance: $c = 0.13$ (iv)

Respiratory quotient (RQ) = $1.4 = d/b$ (v)

Therefore, Eq.(ii) × Eq.(iii) – Eq.(ii), we get,

$$b = 0.2938$$

From Eq.(v), we get,

$$d = 0.4113$$

From Eq.(i), we get,

$$a = 0.2823$$

From Eq.(iii), we get,

$$e = 0.9065.$$

Answer: $a = 0.2823$, $b = 0.2938$, $c = 0.13$, $d = 0.4113$, $e = 0.9065$

EXAMPLE 4.3 Calculate the cell formula for the growth of yeast on sugar when it is observed to give the following balance (assume 1 mole of cell is produced):

$$(100g)\ C_6H_{12}O_6 + (5.1g)\ NH_3 + (46.63g)\ O_2$$
$$\rightarrow (43.23g)\ \text{cell mass} + (41.08g)\ H_2O + (67.42g)\ CO_2$$

If the above reaction produces 0.30017 mole of cell, what will be the cell formula?

Solution: Let us assume the empirical formula of cell mass is $C_mH_pO_nN_g$

100g of substrate (C source) is converted to 43.23 g of cell mass and 67.42 g of CO_2.

From C balance

$$= 3.33 - 1.53 = 1.8$$

From H balance:

$$p = \frac{100\left(\frac{12}{180}\right) + 5.1\left(\frac{3}{17}\right) - 41.08\left(\frac{2}{18}\right)}{1}$$

$$= 6.666 + 0.90 = 3.0$$

From O balance:

$$n = \frac{100\left(\frac{96}{180}\right) + 46.63 - 41.03\left(\frac{16}{18}\right) - 67.42\left(\frac{32}{44}\right)}{16}$$

$$= 3.333 + 2.914 - 2.282 - 3.065 = 0.9$$

From N balance:

$$g = \frac{5.1\left(\frac{14}{17}\right)}{14} = 0.3$$

So, the cell formula becomes

$$C_{1.8}H_{3.0}O_{0.9}N_{0.3}$$

If the reaction produces 0.30017 mole cell mass, the cell mass formula can be obtained by dividing the values computed earlier by 0.30017. Thus,

$m = 1.8/0.30017 = 6$
$p = 3.0/0.30017 = 10$
$n = 0.9/0.30017 = 3$
$g = 0.3/0.30017 = 1$

So, the cell mass formula will be $C_6H_{10}O_3N$.

4.7 ENERGY BALANCE IN FERMENTATION PROCESSES

All fermentation processes need temperature control and this can be achieved by careful addition and removal of heat from the system. For a small bioreactor, control of temperature is easy. But in case of the large bioreactor, heat removal and temperature control pose serious design and operational problems. In the following sections, we will discuss on energy balances and the magnitudes of heat generated in a bioreactor, and the capacity of a bioreactor to remove the additional heat to maintain constant temperature. This understanding is essential for the design of processes and equipment.

4.7.1 Overall Energy Balance

It is known that all biological processes are exothermic in nature and heat is exchanged with the environment causing rise in temperature. To maintain isothermal state in the reactor, heat sources and sinks be dynamically balanced. Mathematical presentation of such balance is given below:

$$mC_p\frac{dT}{dt} = Q_{ox} + Q_{ag} Q_{evp} + Q_{sen} - Q_{ex} \tag{4.1}$$

Here, mC_p = heat capacity of fermentation broth (JK^{-1})
T = temperature of fermentation (°C)
't' = time of operation (h)
Q_{ox} = rate of heat generation by substrate oxidation (Jh^{-1})
Q_{ag} = rate of heat generation by agitation (Jh^{-1})
Q_{evp} = rate of heat loss by evaporation (Jh^{-1})
Q_{sen} = rate of sensible heat gain by equipment or feed (Jh^{-1})
Q_{ex} = rate of heat exchange with the surroundings (Jh^{-1})

Under steady state condition, the left hand side of Eq.(4.1) must be equal to zero in order to maintain a constant fermentation temperature. Q_{ox} is associated with the heat generated due to growth as well as product formation. Q_{ag} is mainly due to mixing by mechanical agitation, which may also be affected by air sparging into the vessel. Q_{evp} and Q_{sen} have normally minor influences, while Q_{ex} includes the influence of surrounding environment. It usually describes the rate of heat gained or lost by temperature control fluids or the air surrounding the fermenter.

4.8 METABOLISM

Metabolism is a sum of complex biochemical reactions in living organisms, which include release of energy by breakdown of complex organic compounds in to simpler forms (known as catabolism) and release energy to form new molecules like synthesis of proteins from amino acids, sugars to polysaccharides (known as anabolism). Both catabolism and anabolism occur in cellular systems concurrently.

Catabolic reactions supply energy required to drive anabolic reactions. This coupling of energy-requiring reactions and electron releasing reactions is made possible through ATP (Adenosine triphosphate) which stores the energy released by catabolic reactions and make it available for anabolic reactions and other cellular works. That is why ATP is known as currency of chemical energy in cells.

Theoretically, the value of standard free energy change ΔG for ATP hydrolysis is 7.3 kcal/mol, however the actual value could be much higher due to higher concentration of ATP than ADP.

$$ATP + H_2O \rightarrow ADP + P_i,\ \Delta G° = 7.3 \text{ kcal/mol}$$

Aerobic catabolism of organic compounds such as glucose consists mainly of four phases:

1. Glycolysis or EMP pathway for fermentation of glucose to pyruvate.
2. Oxidative decarboxylation of pyruvic acid.
3. Krebs, Tricarboxylic acid (TCA) or citric acid cycle for conversion of pyruvate to CO_2 and NADH.
4. Oxidative phosphorylation or terminal oxidation.

4.8.1 Glycolysis

Glycolysis is the process by which glucose or fructose is enzymatically converted into pyruvic acid. It is also known as EMP (Embden-Meyerhof-Paranas) pathway, representing the initials of the three scientists. In eukaryotic cells, cytoplasm is the site where glycolysis occurs, while in prokaryotes membrane-bound enzymes carry out the reactions. Various steps of glycolysis are discussed below (Figure 4.2):

(i) Firstly, one molecule of glucose forms glucose-6-phosphate in presence of hexokinase enzyme.

(ii) This can be isomerized to Fructose-6-phosphate in presence of phosphoglucoisomerase enzyme.

(iii) Fructose-6-phosphate converts into Fructose 1–6 diphosphate by phosphohexokinase enzyme.

(iv) Fructose 1–6 diphosphate breaks down into two smaller compounds—namely one molecule of 3-phosphoglyceraldehyde and another molecule of dihydroxyacetone phosphate. They are inter-convertible in presence of triose-phosphate isomerase enzyme.

(v) 3-Phosphoglyceraldehyde is then oxidized to 1-3 diphosphoglyceric acid, in presence of dehydrogenase enzyme.

$$\text{3-Phosphoglyceraldehyde} + \text{Pi} + \text{NAD} \longrightarrow \text{1, 3 diphosphoglyceric acid} + \text{NADH} + H^+$$

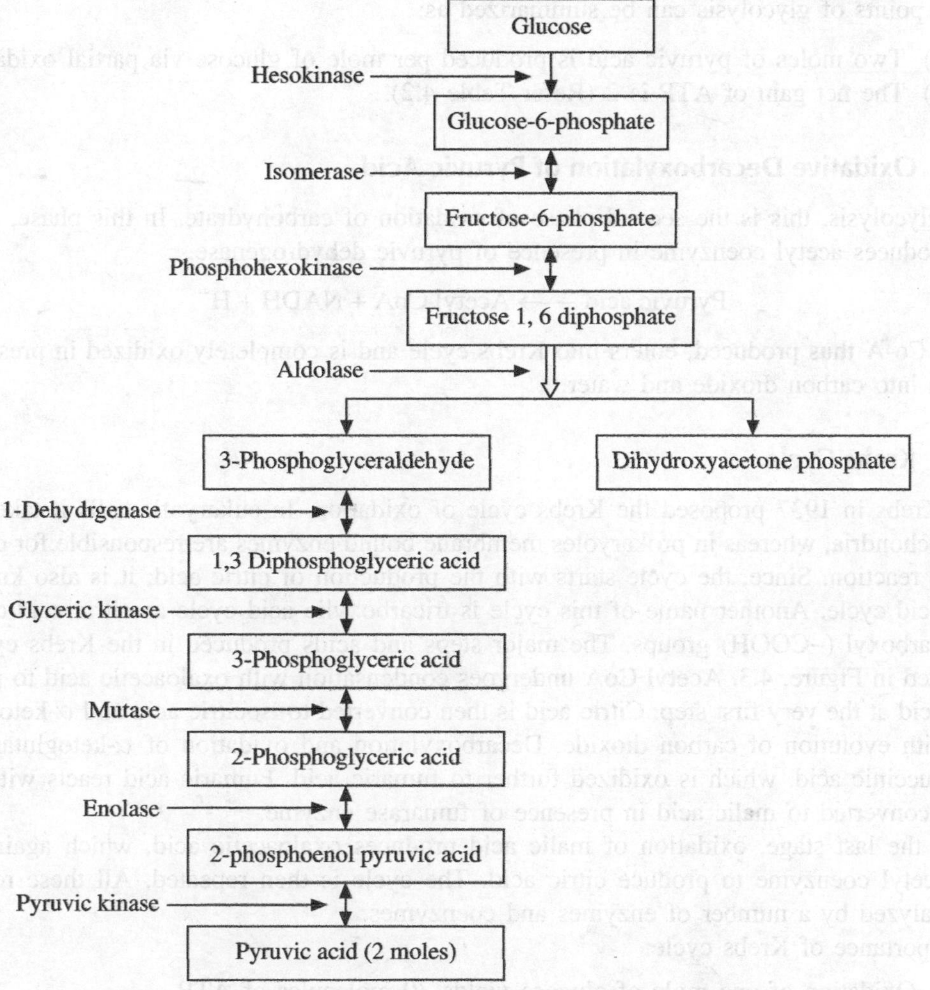

Figure 4.2 Schematic diagram of glycolysis.

(vi) Next 1, 3 diphosphoglyceric acid produces 3-phosphoglyceric acid and ATP in presence of phosphoglycerokinase enzyme.

$$\text{1, 3 diphosphoglyceric acid + ADP} \longrightarrow \text{3-Phosphoglyceric acid + ATP}$$

(vii) 3-Phosphoglyceric acid produces 2-phosphoglyceric acid by enzyme phosphoglyceromutase.

(viii) 2-Phosphoglyceric acid then produces 2-Phosphoenol pyruvic acid, with the help of enolase enzyme:

$$\text{2 Phosphoglyceric acid} \longrightarrow \text{2-Phosphoenol pyruvic acid} + H_2O$$

Finally, at the last stage of glycolysis, 2-phosphoenol pyruvic acid produces pyruvic acid and ATP in presence of enzyme pyruvic kinase. On the whole glycolysis is represented as follows:

$$\text{Glucose + 2ADP + 2NAD}^+ \text{+ 2Pi} \longrightarrow \text{2pyruvate + 2ATP + 2NADH + 2H}^+$$

Salient points of glycolysis can be summarized as:

(i) Two moles of pyruvic acid is produced per mole of glucose via partial oxidation.
(ii) The net gain of ATP is 2 (Refer Table 4.2).

4.8.2 Oxidative Decarboxylation of Pyruvic Acid

After glycolysis, this is the second phase of oxidation of carbohydrate. In this phase, pyruvic acid produces acetyl coenzyme in presence of pyruvic dehydrogenase.

$$\text{Pyruvic acid} \longrightarrow \text{Acetyl CoA} + \text{NADH} + \text{H}^+$$

Acetyl Co-A thus produced, enters into Krebs cycle and is completely oxidized in presence of oxygen into carbon dioxide and water.

4.8.3 Krebs Cycle

Hans Krebs in 1937 proposed the Krebs cycle of oxidation. In eukaryotic cells it takes place in mitochondria, whereas in prokaryotes membrane bound enzymes are responsible for carrying out the reaction. Since, the cycle starts with the production of citric acid; it is also known as citric acid cycle. Another name of this cycle is tricarboxylic acid cycle as citric acid contains three carboxyl (–COOH) groups. The major steps and acids produced in the Krebs cycle are presented in Figure. 4.3. Acetyl CoA undergoes condensation with oxaloacetic acid to produce citric acid at the very first step. Citric acid is then converted to isocitric acid and α-ketoglutaric acid with evolution of carbon dioxide. Decarboxylation and oxidation of α-ketoglutaric acid yield succinic acid, which is oxidized further to fumaric acid. Fumaric acid reacts with water and is converted to malic acid in presence of fumarase enzyme.

At the last stage, oxidation of malic acid produces oxaloacetic acid, which again reacts with acetyl coenzyme to produce citric acid. The cycle is then repeated. All these reactions are catalyzed by a number of enzymes and coenzymes.

Importance of Krebs cycle:

- Oxidation of one mole of glucose yields 30 molecules of ATP.
- Synthesis of important organic acids.
- Helps in nitrogen metabolism and protein synthesis.
- Intermediate compound succinyl CoA is used as the precursor material for chlorophyll, cytochromes, etc.

4.8.4 Oxidative Phosphorylation or Terminal Oxidation

The fourth phase of carbohydrate metabolism is oxidative phosphorylation or terminal oxidation via electron transport chain. It is also called terminal respiration. It occurs in oxysome of mitochondria. Electrons are carried by NADH, H^+ and $FADH_2$ and are transferred to oxygen via a number of electron carriers. As a results, 3 molecules of ATPs are synthesized from each NADH and H^+ and two from each $FADH_2$ in eukaryotes. The reaction in the electron transport chain is given by the following:

$$\tfrac{1}{2}\,O_2 + \text{NADH} + \text{H}^+ \rightarrow H_2O + \text{NAD}^+$$

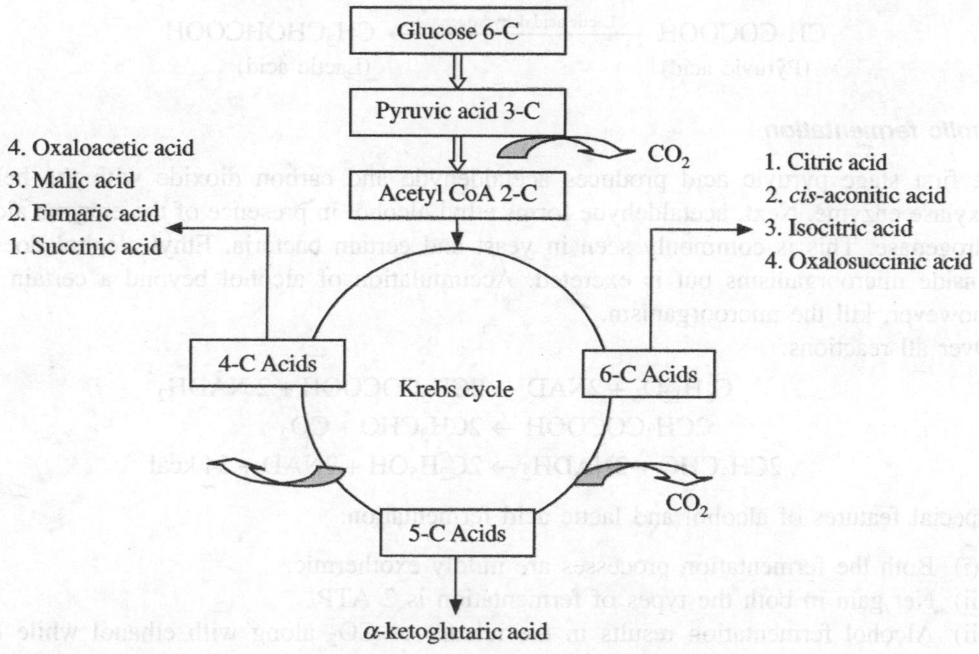

Figure 4.3 Schematic diagram of krebs cycle (TCA cycle).

Summary of ATP generation during aerobic catabolism based on consumption of 1 mole of glucose is presented in Table 4.2.

Table 4.2 ATP generation summary during aerobic catabolism

Process	Number of molecules of ATP generated	Number of molecules of ATP consumed	Net ATP gain
Glycolysis	10	2	8
Oxidative decarboxylation	6	0	6
Krebs Cycle	24	0	24
Total	40	2	38

4.8.5 Fate of Pyruvic Acid

Pyruvic acid is the most important product of glycolysis. It is utilized by cells in three different processes, such as lactic acid fermentation, alcohol fermentation and aerobic respiration. Lactic acid and alcoholic fermentation are briefly outlined below:

Lactic acid fermentation

In the first phase glucose undergoes glycolysis. It is followed by incomplete oxidation of pyruvic acid. In presence of lactic dehydrogenase enzyme, pyruvic acid is converted to lactic acid.

$$\underset{\text{(Pyruvic acid)}}{CH_3COCOOH} \xrightarrow{\text{Lactic acid dehydrogenase}} \underset{\text{(Lactic acid)}}{CH_3CHOHCOOH}$$

Alcoholic fermentation

In the first stage pyruvic acid produces acetaldehyde and carbon dioxide with the help of carboxylase enzyme. Next, acetaldehyde forms ethyl alcohol in presence of the enzyme alcohol dehydrogenase. This is commonly seen in yeast and certain bacteria. Ethyl alcohol does not stay inside microorganisms but is excreted. Accumulation of alcohol beyond a certain limit can, however, kill the microorganism.

Over all reactions:

$$C_6H_{12}O_6 + 2NAD \rightarrow 2CH_3COCOOH + 2\ NADH_2$$
$$2CH_3COCOOH \rightarrow 2CH_3CHO + CO_2$$
$$2CH_3CHO + 2NADH_2 \rightarrow 2C_2H_5OH + 2NAD + 54\ \text{kcal}$$

Special features of alcohol and lactic acid fermentation:

(i) Both the fermentation processes are mildly exothermic.
(ii) Net gain in both the types of fermentation is 2 ATP.
(iii) Alcohol fermentation results in the release of CO_2 along with ethanol while lactic acid fermentation releases lactic acid only.
(iv) The reducing agent NADH and H^+ are re-oxidized to NAD^+ in both the processes.

4.9 METABOLISM OF NITROGEN COMPOUNDS

Nitrogen is the second most important element next to carbon in the living system. It is an essential ingredient of amino acids, enzymes, vitamins, alkaloids and many other important biomolecules. Metabolism of nitrogen involves several reactions such as hydrolysis, deamination, transamination, degradation, etc. Many of these reactions are mediated by enzymes. Proteins get hydrolyzed enzymatically in presence of protease to amino acids which are further converted to organic acids by oxidative or reductive deamination. In transamination, exchange of amino and keto groups of α-keto acids takes place. Final degradation products of protein metabolism are ammonia and carbon dioxide.

4.10 METABOLISM OF HYDROCARBONS

Hydrocarbon metabolism is an important step in bioremediation. Certain bacterial species such as *Pseudomonas*, *Acinetobacter*, etc. certain species of yeast and molds are able to degrade hydrocarbons. These microbes use hydrocarbon as its carbon source for growth. Oxygenation of aliphatic hydrocarbons converts them to alcohols, which on further oxidation in presence of various enzymes form aldehydes and finally to organic acids. Aromatic hydrocarbons are also oxidized by oxygenase enzymes. For example, benzene can be degraded aerobically to acetyl-CoA and succinate via various steps of intermediate formation of cathecol, muconate, etc. In anaerobic metabolism of hydrocarbons C–C bond cleavage occurs with generation of hydrogen and finally methane. However, the conversion steps are more complex.

KEY POINTS TO REMEMBER

- The growth of microorganisms fundamentally involves the process of catabolism and anabolism, collectively known as metabolism.
- The relation between the utilization of substrate (carbon source, nitrogen source, oxygen, minerals, etc.) and production of biomass, biochemicals and carbon dioxide is obtained by performing elemental material balances. C, H, O, N and other elements of biomolecule substrates in a biochemical reaction.
- In aerobic growth, respiratory quotient (RQ) is defined as ratio of the volume of carbon dioxide evolved to that of oxygen consumed.
- Overall growth yield coefficient (or simply yield coefficient) is defined as the ratio of maximum yield of biomass or cell mass to mass of substrate consumed.
- The degree of reductance for organic compounds is denoted as γ and it is defined as the number of equivalent of the available electrons per gram atom carbon.
- Aerobic catabolism of organic compounds such as glucose consists of four phases: Glycolysis or EMP pathway, oxidative decarboxylation of pyruvic acid, Krebs, Tricarboxylic acid (TCA) cycle and terminal oxidation.
- Glycolysis is the process by which glucose or fructose is enzymatically converted into pyruvic acid.
- In oxidative decarboxylation, pyruvic acid produces acetyl coenzyme in presence of pyruvic dehydrogenase.
- In Krebs cycle conversion of pyruvate to CO_2 and NADH takes place.
- Pyruvic acid is utilized by cells in three different processes, such as lactic acid fermentation, alcohol fermentation and aerobic respiration.

QUESTIONS FOR PRACTICE

Multiple Choice Questions

1. Identify the correct statement about glycolysis pathway?
   (a) It can occur both in the presence and absence of oxygen
   (b) It occurs in mitochondria
   (c) It occurs at high ATP concentration
   (d) It occurs in mammalian cells only.
2. In alchoholic fermentation, CO_2 is evolved during
   (a) Decarboxylatin of pyruvic acid (b) Formation of acetaldehyde
   (c) Oxidation of acetaldehyde (d) Both (a) and (b)
3. What is the net gain of ATP during the conversion of glucose to pyruvate?
   (a) 2 ATP (b) 4 ATP
   (c) 6 ATP (d) 1 ATP +1 GTP
4. Which of the following statement about negative free energy change is correct?
   (a) K equilibrium is greater than unity
   (b) Concentration of products become greater than that of reactants

(c) Reaction can occur spontaneously
(d) All are correct

5. RQ (Respiratory Quotient) can be defined as
(a) Volume of carbon dioxide evolved/Volume of oxygen consumed
(b) Volume of carbon dioxide consumed/Volume of oxygen evolved
(c) Volume of oxygen evolved/Volume of carbon dioxide consumed
(d) Volume of oxygen consumed/Volume of carbon dioxide evolved

6. In anaerobic respiration, the Respiratory quotient is
(a) Greater than one (b) One
(c) Zero (d) Infinity

7. Which of the intermediate of the Kreb's cycle is utilised in the formation of amino acids?
(a) Citric acid (b) Malic acid
(c) Isocitric acid (d) α-ketoglutaric acid

8. Acetyl CoA is formed from pyruvate by ________ reaction
(a) Dehydration (b) Reduction
(c) Oxidative decarboxylation (d) Dephosphorylation

9. A single molecule of glucose generates ____________ molecules of acetyl CoA, which enters the Krebs cycle.
(a) 4 (b) 3
(c) 2 (d) 1

10. During the conversion of glucose to pyruvate, two NADH molecules are generated. Which of the following steps generates NADH?
(a) Conversion of fructose-6-phosphate to fructose-1-6-bisphosphate
(b) Conversion of glyceraldehydes-3-phosphate to 1-3-bisphosphoglycerate
(c) Conversion of 3-phosphoglycerate to 2-phosphoglycerate
(d) Conversion of phosphoenolpyruvate to pyruvate

11. What is the degree of reductance of glycerol?
(a) 4.67 (b) 6.13
(c) 3.67 (d) 4.33

12. Which of the following statement about degree of reductance is correct?
(a) A high degree of reductance indicates a high degree of oxidation.
(b) A high degree of reductance indicates a low degree of oxidation.
(c) Degree of reductance has no direct relation with degree of oxidation.
(d) None of the above statements are correct.

Answer Key

1. (a) 2. (d) 3. (a) 4. (d) 5. (a) 6. (d) 7. (d) 8. (c) 9. (c) 10. (b)
11. (a) 12. (b)

Descriptive Questions

1. Calculate the stoichiometric coefficients of the following biological reaction:

$$\text{Hexadecane: } C_{16}H_{34} + a\,O_2 + b\,NH_3 \rightarrow c\,(C_{4.4}H_{7.3}N_{0.86}O_{1.2}) + d\,H_2O + e\,CO_2$$

2. Anaerobic fermentation typically produces a variety of partially oxygenated compounds in addition to cell mass. If the cell molecular formula is given by $C_6H_{10}O_3N$, calculate the unknown coefficients for the following typical equation where coefficients are given as mole quantities and not in mass.

$$5.46\ C_6H_{12}O_6 + NH_3$$
$$\rightarrow \beta\ C_6H_{10}O_3N + \gamma\ C_4H_{10}O + \xi\ (C_4H_6O_4) + 2.20\ C_3H_8O_3 + 3.40\ H_2O + 3.75\ CO_2 + 1.08\ C_2H_5OH$$

3. Arrange the following in the descending order of degree of reductance:

   Oxalic acid, Formic acid, Citric acid, Gluconic acid

4. An empirical formula of a biomass is $CH_\alpha O_\beta N_\gamma$. Calculate the degree of reductance.
5. Define respiratory quotient.
6. Calculate the respiratory quotients for the following reactions
   (i) Aerobic respiration using glucose as carbon source.
   (ii) Anaerobic respiration using glucose as carbon source.
   (iii) Aerobic respiration using malic acid as the carbon source.
7. Name the end products of Glycolysis.
8. Why Krebs cycle is also known as tricarboxylic acid cycle.
9. Name the organic acid, which is formed first during the Krebs cycle.
10. State three importance of Krebs cycle in biochemical processes.
11. What is the net gain of ATP molecules in the following processes?
   (i) Glycolysis
   (ii) Oxidative decarboxylation of pyruvic acid
   (iii) Krebs cycle
12. Write the complete reaction for the hydrolysis of ATP to ADP and inorganic phosphate. Briefly name three chemical differences between reactants and products that explain the high ΔG for the reaction in the cell.
13. The most efficient method of generating ATP for most organisms is aerobic respiration. What molecule is required for aerobic respiration? Some microbes thrive in an anaerobic environment. How do they efficiently generate ATP in these conditions? Pick a specific method/path of respiration such as, fermentation or an alternate pathway of glycolysis.
14. In the absence of oxygen, fermentation must happen; otherwise, glycolysis will stop. This is regardless of whether it is lactic acid fermentation or alcohol fermentation that is used. Explain why.

5 Enzymes and Enzyme Catalysis

Many a time a chemical reaction takes place in presence of a catalyst, where reactants under suitable conditions yield a product. Similarly, in biochemical reaction enzymes are used to act upon a substrate to yield a product. Enzymes are organic catalysts which can speed up biochemical reactions by reducing the energy of activation necessary to carry out reactions and exponentially accelerate the reaction velocity without participating in the reaction. Enzyme can be well considered as nature's catalyst, capable of accelerating a biochemical reaction faster than any man-made catalyst under the mild physico-chemical conditions.

5.1 IDEA OF CATALYSIS BY ENZYMES

The idea of catalysis, a process whereby chemicals in small quantity ensuring facile yield of products without participating in reactions, was propounded by Berzelius in 1836. But the word, 'enzyme' was first coined by a German biologist Kuhne in 1877. The word originates from a Greek word '*enzymos*', meaning leavening of bread. However, much before that enzymes found its early uses in the production of alcohol, cheese, beverage manufacturing and also in meat tenderizing. Earlier it was believed that enzyme is a living entity. But later on during fermentation it was observed that product formation could still continue even if in the absence of any living yeast cells. This was presumably due to the presence of certain enzymes which were produced by microorganisms.

Enzymes are primarily made up of proteins and are present in all living cells. They convert a specific substrate into a specific product just like a catalyst does in chemical reaction. History of enzymes dates back to 1833, with diastase (a mixture of amylases) being the first enzyme to be discovered. Pepsin, invertase and other hydrolytic enzymes were known and reported later.

In absence of enzymes, biological reactions, although being thermodynamically favorable, would require years to yield a product. Scientists have estimated that uncatalyzed cleavage of peptide bonds might take several hundred years at ambient temperature. Same is the case for hydrolysis of phosphomonoester. Had there been no enzymes this would have taken almost a million of years. These examples indicate as to how the action of enzymes could dramatically increase the rate of biochemical reactions. Till date, there have been more than 2000 enzymes of medicinal and industrial uses discovered. It is not uncommon for an enzyme-catalyzed reaction to proceed at a staggering rate is 10^{20} times faster than the uncatalyzed reaction.

5.2 CHEMICAL NATURE OF ENZYME

Enzymes are sophisticated biocatalysts having precise architecture based primarily upon macromolecular proteins and other peripheral components like metal ions, carbohydrates and nucleotides. They are large molecular weight compounds comprised primarily of a series of amino acids joined mutually by peptide bonds. Enzymes undergo denaturation and precipitation in presence of certain reagents like acetone, methanol, tricholoro-acetic acid, etc. Their molecular weight ranges from 10^3 to 2×10^6 Da. Even though enzyme itself remains unchanged in course of its reaction with substrate, the amino acid residues of enzyme may get fragmented or form some temporary bonds with substrate. Additionally these fragmented residues may undergo further internal rearrangement and more enzyme-substrate compounds are formed.

Sometimes, whole cells of intact living microorganisms may be used to provide catalyst materials (enzymes), this is known as '*in vivo*'. If the cells are disrupted to extract enzyme and purify it is called '*in vitro*'. However, there are chances of enzyme loss or deactivation in '*in vitro*' process.

Protein parts of enzymes are called apoenzyme while non-protein parts are called cofactor or coenzyme. Generally a cofactor is a metal ion and coenzyme is an organic molecule.

According to Holum, the cofactor may be:

(i) Coenzymes are loosely attached small organic molecules. They are easily dialyzable, and relatively stable towards thermal degradation. Examples of coenzymes include vitamins or its derivatives, ATP (Adenosine triphosphate), NAD (Nicotinamide adenine dinucleotide), FAD (Flavine adenine dinucleotide), etc. Often these cofactors exist in equilibrium between enzyme, apoenzyme and cofactor.

(ii) Sometimes a stable organic compound is hooked irreversibly with apoenzymes. These are called prosthetic groups. Although prosthetic groups have functional similarity with cofactors, the former is much stable as compared to the latter. They can be released only under extreme conditions of denaturation.

(iii) Generally metal cations such as Fe^{3+}, Cu^{2+}, Co^{2+}, Zn^{2+}, Ca^{2+}, etc.

Examples:		
	Ca^{2+}	α-amylase (swine pancreas)
	Co^{2+}	glucose isomerase
	Na^+	plasma membrane ATPase
	Fe^{2+}/Fe^{3+}	catalase, peroxidase, etc.
	Zn^{2+}	carboxypeptidase

5.3 NAMING AN ENZYME

Naming an enzyme is largely based on what the enzyme does rather than what the enzyme is. Suffix 'ase' is added either to substrate name (urease $\rightarrow$ catalyzes urea decomposition; sucrase $\rightarrow$ catalyzes breaking down of sucrose; lipase $\rightarrow$ fat hydrolyzing enzyme) or to the reaction which is catalyzed (alcohol dehydrogenase $\rightarrow$ oxidative dehydrogenation of alcohol; polymerase

→ aids in polymerization; dehydrogenase → removal of H atoms). The International Union of Biochemistry (IUB) introduced the norms for enzyme nomenclature. This is based on both the nature of reaction and the name of substrate. For example, enzyme responsible for the oxidative dehydrogenation of alcohol is named as alcohol dehydrogenase. On the other hand, enzyme which acts on the substrate urea is called urease or the one acting on protein is called protease and so on. However, there are exceptions to the above convention also, like pepsin, trypsin, ptyalin (human digestive enzymes), renin (used in cheese making), etc. Apart from this enzymes are also named based on its original source. For example, papain is extracted from papaya. The names of some enzymes end with an 'in' denoting that they are inherently proteins. For example, pepsin, trypsin, etc.

Enzymes are heat sensitive. They double their activity for each 10°C rise in temperature; but they very much susceptible to denaturation by heat, wide pH fluctuations and in contact with strong acids or alkalis. They ideally remain intact in the process of product formation and take up many molecules of substrate to convert into products.

5.4 TYPES OF ENZYME

All enzymes are categorized into six distinct types based on the kinds of reactions that they catalyze. These six major groups of enzymes are discussed below:

5.4.1 Oxidoreductase

As the name suggests, these enzymes support oxidation and reduction type of reactions carried out by electron transfer. They also frequently involve one of the coenzymes such as NAD^+, NADP or FAD during reaction. This group can further be subdivided into 4 sub-groups:

(a) *Dehydrogenase*: In this a hydrogen atom is shifted from a substrate to NAD^+ cofactor. Lactate dehydrogenase catalyzes conversion of lactate to pyruvate:

$$\text{Lactate} + NAD^+ \rightarrow \text{Pyruvate} + NADH + H^+$$

(b) *Oxidase*: These types of enzymes catalyze a reaction in which molecular oxygen acts as electron receiver. As a result C–N and C–O bonds are oxidized with the formation of hydrogen peroxide.

$$\beta\text{-D-glucose} + O_2 \rightarrow \text{Gluconolactone} + H_2O_2$$

(c) *Peroxidase*: This type of enzymes uses hydrogen peroxide as electron donor to catalyze substrate oxidation. A model example of peroxidase enzyme is horseradish peroxidase responsible for carrying out oxidation of dyes, amines, hydroquinones, etc., with simultaneous reduction of H_2O_2. Peroxidase catalyzes the oxidation of a number of different reducing substances such as dyes, amines, hydroquinones, etc., and the simultaneous reduction of hydrogen peroxide. A representative reaction of the oxidation of neutral ferrocene to ferricenium in presence of hydrogen peroxide is shown below:

$$2\,[Fe\,(Cp)_2] + H_2O_2 + 2H^+ \rightarrow 2[Fe\,(Cp)_2]^+ + 2H_2O$$

(d) *Oxygenase*: These enzymes carry out substrate oxidation where one 'O' atom of molecular oxygen is accommodated into the substrate while the other is reduced by forming water after combining with hydrogen. For example, lactate-2-monooxygenase brings about the oxidation of lactate into acetate with liberation of CO_2.

$$\text{Lactate} + O_2 \rightarrow \text{Acetate} + CO_2 + H_2O$$

5.4.2 Transferase

This type of enzymes facilitate relocation or shifting of a particular group such as alkyl, acyl, aldehyde, amino, phosphate or glucosyl from one substrate to another. Examples: Transaminase, transmethylases, transaldoses, transketolases, acyltransferases, etc. Alanine aminotransferase, rearranges the alpha amino group between alanine and aspartate.

$$\text{Alanine} + \text{Oxaloacetate} = \text{Pyruvate} + \text{Aspartate}$$

Other types of transferases transfer phosphate groups between ATP and other compounds, sugar residues to form disaccharides and so on.

5.4.3 Hydrolase

These enzymes catalyze hydrolysis via hydrolytic cleavage of bond or the reverse fragment condensations. Examples of this class of enzymes include cholesterol esterage, alkaline phosphatase, cutinase, acylase, lipase, glucoamylase, etc. For example, phosphatases break the oxygen-phosphorus bond of phosphate ester. Some other hydrolases help in protein digestion.

5.4.4 Isomerase

This class of enzymes facilitates isomerization reaction by intramolecular rearrangement. These are further subdivided into racemases, epimerases, mutases and *cis-trans*-isomerases leading to the formation of racemic mixtures, epimers, etc. In many of these reactions substrate and products become isomers to each other. An example of this class of enzyme is glucose isomerase which catalyzes the isomerization of glucose to fructose. Triose phosphate isomerase carries out the rearrangement reaction of converting glyceraldehyde-3-phosphate to dihydroxyacetone phosphate.

5.4.5 Lyases

This class of enzymes eliminates functional groups from their substrates along with the simultaneous formation of double bonds or even are capable of adding new groups across double bonds such as amino groups. For example, decarboxylase removes CO_2 from α or β-keto acids. Dehydratase removes water as in fumarase (fumarate hydratase); deaminases remove ammonia, for example, in the removal of amino groups from amino acids.

$$\text{Serine} \longrightarrow \text{Pyruvate}$$

5.4.6 Ligases

This class of enzymes (also called synthetases) catalyzes joining of two molecules by the formation of new bonds. Synthetase is a subclass of ligase that uses the hydrolysis of ATP to drive this formation. For example, amino-acyl transfer RNA synthetases join amino acids to their respective transfer RNAs in preparation for protein synthesis. Major types of enzymes and their salient functions are presented in Table 5.1.

Table 5.1 Major enzyme types and their functions at a glance

Class	Designation	Function
EC1	Oxidoreductase	Catalyzes oxidation and reduction reaction
EC2	Transferase	Transfers a functional group (e.g. methyl or phosphate group)
EC3	Hydrolase	Catalyzes the hydrolysis of various bonds
EC4	Lyase	Cleaves various bonds by means other than hydrolysis and oxidation
EC5	Isomerase	Catalyzes isomerization within a molecule
EC6	Ligase	Joins two molecules by covalent bonds.

5.5 SALIENT FEATURES OF AN ENZYME CATALYZED REACTION

Enzymes catalyzed reactions are unique on many counts. These are briefly summarized below:

5.5.1 High Catalytic Activity

Enzyme catalyzed reactions are more efficient in terms of high turnover numbers. Turnover number indicates the maximum number of substrate molecules per catalytic active site per unit time for a given enzyme concentration at a given temperature. Enzymes have high turnover numbers at room temperature compared to industrial chemical catalysts.

Catalase with turnover number 40 million is one of the most effective enzymes reported so far. According to Braunschweig Enzyme Database (BRENDA), one molecule of catalase can convert over 2.8 million molecules of hydrogen peroxide to water and oxygen per second. Graying of hair with age is attributed to the fall of catalase activity.

Some other enzymes such as lactate dehydrogenase can have the turnover number as high as 1000, while carbonic anhydrase has the turnover number in the range of 6×10^5. Commonly turnover number of enzymes range between 1 to 10,000. Some turnover numbers (N) for enzyme and solid catalyzed reactions are presented in the Table 5.2.

Table 5.2 Examples of turnover numbers of a few enzymes and solid catalysts

Enzymes	Reaction	Range of reported N values at 0 to 37°C
Ribonuclease	Transfer phosphate of a polynucleotide	$2 - 2 \times 10^3$
Trypsin	Hydrolysis of peptide	$3 \times 10^{-3} - 1 \times 10^2$
Papain	Hydrolysis of peptides	$8 \times 10^{-2} - 1 \times 10^1$
Fumarate hydratase	Interconversion of malate and fumerase	1×10^3

(Contd.)

Solid catalysts	Reaction	N at (°C)
SiO_2–Al_2O_3 (impregnated)	Cumene cracking	3×10^{-8} (25°C)
V_2O_3	Cyclohexane dehydrogenation	7×10^{-11} (25°C)

The definition of turnover number is (moles substrate) (mole enzyme)$^{-1}$s^{-1}. The units of turnover number are s^{-1}.

Estimation of turnover number of enzyme

Turnover number can be estimated for any enzyme catalyzed reaction. For this the reaction should follow zero-order kinetics and occur at maximum velocity *V*. This is true, when concentration of enzyme-substrate complex [*ES*] is maximal, i.e. when all the enzymes in a system are bound by substrate. Under these conditions:

$$V = k_{cat}[ES] = k_{cat}[E] \tag{5.1}$$

Or

$$k_{cat} = \frac{V}{[E]} \tag{5.2}$$

where, by convention, *V* = moles of substrate converted per minute and [*E*] = moles of enzyme taking part in the reaction.

When there is more than one catalytic centre on the enzyme molecule, the equation will be:

$$k_{cat} = \frac{V}{\text{Number of active sites} \times [E]} \tag{5.3}$$

which gives the moles of substrate converted per mol active site of the enzyme. Unfortunately, the number of active sites per molecule of enzyme is known for only a few enzymes, and the more approximate equation $k_{cat} = \frac{V}{[E]}$ must be used. A limitation of the above approach of predicting the turnover number of enzymes lies with the fact that it is difficult to estimate the number of active sites of an enzyme accurately. Using previous equation, turnover number of catalase was estimated to be 5×10^6. But actually it should have been one fourth of 5×10^6 as it has four active sites. Similarly turnover numbers of other enzymes were also not calculated precisely due to the lack of information about the exact number of active sites in those enzyme molecules. Hence turnover numbers can only give an approximate idea about the enormity of enzyme catalyzed reactions.

5.5.2 Specificity and Selectivity

Probably the most distinguishing features of enzymatic reactions are their specificity and selectivity by nature (Figure 5.1). Even sometimes enzymes have absolute discrimination (or specificity) for only one type of molecules, that they were designed to act on, and ignore everything else. Enzyme specificity implies that an enzyme is extremely specific in its choice of a substrate. Some notable types of specificity exhibited by the enzymes include absolute specificity, functional group specificity, stereochemical specificity, product and substrate specificity.

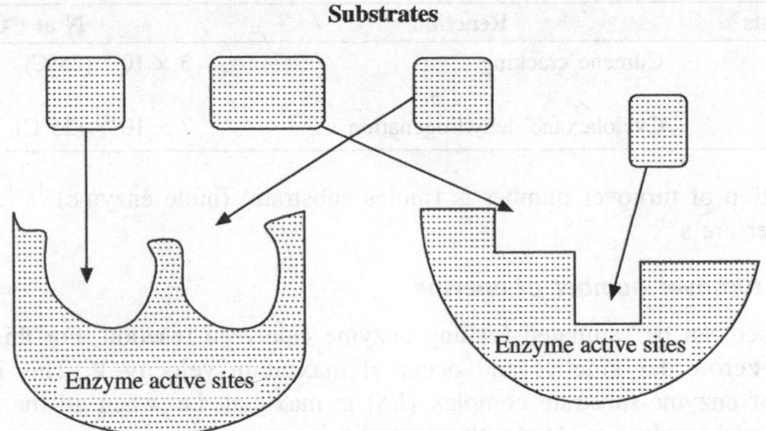

Figure 5.1 Pictorial representation of substrate specificity of enzymes.

Selectivity implies the extent to which a desired substrate is complexed with enzyme and reacts by the action of a particular enzyme. The term selectivity comes from the choice an enzyme makes in the presence of multiple substrates. If a substrate molecule has two stereoisomers such L and D isomers, enzyme can form complex with only one of the two. For example, glucose oxidase will oxidize D (+) glucose only and no other hexose.

Stereo specificity is related to the recognition of a particular optical isomer by the enzyme. On the other hand group specificity is related to the recognition of a particular functional group. Stereochemical specificity of enzymes is a function of the orientation of the active site or active centers of the enzyme and their proximity. Alcohol dehydrogenase which catalyzes the oxidation of a variety of alcohols is an example of group specific enzyme. In this case, the oxidation or dehydrogenation occurs with the alcoholic –OH group irrespective of the types of alcohol. Active sites of enzymes are stabilized by the presence of several molecular forces such as charge transfer, hydrophobic-hydrophilic interaction, and hydrogen bonding.

5.5.3 Versatility

The versatility of enzyme catalyzed reactions is exhibited by the variety of reactions that can be catalyzed. A large number of enzymes are being commercially used in the production of biochemicals and various other value added and innovative products, formulations via organic synthesis route. A few examples of widely used enzymes include amylase, cellulase, protease and so on. Enzyme markets are ever expanding with growing demands.

5.5.4 Controlled Expression

Sometimes enzymes may contain additional substrate binding sites over and above the usual active sites. In these additional sites small molecules known as effector or regulator or modulators can be attached. These enzymes are termed as allosteric enzymes. However, these effector molecules could have either positive or negative effects. Allosteric regulators may bring about conformation changes thereby rendering the enzymes more active or less active. In some cases substrate molecules can act as modulator and in some other cases modulator is different from the substrate. The former types are known as homotropic allosteric enzymes and the latter is called heterotrophic allosteric enzymes.

5.5.5 Stability

Mostly proteins in enzymes have tertiary structure stabilized by weak molecular forces. As a consequence enzymes are less stable biomolecules. Moreover, active sites of enzymes are highly reactive in nature and they can hardly exist free or uncomplexed. This means active sites always tend to form complex with either substrates or even look alike substance (substrate analog) to get stabilized. They also need to get stabilized against different denaturing conditions, namely against their inactivation caused by unfolding of proteins. Immobilization of enzymes is one of the important methods enzyme stabilization. This will be discussed later.

5.6 FACTORS AFFECTING ENZYME ACTIVITY

There are several physico-chemical factors that can influence the catalytic performance of enzymes in biochemical reaction, primarily by affecting the enzyme's structural or chemical state. These are discussed in brief below:

5.6.1 Temperature

Generally increase in temperature complement the rate of chemical reactions. But in case of enzyme catalyzed biochemical reactions rate of reaction increases initially up to certain temperature, attains a maxima but thereafter the rate decreases. Since enzymes are basically protein compounds, unfolding of their globular structure takes place at high temperature. This results in enzyme deactivation. Spatial structure of enzyme has a great effect on the rate of biochemical reactions.

As the enzyme structure becomes distorted, due to protein denaturation at elevated temperature, reaction rate drops. The Figure 5.2 indicates the change of reaction rate with change in temperature. It also suggests that there must be an optimum temperature for each enzyme to function at a maximum possible rate without getting deactivated.

The rising part of the plot (Figure 5.2) is known as temperature activation. In this region rate varies according to the Arrhenious equation

$$k = Ae^{-E_a/RT} \tag{5.4}$$

Where E_a = activation energy
R = gas law constant
A = frequency factor
T = absolute temperature

The falling part of the figure is known as temperature inactivation or thermal denaturation. For many proteins, denaturation begins to occur at 45 to 50°C and is severe at 55°C. With the rise in reaction temperature enzyme molecules tend to acquire more thermal energy and overcome the weak biding force of globular structure. Consequently they experience thermal denaturation and deactivation.

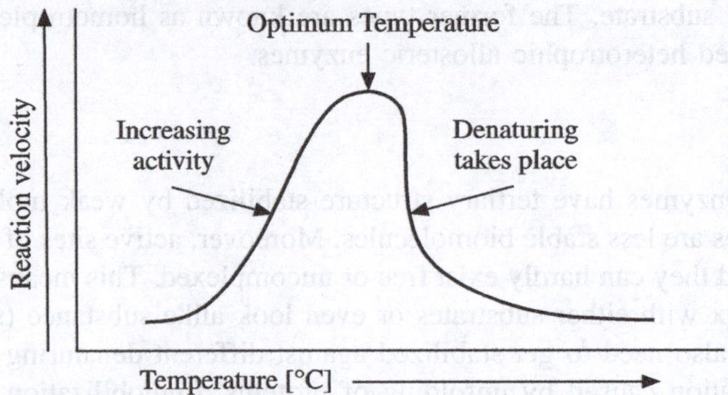

Figure 5.2 Effect of temperature on the rate of enzymatic reactions (with kind permission from Worthington Enzyme Manual; Worthington, K., and Worthington, V (2011). Worthington Biochemical Corporation. (http://www.worthington-biochem.com/pap/default.html).

5.6.2 pH

Since enzyme proteins contain both amino and carboxylic groups (of amino acid) their ionization states change with the change in pH. Acidic amino acids contain carboxylic acid group while basic amino acids contain amino group on their side chains. Alteration of pH may also lead to structural alteration of the enzyme leading to its partial or complete deactivation. This is due to the fact that an incompatible active site proves unsuitable for a substrate to bind. Most of the physico-chemical processes occur in the neutral range of pH and many enzymes get inactivated at pH values outside the range of 5–9. Optimum pH of the enzymes like catalase, urease, amylase are 6.5 to 7.0; while that for pepsin is 1.5 and invertase, 4.5. Optimum pH of lipase (pancreas) is 8.0 (Figure 5.3).

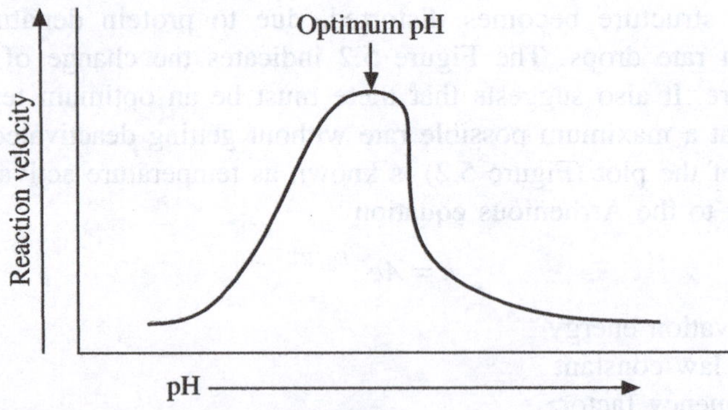

Figure 5.3 Effect of pH on the rate of enzymatic reactions (With kind permission from Worthington Enzyme Manual; Worthington, K., and Worthington, V.(2011). Worthington Biochemical Corporation. (http://www.worthington-biochem.com/pap/default.html).

The maximum reaction rate, K_m (Michaelis constant) is a function of pH and the stability of enzyme. In any event pH effect on K_m are relatively insignificant, however if the substrate does not have different ionization state with different affinity for free enzyme and if the formation of enzyme-substrate complex (ES) does not influence ionization equilibrium constants.

5.6.3 Mechanical Forces

Drag force, viscosity, shear stress, interfacial tension, hydrostatic pressure, etc. caused by fluid has direct impact on the enzyme activity. Under the influence of such hydrodynamic forces deformation of active sites might occur. Distorted active sites lead to enzyme deactivation. Cytoplasmic fluid has surface tension of 1 dyne/cm^2 at ambient temperature—an optimum condition for most enzymatic reactions. Since, the surface tension of the interface between air and pure water is 80 dyn/cm, foaming or frothing in protein solutions commonly causes denaturation of protein adsorbed at the air-water interface.

5.6.4 Chemical Agents

Presence of some chemical compounds like detergent, urea solution, alcohol, hydrogen peroxide, etc. may also cause danaturation of protein and hence can cause enzyme deactivation.

5.6.5 Irradiation (Light, Sound, Ionizing Radiation)

Presence of light, sound, and ionizing radiation may also cause the deformation of active sites of the enzyme and thereby resulting in enzyme deactivation.

5.7 VARIOUS HYPOTHESES FOR ENZYME SUBSTRATE REACTION

In an enzymatic reaction, enzymes bind to the substrate to form a transient-state species, commonly known as enzyme-substrate complex, which subsequently breaks in to product with making the enzyme free. A substrate joins with an active site of enzyme to form an intermediate enzyme-substrate complex. Enzyme-substrate complex is stabilized by various bonds such as electrostatic, hydrogen, non-covalent and by van der Waals interaction. Active sites are 'pit' or crevice like structure and are generally located on the surface of the enzyme molecule. Affinity, complementarity, and mutual flexibility are some of the important prerequisites for the formation of appropriate enzyme-substrate complex.

5.7.1 Lock and Key Model

The 'Lock and Key Model' of enzyme-substrate reaction was put forward by Emil Fisher in the year 1894. It is also known as template hypothesis. According to this hypothesis an enzyme is imagined as a lock and the substrate as key (Figure 5.4). Just as a key with appropriate size, shape and orientation can fit into a lock to open it, a substrate having matching size and shape with an active site of the enzyme can form an enzyme-substrate complex. This enzyme-substrate complex formation constitutes the essential precondition of an efficient enzymatic reaction to yield a product.

Once product is formed, it can no longer be attached with the enzyme site and is released. The enzyme active site becomes free again to take up another molecule of substrate. This process then continues. The existence of enzyme-substrate complexes are substantiated by various

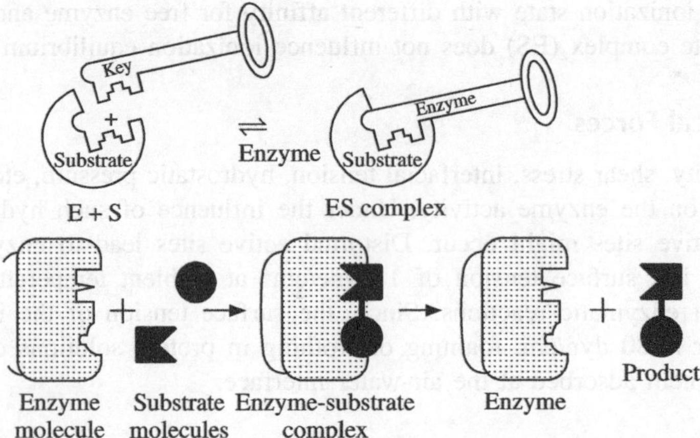

Figure 5.4 Schematic representation of Lock and Key Model (reproduced with permission from Zigya Technology Labs Private Limited, New Delhi).

techniques particularly X-ray and Raman spectroscopy studies. The Lock-and-Key model has been successful in explaining many features of the enzyme specificity; but it could not explain some of the conformational changes taking place in the enzymes when they are in close vicinity with the substrates.

5.7.2 Induced Fit Model

Induced fit model can be viewed as an extension of Lock and Key model with the provision of more dynamic and flexible interaction between enzyme and substrate. In induced fit model enzyme active site accepting the substrate can slightly change its shape to accommodate the substrate by undergoing a dynamic conformational change (Figure 5.5). Major takeaways from the induced fit theory as proposed by Koshland are as follows:

(i) Exact and facile orientation of active sites.
(ii) Substrate binding at the active site resulting in change enzyme structure.
(iii) Such structural changes bring about desired orientation for an enzyme-substrate reaction to take place.

A good example to understand induced-fit model is to compare the same with hands-in-glove. While wearing gloves on our hands, it becomes difficult to insert any finger at first into the proper place, but if it is done the other fingers can be adjusted easily.

$$\text{D-hexose} + \text{ATP} \leftrightarrow \text{D-hexose-6-phosphate} + \text{ADP}.$$

This reaction is catalyzed by yeast hexokinase enzyme, is a typical example of induced fit hypothesis. Both lock and key model and induced fit model assume great importance in the applications leading to drug design, using the idea of complementarity.

5.7.3 Proximity and Orientation Effects

Rate of enzyme catalyzed bioreactions increase due to proximity and orientation effects. The starting point of enzyme catalyzed reactions is the attachment of substrate to the enzyme active

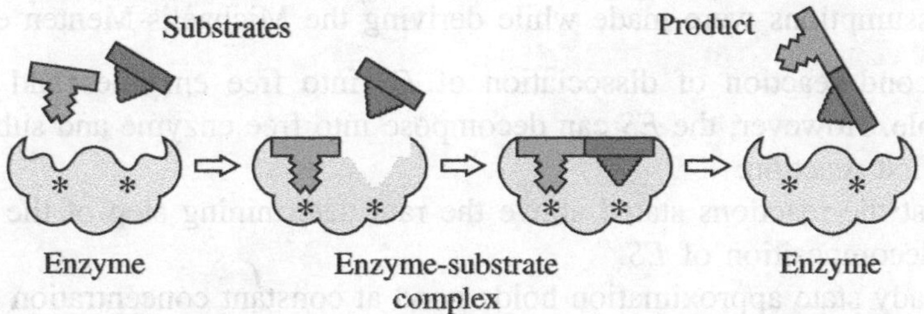

Figure 5.5 Induced fit model of enzyme kinetics (*Image courtesy: Prof. Steven M. Carr, Memorial University of Newfoundland, Canada*).

site, which is the prime functional region of an enzyme. Under the influence of proximity effect substrates come close to each other in a favorable position near to the active sites. In other words active site of an enzyme gathers substrate to its vicinity. This also helps in reduction of total entropy of the reactant species while reacting by combing with substrates.

In orientation effect substrate molecules get aligned in the right positions considered to be most favorable to combine with the active sites of the corresponding enzyme. This is generally carried out by orbital steering. Both these methods help in formation of the transition state enzyme-substrate complex, which further dissociates into desired products.

5.8 EXPRESSION OF ENZYME ACTIVITY

As compared to substrate concentration, enzymes are required in very low concentration in a reaction to occur. Enzyme concentration is directly proportional to the initial reaction velocity. The ratio of reaction velocity to enzyme concentration is a measure of enzyme activity. A commonly used term for expressing enzyme action is either 'enzyme unit' or 'international unit' or simply as 'unit'. If 1 μmol of product is produced by conversion of 1 μmol of substrate in 1 min, it is considered to be 1 enzyme unit. For a known molecular weight of enzyme it can also be expressed in terms of molecular activity. Accordingly I enzyme unit is defined as the moles of products formed or substrate consumed per mol of enzyme per min. Katal (short form Kat) is also used to express enzyme activity. 1 Katal is defined as the amount of enzyme required for the transformation of 1 mol of substrate into product per second. The relationships between Katals and Units are:

$$1 \text{ kat} = 60 \text{ mol.min}^{-1} = 6 \times 10^7 \text{ Units}$$

$$1 \text{ Unit} = 1 \text{ micro mol per min} = 16.67 \text{ nano katal}$$

5.9 ENZYME KINETICS

5.9.1 Michaelis-Menten Equation

Let us consider a simple reaction taking place between an enzyme E and a substrate S according to the following reaction

$$E + S \xleftrightarrow{K_{f1}} ES$$

$$ES \xrightarrow{K_{f2}} E + P$$

Following assumptions were made while deriving the Michaelis-Menten equation:

(i) The second reaction of dissociation of *ES* into free enzymes and products is not reversible. However, the *ES* can decompose into free enzyme and substrate according to the first reaction.
(ii) Amongst the reactions stated above the rate determining step of the above reactions is the decomposition of *ES*.
(iii) The steady state approximation holds good at constant concentration of *ES* and when total enzyme concentration is negligibly small compared to substrate concentration, i.e. $E_{tot}S$. This assumption holds good in most of the cases over a long period of times.
(iv) Isothermal conditions are maintained.

Since, the concentration of *ES* is constant, $\frac{d}{dt}(ES) = 0$ (5.5)

Enzyme-substrate complex *ES* is produced by a single reaction but dissociates by two pathways, the one being reversal of formation and the other is product formation. Since all enzymes are present either in free (*E*) or complexed (*ES*) form, at any time the total enzyme concentration of the system would be

$$E_{tot} = E + ES \tag{5.6}$$

or, we can write

$$\frac{d}{dt}(ES) = K_{f1}.E.S - K_{r1}.ES - K_{f2}.ES = 0$$

$$= K_{f1}.E.S - ES(K_{r1} + K_{f2}) = 0 \tag{5.7}$$

$$K_{f1}.E.S = ES(K_{r1} + K_{f2})$$

or

$$E = \frac{ES(K_{r1} + K_{f2})}{K_{f1}S} \tag{5.8}$$

$$E_{tot} = \frac{ES(K_{r1} + K_{f2})}{K_{f1}S} + ES = ES\left[\frac{(K_{r1} + K_{f2})}{K_{f1}S} + 1\right] = ES\left[\frac{(K_{r1} + K_{f2}) + K_{f1}S}{K_{f1}S}\right]$$

$$= ES\left[\frac{(K_{r1} + K_{f2})/K_{f1} + S}{S}\right] \tag{5.9}$$

$$ES = \frac{E_{tot}.S}{\frac{(K_{r1} + K_{f2})}{K_{f1}} + S} \tag{5.10}$$

Initial rate of reaction can be written as

$$V = K_{f2}.ES = \frac{K_{f2}.E_{tot}.S}{\frac{(K_{r1} + K_{f2})}{K_{f1}} + S} \tag{5.11}$$

The term $\frac{K_{r_1} + K_{f_2}}{K_{f_1}}$ can be considered as a constant K_m (Michaelis constant)

Substituting K_m in the above equation we get,

$$V = \frac{K_{f_2}.E_{tot}.S}{K_m + S} \tag{5.12}$$

If $V_{\max}$ is the maximum reaction rate (under this condition it is assumed that all enzymes are consumed in the reaction).We can write

$$V_{\max} = K_{f_2}.E_{\text{tot}} \tag{5.13}$$

Therefore,

$$V = \frac{V_{\max}.S}{K_m + S} \tag{5.14}$$

Eq.(5.14) is known as the Michaelis–Menten Equation for enzymatic reaction.

5.9.2 Evaluation of Kinetic Parameters of M-M Equation

The evaluation of kinetic parameters $V_{\max}$ and K_m of Michelis-Menten equation involves graphical transformation of the equation to make it a linear form. The rearrangement of the M-M equation is done by various methods. A few popular methods are discussed below:

(i) Lineweaver-Burk plot: The Lineweaver-Burk plot inverts the M-M equation in to a double reciprocal form. Hans Lineweaver and Dean Burk in 1934 described this double reciprocal equation. It is so called because manipulation is done using the reciprocal of the values of both the velocity and the substrate concentration. M-M equation can be written as

Reciprocal form of Eq.(5.14) can be written as:

$$\frac{1}{V} = \frac{K_m}{V_{\max} S} + \frac{S}{V_{\max}} \frac{1}{S} \tag{5.15}$$

or

$$\frac{1}{V} = \frac{K_m}{V_{\max}} \frac{1}{S} + \frac{1}{V_{\max}} \tag{5.16}$$

A plot of $1/V$ versus $1/S$ (Figure 5.6) gives a straight line with an intercept of $\frac{1}{V_{\max}}$ and slope $\frac{K_m}{V_{\max}}$

Although this kind of reciprocal transformation provides a good estimate of $V_{\max}$ but for K_m it does not. Experimental data points for low and high substrate concentrations have impact on slope as well as intercept.

(ii) Eadie-Hofstee plot: The Eadie–Hofstee plot seeks to improve the values of constants as obtained in the Lineweaver–Burk plot in that the experimental points are usually more equally spaced. If Lineweaver–Burk double reciprocal equation is multiplied by $V_{\max}$ throughout, we can get,

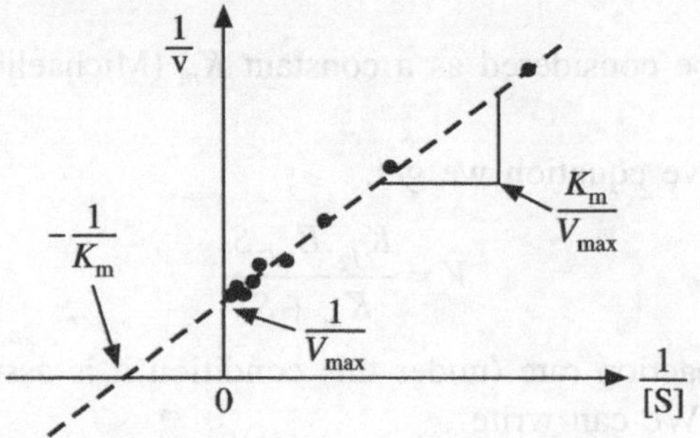

Figure 5.6 Lineweaver-Burk Plot.

$$\frac{1}{V}V_{max} = \frac{K_m}{V_{max}}\frac{1}{S}V_{max} + 1 \tag{5.17}$$

or
$$V_{max} = \frac{V}{S}K_m + V \quad \text{or} \quad V = -K_m\frac{V}{S} + V_{max} \tag{5.18}$$

The plot of V against $\frac{V}{S}$ will be linear (Figure 5.7). Intercepts on V and $\frac{V}{S}$ axes will be V_{max} on the and $\frac{V_{max}}{K_m}$ respectively.

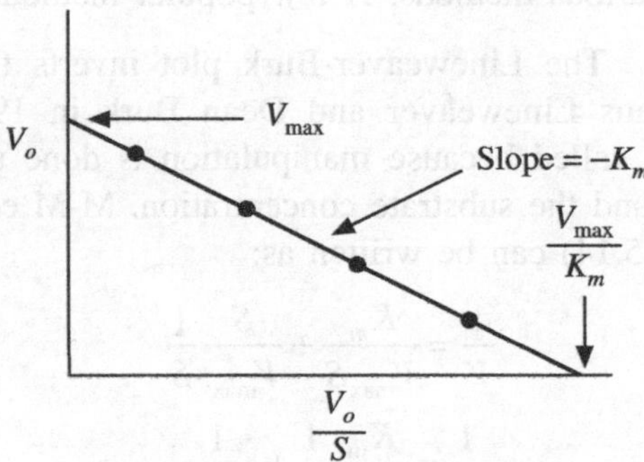

Figure 5.7 Eadie-Hofstee Plot.

Since both V_o and V_o/S are velocity dependent variables, the plot is not entirely free from deviation. There are chances of experimental errors.

In the Michaelis-Menten equation $V = \frac{V_{max}.S}{K_m + S}$, the term '$V$' can be interpreted as the rate of substrate utilization. Therefore,

$$-\left(\frac{ds}{dt}\right) = V = \frac{V_{max}\,S}{K_m + S} \tag{5.19}$$

or
$$-K_m.\frac{dS}{S} - S\frac{dS}{S} = V_{max}dt \tag{5.20}$$

or $$-K_m \int_{S_o}^{S} d\ln S - \int_{S_o}^{S} dS = V_{\max} \int_0^t dt \quad (5.21)$$

or $$-K_m \ln\frac{S}{S_o} - (S - S_o) = V_{\max}.t \quad (5.22)$$

where K_m, S_o and V_{max} are constants.

5.10 INHIBITION OF ENZYME ACTIVITY

Any substance or chemical that can retard the rate of enzymatic reaction is termed as an inhibitor. An inhibitor blocks the active sites of the enzyme in such a way, that it becomes unavailable to provide space for the actual substrate for combination. In general, inhibitors could be of two types, reversible and irreversible, depending on the nature of its binding with the enzyme active sites. Reversible inhibitors are loosely attached to the active sites and can be easily retrievable and reused upon dilution. On the other hand, irreversible inhibitors are strongly bonded with active sites via formation of covalent bonds or some other strong interaction and are difficult to dislodge. For example, certain insecticides exhibit neurotoxic effects as a result of their irreversible binding with the active site of acetylcholinesterase enzyme.

The concept of enzyme inhibition has been widely exploited in the pharmacological studies of drug applications and also in understanding various metabolic reaction pathways. Some the processes of chemotherapy used for strategic destruction of cancer cells depend on inhibition of a particular critical enzyme of the infecting cell so as to prevent its proliferation. Some examples of enzyme inhibitors are presented in Table 5.3.

Table 5.3 Some examples of enzyme inhibitors

S. No	Enzymes	Inhibitors
1	Succinic dehydrogenase	Malonate, hematin, Se
2	Lipase	Benzaldehyde, some metal ions
3	Carboxylase	Formaldehyde acetaldehyde, benzaldehyde
4	Alkaline phosphatase	Arsenate, Fluoride ions, Chelating compounds
5	Urease	o-quinone, heavy metals
6	Carboxy peptidase	Iodoacetate, sulphide and cyanide ions
7	Pyrophosphatase	Formaldehyde, Idoacetate

There are three different types of inhibition occurs in enzyme catalyzed reactions. These are briefly discussed below:

5.10.1 Competitive Inhibition

In competitive inhibition, an inhibitor and a substrate compete for the same active site of enzyme for binding. In most of the cases the inhibitor is a look-alike of the substrate or it is also known as substrate analog. These inhibitors although can fit into the enzyme active site, but no further product formation takes place and the site is blocked.

$$(CH_2)_2\,(COOH)_2 \longrightarrow HOOC\text{–}CH=CH\text{–}COOH$$

This reaction is catalyzed by succinic dehydrogenase. But in presence of malonic acid (CH_2) $(COOH)_2$ this reaction is completely inhibited. Malonic acid can form a complex with succinic dehydrogenase but it does not react.

Sulpha drug or sulphanilamide which has structural similarity with p-amino benzoic acid—an important vitamin for many bacteria can act as a competitive inhibitor for p-amino benzoic acid. Another example is the inhibition of alcohol dehydrogenases by ethanol in methanol poisoning.

Competitive inhibitors find their uses in making pharmaceutical products, particularly drugs for chemotherapy. For example, methotrexate, a chemotherapy drug having similarity with coenzyme folate, acts as competitive inhibitor. It blocks the active site of the dihydrofolate reductase enzyme.

Succinic acid and malonic acids have a considerable structural similarity, although their molecular weights are different. Thus, malonate can bind to the active site of the enzyme succinate dehydrogenase, thereby posing as an inhibitor. This could be a preventive reaction to stop undesirable metabolic reactions in pathogenic organisms.

When a competitive inhibitor I is present, the reaction rate V is given by:

$$V = \frac{V_{\max}[S]}{[S] + K_m\left(1 + \frac{[I]}{K_i}\right)} \tag{5.23}$$

where $[S]$ is the substrate concentration (mol/L); $[I]$ is inhibitor concentration (mol/L); K_i is the inhibition coefficient; $V_{\max}$ is the maximum enzymatic reaction rate (t^{-1}) and K_m is the Michaelis constant.

Effect on $\boldsymbol{V_{\max}}$**:** Increase in substrate concentration may nullify the effect of competitive inhibition. If the substrate concentration is considerably high, $V_{\max}$ could be high and remain unaffected by inhibitor.

Effect on $\boldsymbol{K_m}$**:** For a specific substrate, K_m increases with competitive inhibition. It suggests that in case of competitive inhibition, substrate concentration should be more to achieve ½ $V_{\max}$.

Competitive inhibitors are used in arresting the growth and propagation of pathogenic bacteria. Folic acid, an essential factor of cell growth, is usually unavailable to cells by normal transport processes. Hence, it is required to be synthesized from p-amino benzoic acid.

5.10.2 Non-competitive Inhibition

In non-competitive inhibition, the inhibitor can combine with either the free enzyme or the enzyme-substrate complex. In either of the cases the reaction will cease to occur. It is irreversible in nature. Examples of such type of inhibitors include ions of heavy metals, cyanides, penicillin, glucose-6-phosphate, etc.

Effect on $\boldsymbol{V_{\max}}$**:** Increasing substrate concentration does not result in mitigating non-competitive inhibition. $V_{\max}$ of the reaction decreases in such inhibition.

According to Maichaelis-Menten kinetics, the reaction rate is given by:

$$V = \frac{V_{\max}[S]}{(K_m + [S])\left(1 + \frac{[I]}{K_i}\right)} \tag{5.24}$$

***Effect on K_m*:** In non-competitive inhibition K_m remains unchanged.

5.10.3 Uncompetitive Inhibition

In uncompetitive inhibition the inhibitor does not interact with free enzyme (E), rather it combines with the enzyme–substrate complex (ES) to form another complex i.e. enzyme-substrate-inhibitor ESI. Here the inhibitor does not compete with the substrate to occupy the active site. The reaction velocity in uncompetitive inhibition can be given by:

$$V = \frac{V_{\max}[S]}{K_m + [S]\left(1 + \frac{I}{K_i}\right)} \tag{5.25}$$

In this type of inhibition dissociation of products from the enzyme-substrate complex becomes difficult. Both K_m and $V_{\max}$ decreases in such inhibition. Uncompetitive inhibitors find its application in repressing or preventing propagation of various diseases like cancer, tumors in biological systems. Human alkaline phosphatase, an enzyme found to be over expressed in certain cancer cells, functions via uncompetitive inhibition.

5.11 INDUSTRIAL USES OF ENZYMES

Today, enzymes are widely used in the dairy, food, pharmaceutical, pulp and paper, textile, leather, specialty chemicals and different other allied chemical process industries. Many of the enzymatic processes are already well established and have been successfully commercialized. Driven by the philosophy of sustainable development via application of green chemistry principles in the synthesis of chemicals and petrochemicals, enzymatic production processes are adopted by the industry and the research organizations. Multiple benefits of bio-catalysis include mild physico-chemical conditions, enhanced selectivity, less pollution potential, simplified synthesis schemes and improved chemical process economics.

5.11.1 Pharmaceuticals Industry

Use of enzymes in pharma industries helps reducing the process steps in the system thereby improving the process economics. Due the high selectivity of enzymes high degree of product purity could be achieved. Streptokinase, urokinase, lipase, α-amylase and many other enzymes are successfully being used in various diagnostic as well as therapeutic applications. Enzymes are used not only in diagnosis but also in treatment and therapy. In the antibiotic industry, penicillin acylase is used as precursor of making a wide variety of penicillin derivatives.

5.11.2 Food and Beverage Industry

Due to toxicity and hazardous nature of many chemical reactants and extremities of physico-chemical conditions, classical production process routes are less viable for food processing industries. Enzymes are used in wine and cheese making even in ancient times. They are used in food and beverage industries for quality upgradation and preservation. Enzymes are commonly used in beer fermentation, juice debittering, bread baking, conversion of starch to glucose and fructose, faster cheese ripening, production of high fructose syrup and many more processes.

In beverage industry the enzymes are used not only in the juice clarification processes, these are also exploited to both increase the juice yield and decrease energy consumption by enzymatic breakdown of pectins and other compounds of plant cell wall. In the wine industry, pectinase is used to improve juice yield while β-glucanase, as a clarification and filtering aid. Pectolytic enzymes (like pectinol, mylase, etc.) are used in the clarification of fruit juices for making ready-to-serve beverages. These enzymes are also used to clarify wines made out of fruit juices. Chill-proofing of beer is another very useful industrial application. Beers produced from various cereal grains show a certain degree of haziness, which is required to be removed. This is known as Chill-proofing of beer and is done by an enzyme papain (extracted from the latex of raw papaya fruits).

5.11.3 Paper and Pulp Industry

The major applications are boosting of bleaching, fibre modification, pitch control, delignification, deinking and so on. Cellulase, xylanase, laccase and lipase are the ideal enzymes fulfilling multiple requirements of pulping and paper making process. Xylanase enzyme is extensively used for the pretreatment of pulp obtained in Kraft process. Besides, the enzymatic pitch control method by lipase from *Candida antarctica*; enzymatic deinking by amylase or cellulase from *Bacillus licheniformis* and laccase mediated pulp bleaching are also important enzymatic processes. Enzymatic debarking, enzymatic beating, retting of flax fibres, and selective removal of xylan are some of the processes which are still in the developing stage.

5.11.4 Textile Industries

Textile industry is a major user of enzymatic processes in recent times. Around thirty different enzymes are used by textile and paper industries at various levels of processing. Some important enzymes include amylase, pectinase, cellulose, peroxidase, glucose oxidase and so on. Amylase is extensively used in textile processing and finishing for the removal of starch-based sizes. This process is commonly known as enzymatic desizing. Amylase facilitates hydrolytic degradation of starch into simple sugars and smaller molecules. Enzymatic desizing can be accomplished at a relatively low temperature (30–60°C) and pH of 6 ± 0.5. The process becomes less energy consuming. Pectinase can also digest sizing materials such as pectin by rupturing the cuticle structure of cotton fibres. The enzyme cellulase is also responsible for solubilizing cell wall. These enzymes are used in combinations depending on the process requirement. Suitable enzymes can even replace the chemical bleaching agents like hydrogen peroxide. Peroxidase enzymes are used in the rinsing process after dyeing of fibres. Different enzymes are also used in bleaching and lignin degradation. Enzymatic treatment of textile materials results in less detrimental wastewater as compared to chemical treatment.

5.11.5 Leather Industries

Important steps in the processing of leather include soaking, dehairing, bating, degreasing and post-tanning operations. Different enzymes such as lipase, alkaline protease, carbohydrase, trypsin, etc. are used these steps. Soaking is the very first step of leather processing where hides or skins undergo rehydration to make it soft. Enzymatic processes are gradually replacing chemical processing steps in tanneries. Alkaline protease and lipase used in soaking facilitates breakdown of protein in the hide and make them more soluble. Use of enzymes can also reduce the soaking time. In addition, protease helps in effective dehairing, thereby eliminating the use of lime, sulphides and other chemicals. Lipases are used in degreasing of leather and also impart waterproofing quality of leather.

There are many more applications of enzymes in detergent, cosmetics, organic synthesis of several specialty chemicals, polymer processing and so on. Some of these applications are briefly tabulated in Table 5.4.

Table 5.4 Brief summary of important enzymes and their industrial applications for various purposes

Types of industry	Enzymes used	Purpose
Dairy and food processing	Rennin from the stomach of young calves	To coagulate proteins during manufacture of cheese
	Patalase	To enhance cheese flavour
	Lipozyme	Inter-esterification of vegetable oil.
	Pectinase and protease	To clear haziness from bottled juice
	Neutral proteinase	Maturation of cheese (ripening), treatment of fruit juice to remove bitterness.
	Lipase	Faster cheese maturation, flavor customized cheese
	Aminopeptidase	Maturation of cheese
	Catalase	Treatment of cheese
	Transglutaminase	Joining of proteins by cross linking
Brewing and Beverage	Amylase	To digest polysaccharides in malt
	Protease	To digest protein in malt
	Pectinase	Depectinization
	Glucose oxidase	Deoxygenation of beer
	Cellulase	Fruit liquefaction
	α-Amylase	Solubilization of starch
	β-Amylase	Solubilization of starch
	β-Glucanase	Reduction of haze formation
	protease	Reduction of haze formation
	Pullulanase	Starch saccharification
	Naringinase; limoninase	Debittering

(Contd.)

Types of industry	Enzymes used	Purpose
	Arabinanase	Removal of post-maceration haze
Pharmaceutical and healthcare	Streptokinase	Anti-inflammatory agent
	Urokinase	Prevention and removal of blood clots
	Asparaginase	Lymphatic cancer therapeutic.
	Penicillin amylase	Synthetic antibiotic production
	Lipase	Synthesis of chiral compounds.
	α-Amylase, lipase	Treatment of digestive disorder
	Collagenase	Treatment of skin ulcer
	β-Lactamase	Treatment of antibiotic resistance
Leather processing	Alkaline and neutral Protease	Dehairing of hides, bating and soaking
	Amylase	Fibre splitting
	Keratinase	Hydrolysis of keratins
	Acid or fungal lipase	Degreasing of animal skins
Paper and pulp	Cellulase	To breakdown complete starch to smaller molecules and to make smoother paper.
	Lipase	Controlling pith particle deposits
	Protease	Removal of film formed by microbial cells
	Amylase	Deinking, drainage improvement
	Xylanase	Enhancement of bleaching
	Laccase	Non-chlorine bleaching, delignification
Detergent	Protease	Used to breakdown precise in stains, in dishwashers to remove food residue
	Subtilisin	Laundry detergent
	Amylase	Removal of smudge due to carbohydrate
	Lipase	Stain removal due to fat
	Protease	Protein stain removal
Textile	Laccase	Non-chlorine bleaching
	Cellulase	Biofinishing, biopolishing, brightness improvement, etc.
	Pectate lyase	Bio-scouring
	Catalase	Bleach termination
	Lipase	Removal of size lubricants, denim finishing,
	Amylase	Desizing
	Ligninase, Collagenase	Processing of raw wool
	Esterase	Development of polymer film
Cosmetics	Superoxide dismutase	Skin care applications.
	Protease	Dead skin scavenging
	Endoglycosidase	Treatment of teeth and gums
	Laccase	Hair dying
	Lipase	Skin care

(*Contd.*)

Types of industry	Enzymes used	Purpose
Polymer processing	Lipase	Polycondensation, ring-opening polymerization of lactones, carbonates
	Laccase	Synthesis of bisphenol A polymer
	Glucose oxidase	Synthesis of polyaniline
	Transglutaminase	Crosslinking of protein
	Tyrosinase	Synthesis of lignin and chitosan polymers

5.12 ENZYME IMMOBILIZATION

Immobilization of enzymes indicates that their mobility is arrested and they are confined and localized so that they can not only be used continuously but the stability of the enzyme will be increased to a great extent.

5.12.1 Merits and Demerits of Immobilization

Merits of immobilization:

(i) Overcomes the problem of enzyme removal from end products.
(ii) Reuse of enzymes for a long period of time.
(iii) Greater turnover number.
(iv) Impart increased stability of enzyme.
(v) Reduction in contamination and destruction.
(vi) Deactivation due to higher temperature and range of pH is minimized.
(vii) Prolongation of the half-life or activity of the enzyme and improved storage quality.
(viii) Immobilized systems are increasingly used in diagnostics and therapeutic applications.

Demerits of immobilization

(i) Substrate and product diffusivity may become a great problem, particularly in insoluble substrate.
(ii) Partial steric hindrance of the active site of the enzyme.
(iii) Due to insolubility imparted, agitation and stirring may become essential.
(iv) All types of immobilization principles may not be applicable to all systems.

5.12.2 Immobilization Techniques

Commonly used methods of enzyme immobilization are broadly categorized into two major types—surface immobilization and entrapment.

Surface immobilization

In surface immobilization enzyme molecules are bound on the surface of supporting carrier or media. These are held together under the influence of several interactions such as covalent, van der Waals, physical force like dispersion, etc. These are mainly of two types, adsorption and covalent binding.

In adsorption enzymes are adsorbed onto the surface of a suitably selected adsorbent material. Binding of enzymes takes place as a result of weak physical forces as mentioned earlier. Examples of support materials for adsorption immobilization are activated carbons, silica gel, clay, resins, collagen, porous glass, diatomaceous earth, carboxyl methyl cellulose (CMC), other cellulosic materials, sephadex and so on. To ensure stable attachment on the adsorbent site, crosslinking of enzymes are often cross linked with multi-functional cross linking agents like glutaraldehyde, *bis*-biazobenzidine, etc. Sometimes, pretreatment of support materials is required for efficient and durable attachment. This can be done by surface activation or modification of the materials by coating with organic functional groups using silane reagent or by attaching spacer arms or moieties like *n*-propyl amine. Schematic diagram of various immobilization methods is presented in Figure 5.8.

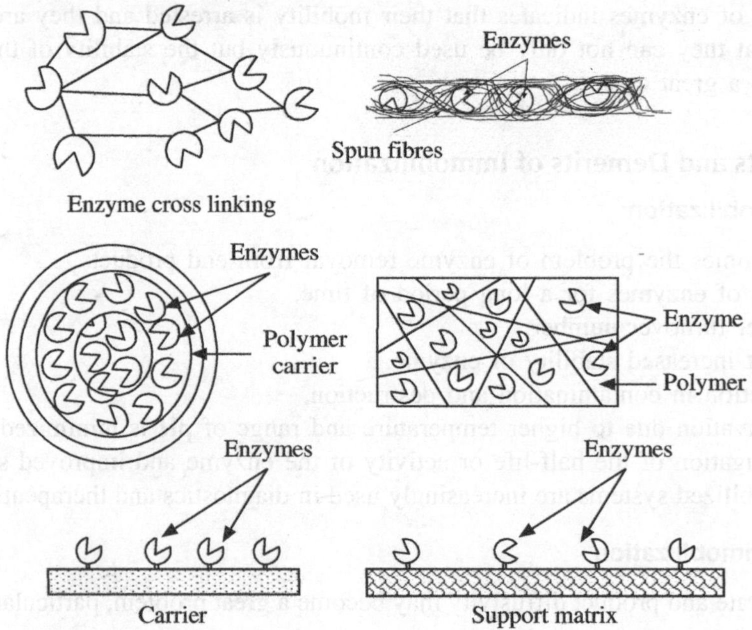

Figure 5.8 Schematic diagram of immobilization methods.

In covalent binding enzymes form covalent bonds with the support materials through certain functional groups such as amino, hydroxyl, carboxyl, suphydryl, etc. These groups are chemically activated using certain reagents like glutaraldehyde, cyanogen bromide, carbodiimide, etc. Two examples are illustrated in Figure 5.9.

Some examples of water insoluble support materials are acrylamide, maleic anhydride, styrene-based polymers, cellulose, dextran, agarose (sepharose) and so on.

Entrapment

The entrapment method of enzyme immobilization involves the localization and physical enclosure of an enzyme within the matrix or capsule of a suitably selected carrier. Entrapment is of two types, lattice type or microencapsulation. Generally the matrix is made of water-insoluble polymeric materials, both synthetic and natural, such as polyacryl amide, polyvinyl alcohol, polystyrene, starch, agar, calcium alginate, collagen, etc. The matrix for entrapment could be even some inorganic materials such as porous ceramics, activated carbon, diatomaceous earth, etc. Physically the matrix should be a particle, membrane or fibre having interstitial space to accommodate and retain the enzyme.

In microencapsulation the enzyme is encapsulated into a microscopic hollow sphere enclosed by a porous polymeric membrane. The sphere holds the enzyme phase. This is accomplished by interfacial polymerization method where enzyme solution is mixed with the hydrophilic monomers. As a result the enzyme in its aqueous phase gets enclosed in a polymer membrane. In addition, dispersion of enzymes into a water-in-oil type emulsion may lead to formation of microdrops. Hollow fibre membranes made of nylon, cellulose, polyamide, polysulfone are widely used for membrane entrapment. Selection of support materials include several considerations such as chemical and thermal stability, insolubility under reaction condition, mechanical strength and dimensional stability to protect the enzyme structure, high affinity to enzymes, high available surface area, biocompatibility, and economic viability.

Disadvantage of non-porous solid matrix

As the nonporous materials have very low surface area, enzyme attachment onto these materials becomes less effective. Use of finely grained materials could be a partial solution to enzyme loading problem.

The particles are difficult for both to remove from solution and to employ in continuous reactor, as they lead to high pressure drop and limited flow rates.

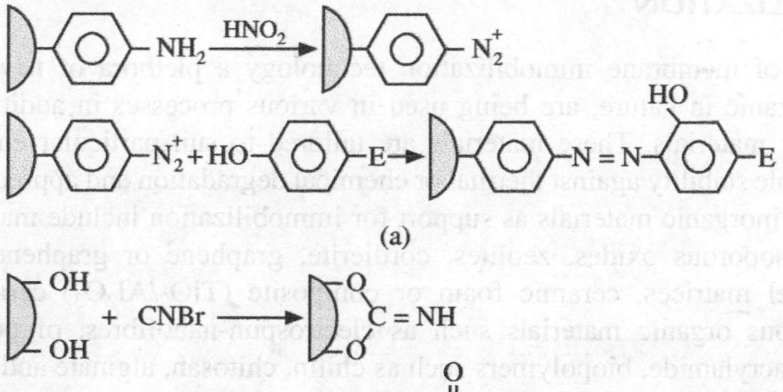

$$\text{Support}-C_6H_4-NH_2 \xrightarrow{HNO_2} \text{Support}-C_6H_4-N_2^+$$

$$\text{Support}-C_6H_4-N_2^+ + HO-C_6H_4-E \longrightarrow \text{Support}-C_6H_4-N=N-C_6H_3(HO)-E$$

(a)

$$\text{Support}(-OH)_2 + CNBr \longrightarrow \text{Support}(-O-)_2C=NH$$

$$\text{Support}(-O-)_2C=NH + ENH_2 \longrightarrow \text{Support}(-OH)(-O-\overset{O}{\overset{\|}{C}}-NHE)$$

(b)

Figure 5.9 Some examples of covalent attachment methods of enzyme immobilization (a) Diazo method (b) Cyanogen bromide method (E indicates enzyme) (Adapted from Biochemical Engineering Fundamentals, J.E. Bailey and D.F. Oillis, 2nd Edition, 1986, McGraw Hill Book Company).

Advantage of non-porous solid matrix

In non-porous enzymes are fixed to outer surface and are in readily in touch with substrate. Therefore, diffusion effect is minimum and large substrate conversion is expected.

Advantages and disadvantages of porous carrier

Disadvantage:

(i) The internal structure of porous carriers might be unavailable for larger size enzyme molecules to penetrate during to steric effect.
(ii) Sometimes internal surface of porous structure acquires negative charge. Thus any substrate having negative charge would experience repulsion leading to inefficient attachment.
(iii) Broad pore distribution may also reduce substrate-enzyme interaction.
(iv) Entrapped or encapsulated enzymes are severely limited by diffusional effects.

Advantage:

(i) Enhanced surface area for immobilization
(ii) Enzymes attached to the internal surface are immune to damage due to external turbulence.
(iii) Positively charged substances can effectively attach to the negatively charged surfaces.
(iv) Large enzyme loading is ensured in case of encapsulation or entrapping.
(v) Elastic carrier can be used in thin membrane form and it implies a small diffusional path and optimum available surface for reaction.

5.13 NEW AND EMERGING SUPPORT MATERIALS FOR ENZYME IMMOBILIZATION

With the advent of membrane immobilization technology a plethora of new materials, both organic and inorganic in nature, are being used in various processes in addition to traditional support or carrier materials. These materials are tailored to suit particular enzymes, and they exhibit considerable stability against thermal or chemical degradation and appreciable mechanical strength. Various inorganic materials as support for immobilization include magnetic iron oxide nanoparticle, mesoporous oxides, zeolites, cordierite, graphene or graphene oxides, carbon nanotubes, sol-gel matrices, ceramic foam or composite (TiO_2/Al_2O_3) ceramic membranes and so on. Various organic materials such as electrospun-nanofibres, of polyvinyl alcohol, polystyrene, polyacrylamide, biopolymers such as chitin, chitosan, alginate and cellulose, hybrid and composite materials and others. The major advantage of the new materials is that they are available in various geometrical shapes thereby facilitating improved catalytic efficiency and reducing the diffusional limitations. In addition, there are hybrid support materials which are synthesized by combining organic or inorganic precursors. These materials exhibit appreciable thermochemical stability and wear resistance and usually long-lasting binding with the enzyme.

5.14 ENZYMES REQUIRED FOR RECOMBINANT DNA TECHNOLOGY

Recombinant DNA technology makes extensive use of a wide variety of enzymes with which one can manipulate the DNA molecule to one's will. At each step of such experiments one or more kinds of these enzymes are required. The most common enzymes of *r*DNA technology are restriction enzymes (type I, II and III). These are used endogenously by bacteria to restrict the growth of foreign DNA especially viral DNA, within itself, hence the name restriction enzyme. Other modifying enzymes include DNA polymerase, RNA polymerase, reverse transcriptase, polynucleotide kinase, terminal transferase, etc.

5.15 PERFORMANCE OR DESIGN EQUATION FOR BIOREACTORS FOR ENZYMATIC REACTIONS

Various enzymatic conversions are carried out using bioreactor systems. In the following section we will incorporate the Michaelis-Menten equation to develop design equation for different types of reactor. Assuming the reaction mechanism proposed by Michaelis and Menten for an enzyme catalyzed reaction of a single substrate (A):

Let C_{A_0} = initial concentration of the substrate (at $t = 0$)
C_A = concentration of substrate at any time $t = t$
P = concentration of the complex EA at any time $t = t$
e = concentration of the total enzyme at $t = t$
$C_{E:}$ concentration of E at time $t = t$

By enzyme balance

$$e = C_E + p \tag{5.26}$$

or

$$C_E = e - p \tag{5.27}$$

The rate of enzyme complex reaction:

$$\frac{dp}{dt} = K_1 C_A C_E - K_2 p - K_3 p \tag{5.28}$$

Substituting Eq.(5.27) into Eq.(5.28):

$$\frac{dp}{dt} = K_1 C_A (e - p) - (K_2 + K_3) p \tag{5.29}$$

Under steady state conditions, $\frac{dp}{dt} = 0$, gives:

$$p = \frac{eC_A}{\frac{K_2 + K_3}{K_1} + C_A} = \frac{eC_A}{K_m + C_A} \tag{5.30}$$

Where, $K_m = \frac{K_2 + K_3}{K_1}$

The rate of enzyme reaction:

$$-r_A = -\frac{1}{V}\frac{dN_A}{dt} = K_3 p \tag{5.31}$$

For liquid phase reaction:

$$-r_A = -\frac{1}{V}\frac{dN_A}{dt} = -\frac{dC_A}{dt} \tag{5.32}$$

Hence, $$\frac{dC_A}{dt} = K_3 p = K_3 e \frac{C_A}{K_m + C_A} \quad (5.33)$$

5.15.1 Performance Equation by Agitated Batch Reactor

The performance or design equation for different bioreactors is developed from the material balance of the limiting component. In general this may be written as:

Input = Output + Disappearance + Accumulation by chemical reaction (5.34)

Dimension of each term must be either mol/time or mass/time.
In an agitated batch reactor,

Input = 0

Output = 0

Disappearance by chemical reaction = $(-r_A)\ V_{mol}$/time

$$\text{Accumulation} = \frac{dN_A}{dt}\frac{\text{mol}}{\text{time}}$$

So the balance equation will take the form:

$$-\frac{dN_A}{dt} = (-r_A)V \quad (5.35)$$

or $$dt = -\frac{dN_A}{(-r_A)V} \quad (5.36)$$

Now,

$N_A = N_{A_0}(1 - X_A)$, where X_A = fractional conversion of $\dfrac{N_{A_0} - N_A}{N_{A_0}}$

$$-dN_A = N_{A_0} dX_A \quad (5.37)$$

Hence from Eq.(5.36) and Eq.(5.37)

$$\int_0^t dt = \int_0^{X_A} \frac{N_{A_0}\, dX_A}{(-r_A)V}$$

For liquid phase reaction, $\dfrac{N_{A_0}}{V} = C_{A_0}$

So, $$t = C_{A_0} \int_0^{X_A} \frac{dX_A}{(-r_A)} \quad (5.38)$$

This is the performance equation of agitated batch bioreactor. Substituting the value of $-r_A$ from Eq.(5.33) in Eq.(5.38)

$$t = C_{A_0} \int_0^{X_A} \frac{dX_A}{K_3 e \dfrac{C_A}{K_m + C_A}} \quad (5.39)$$

Now for the liquid phase reaction:

$$C_A = C_{A_0}(1 - X_A) \tag{5.40}$$

Then,
$$t = \frac{C_{A0}}{K_3 e}\int_0^{X_A}\frac{(K_m + C_A)dX_A}{C_A}$$

$$= \frac{C_{A0}}{K_3 e}\left[\int_0^{X_A}\frac{K_m}{C_{A_0}(1 - X_A)}dX_A + \int_0^{X_A} dX_A\right] \tag{5.41}$$

or
$$K_3\, et = C_{A_0}\left[X_A - \frac{K_m}{C_{A0}}\ln(1 - X_A)\right]$$

$$= C_{A_0}X_A - K_m \ln(1 - X_A) \tag{5.42}$$

The term $K_3 et$ is known as reaction capacity and is denoted by Rearranging Eq.(5.42) we get

$$C_{A_0}X_A = K_m \ln(1 - X_A) + C \tag{5.43}$$

It is noticed that a plot of $C_{A_0}X_A$ vs. ln $(1 - X_A)$ should yield a straight line of slope K_m and intercept C.

5.15.2 Mixed Flow Reactor (MFR) or Continuous Stirred Tank Reactor (CSTR)

Let F_{A_0} = molar flow rate of A, mol/time
F_A = molar flow rate of A leaving the reactor, mol/time = F_{A_0} $(1 - X_A)$

Applying general material balance Eq.(5.34)

Input = F_{A_0} mol/time
Output = F_A, mol/time = $F_{A_0}(1 - X_A)$

Disappearance by chemical reaction = $(-r_A)$ V_{mol}/time
Accumulation = 0 (Since it is a steady state process)

Hence,
$$F_{A_0} = F_A + (-r_A)V$$

$$= F_{A_0}(1 - X_A) + (-r_A)V \tag{5.44}$$

or
$$\frac{V}{F_{A_0}} = \frac{X_A}{(-r_A)} \tag{5.45}$$

Defining τ_m = space time of MFR = $\frac{V}{V_0}$, where V_0 = volumetric flow rate and noting that

$$F_{A_0} = V_0 C_{A_0} \tag{5.46}$$

$$\tau_m = \frac{C_{A_0}X_A}{(-r_A)} \tag{5.47}$$

Eq 5.45 and Eq 5.46 are the performance equations of MFR or CSTR. Now substituting Eq.(5.33) in Eq.(5.47), we get

$$\tau_m = C_{A_0} \frac{V_A}{K_3 e \frac{C_A}{K_m + C_A}}$$

$$K_3 e \tau_m = \frac{C_{A_0}(K_m + C_A)X_A}{C_A} = \frac{C_{A_0} K_m X_A}{C_{A_0}(1 - X_A)} + C_{A_0} X_A$$

or $$K_3 e \tau_m = K_m \frac{X_A}{1 - X_A} + C_{A_0} X_A \tag{5.48}$$

A plot of $C_{A_0}X_A$ vs $\dfrac{X_A}{1 - X_A}$ should yield a straight line having slope of $-K_m$.

5.15.3 Packed-bed Reactor or Tubular Reactor Considering Ideal Plug Flow Behavior

In this case material balance has to be set up around a differential volume dv since the concentration of A is changing with position. Let us consider that input to the differential volume element $(dv) = F_A$, mol/time, output from the volume element $= F_A + dF_A$, mol/time

Disappearance by chemical reaction $= (-r_A)_{dv}$

Accumulation = 0, steady state reactor

From the material balance equation (Eq. 5.34), we can write,

$$F_A = F_A + dF_A + (-r_A)dv \tag{5.49}$$

Noting that $F_A = F_{A_0}(1 - X_A)$ or $dF_A = -F_{A_0}\, dX_A$

$$F_{A_0} dX_A = (-r_A)V \text{ or } \frac{V}{F_{A_0}} = \int_0^{X_A} \frac{dX_A}{(-r_A)} \tag{5.50}$$

Defining space time $\tau_p = V/V_0$

From Eq.(5.14) and Eq.(5.17):

$$\tau_m = C_{A_0} \int_0^{X_A} \frac{dX_A}{-r_A} \tag{5.51}$$

Eq.(5.50) and Eq.(5.51) are performance equations of plug flow reactor. Now substituting M-M equation, we get

$$\tau_p = C_{A_0} \int_0^{X_A} \frac{dX_A}{K_3 e \frac{A}{K_m + C_A}} \tag{5.52}$$

or $$K_3 e \tau_p = C_{A_0} \int_0^{X_A} \frac{K_m + C_A}{C_A} dX_A = C_{A_0} \left[\int_0^{X_A} \frac{K_m dX_A}{C_A} + \int_0^{X_A} dX_A \right]$$

$$= C_{A_0} \left[\int_0^{X_A} \frac{K_m dX_A}{C_{A_0}(1 - X_A)} + X_A \right]$$

$$= K_m \ln(1 - X_A) + C_{A_0} X_A \tag{5.53}$$

or

$$C_{A_0} X_A = K_m \ln(1 - X_A) + K_3 e \tau_p \tag{5.54}$$

In a packed-bed reactor, volume of packed section (V_R):

$$\varepsilon V_R = V$$

where ε = bed voidage, Thus Eq.(5.54) becomes:

$$C_{A_0} X_A = K_m \ln(1 - X_A) + \frac{K_3 e \varepsilon V_R}{V_0} \tag{5.55}$$

SOLVED NUMERICAL PROBLEMS

EXAMPLE 5.1 Substrate '*A*' decomposes according to the following in presence of an enzyme (E):

$A \rightarrow 2R$; Initial concentration of A = 1.0 mol/dm^3 and initial concentration of enzyme = 0.01 mol/dm^3. It is desired to produce 100 kg R/day in an agitated batch reactor. The time of filling, discharging and cleaning is 15 minutes. The final concentration of '*A*' is 0.1 mol/dm^3 and the molecular weight of R is 100. Find the volume of the reactor. The rate equation:

$$-r_A = \frac{50 C_{E_0} C_A}{1 + C_A} \frac{\text{mol}}{(\text{h})(\text{dm}^3)}$$

Solution: Here C_{A_0} = 1.0 mol/dm^3; C_A = 0.1 mol/dm^3

$$X_A = \frac{C_{A_0} - C_A}{C_{A_0}} = \frac{1.0 - 0.1}{1.0} = 0.9$$

From rate equation, $K_3 = 50\ \text{h}^{-1}$; C_{E_0} = 0.01 mol/dm^3; K_m = 1.0 mol/dm^3
Time of reaction (t) from Eq.(5.42)

$$K_3 C_{E_0} t = C_{A_0} X_A - K_m \ln(1 - X_A)$$

or

$$50 \times 0.01 t = 1.0 \times 0.9 - (1) \ln(1 - 0.9)$$

or

$$t = 6.4 \text{ h}$$

Total time of processing = 6.4 + 15/60 = 6.65 h
Number of batches per day: 24/6.65 = 3.61
K moles of R produced per m^3 of batch = $2C_{A_0} X_A = 1.8$
K moles of R produced/m^3 day = 1.8 × 3.61 = 6.5
Kg R formed per m^3 per day 6.5 ×100 = 650
Thus volume of the reactor needed 1000/650 = 1.54 m^3

EXAMPLE 5.2 The following data were obtained for the hydrolysis of carbobenzoxyglycyl-L-tryptophan catalyzed by the enzyme carboxypeptidase. The reaction in general can be written as:

$$\text{Carbobenzoxyglycyl-L-tryptophan} + H_2O \rightarrow \text{Carbobenzoxyglycine} + \text{L-tryptophan}$$

Plot these results using the Lineweaver-Burk method, and determine values for K_M and V_{max}. The symbol mM represents millimoles L^{-1}; 1 mM = 1×10^{-3} moles L^{-1}:

Substrate concentration (mM)	Velocity (mM s^{-1})
2.5	0.024
5.0	0.036
10.0	0.053
15.0	0.060
20.0	0.064

Solution: Taking the reciprocal of substrate concentration and of velocity gives the following results:

$1/[S]$ (mM^{-1})	$1/V$ (mM s^{-1})$^{-1}$
0.400	41.667
0.200	27.778
0.100	18.868
0.067	16.667
0.050	15.625

Plotting the results gives a straight line; the best fit to the experimental points is $1/V = 75.431(1/[S]) + 11.8$. The reciprocal of the y intercept is V_{max} and the slope is K_M/V_{max}. Hence $V_{max} = 0.0847$ mM s^{-1}, $K_m = 6.39$ mM.

KEY POINTS TO REMEMBER

- Enzymes are sophisticated biocatalysts based primarily upon macromolecular proteins and other peripheral components like metal ions, carbohydrates and nucleotides.
- Enzymes increase the rate of biochemical reactions reducing the energy of activation.
- Protein parts of enzymes are called apoenzyme while non-protein parts are called cofactor or coenzyme. Generally a cofactor is a metal ion and coenzyme is an organic molecule.
- According to the Enzyme Commission all enzymes are categorized into six distinct types based on the kinds of reactions that they catalyze. These are oxidoreductase, transferase, hydrolase, isomerase, lyase, and ligase.
- Turnover number indicates the maximum number of substrate molecules per catalytic active site per unit time for a given enzyme concentration at a given temperature. Enzymes have high turnover numbers at room temperature compared to industrial chemical catalysts.

- Enzymes are highly specific and selective in their reaction with substrates.
- Factors affecting enzyme activity include temperature, pH, mechanical forces, presence of chemical agents, etc.
- In an enzymatic reaction, enzymes bind to the substrate to form a transient-state species, commonly known as enzyme-substrate complex, which subsequently breaks in to product with making the enzyme free.
- Two important hypotheses of enzyme-substrate complex formation are lock and key model and the induced fit model.
- Michaelis-Menten equation can be used to study the rate of enzyme catalyzed reaction.

  $V = \frac{V_{max}.S}{K_m + S}$, V_{max}: maximum reaction rate; S: substrate concentration; K_m: Michaelis Constant; V: reaction velocity.
- There are three types of enzyme—inhibition, competitive, uncompetitive and non-competitive.
- Immobilization of enzymes indicates the arresting or confinement of their free movements, so that they are used continuously with more stable performance.
- Commonly used methods of enzyme immobilization are broadly categorized into two major types—surface immobilization and entrapment.
- In surface immobilization enzyme molecules are bound on the surface of supporting carrier such as silica gel, clay, resins, collagen, porous glass, carboxyl methyl cellulose (CMC), sephadex, etc., by adsorption or covalent bonding.
- Entrapment involves the physical enclosure of an enzyme within the matrix or capsule of a suitably selected carrier or lattice made of made of water-insoluble polymeric materials.

? QUESTIONS FOR PRACTICE

Multiple Choice Questions

1. Which of the following lists represent bonding interactions in their general order of strength on a "per bond" basis from lowest to highest?
   (a) Hydrogen, ionic, hydrophobic, covalent
   (b) Hydrogen, hydrophobic, ionic, covalent
   (c) Ionic, hydrophobic, hydrogen, covalent
   (d) Hydrophobic, hydrogen, ionic, covalent
2. Which of the following factors does not affect the shape of the active site of an enzyme?
   (a) A drop in temperature (b) Non-competitive inhibitor
   (c) A change in pH (d) Binding of substrate
3. Which of the following is not true about enzymes?
   (a) They are consumed by the reactions they catalyze
   (b) They are usually made of amino acids
   (c) They lower the activation energy of chemical reactions
   (d) Each one is specific to the particular substrate, or substrates, to which it binds

4. Which of the following analogies best describe the induced-fit model of enzyme-substrate binding?
   (a) A hug between two people
   (b) A key fitting into a lock
   (c) A square peg fitting through the square hole and a round peg fitting through the round hole of a children's toy
   (d) The fitting together of two jigsaw puzzle pieces
5. What is the function of enzymes?
   (a) To increase the free energy of reactions
   (b) To increase the sum of the heats of formation
   (c) To lower the randomness of the chemicals in the reaction
   (d) To lower the activation energy of a reaction
6. What happens if an enzyme is not functioning in a chemical reaction in a living organism that needs it?
   (a) The reaction stops
   (b) The reaction proceeds, but much more slowly
   (c) The reaction proceeds faster without the interference
   (d) There is no change in the reaction rate
7. An enzyme that joins the ends of two strands of nucleic acid is:
   (a) Polymerase (b) Ligase
   (c) Synthetase (d) Helicase
8. Which of the following is produced with the combination of apoenzyme and coenzyme?
   (a) Holoenzyme (b) Enzyme substrate complex
   (c) Prosthetic group (d) Enzyme product complex
9. Which one among them is the example of competitive inhibition of an enzyme?
   (a) Succinic dehydrogenase by malonic acid
   (b) Cytochrome oxidase by cyanide
   (c) Hexokinase by glucose-6-phosphate
   (d) Carbonic anhydrase by carbon dioxide
10. Blocking of enzyme action by blocking its active site is called as:
    (a) Allosteric inhibition (b) Feedback inhibition
    (c) Competitive inhibition (d) Non-competitive inhibition
11. Enzyme catalyzing rearrangement of atomic grouping without altering molecular weight or number of atom is:
    (a) Ligase (b) Isomerase
    (c) Oxidoreductase (d) Hydrolase
12. Enzyme complex involved in alcoholic fermentation is:
    (a) Zymase (b) Invertase
    (c) Lipase (d) Amylase
13. The enzyme which hydrolyses starch to maltose is:
    (a) Protease (b) Amylase
    (c) Lactase (d) Maltase

14. How do enzymes catalyze the biochemical reactions in our bodies?
(a) They lower the activation energy required to start a chemical reaction.
(b) They raise the temperature required to start a chemical reaction.
(c) They lower the pH required to start a chemical reaction.
(d) They decrease the amount of products produced by a chemical reaction.

15. When an inhibitor binds to an enzyme at a place other than the active site, but only when the enzyme and substrate are already bound in complex, which of the following best represents the situation?
(a) Competitive inhibition (b) Allosteric inhibition
(c) Uncompetitive inhibition (d) Non-competitive inhibition

16. In a chemical process, which of the following modifications cannot be achieved by an enzyme?
(a) Change in enthalpy (b) Change in reverse reaction rate
(c) Change in activation energy (d) Change in forward reaction rate

17. Which of the following best describes the class of enzymes that break bonds by generating a new double bond or ring structure rather than by hydrolysis or oxidation?
(a) Ligases (b) Isomerases
(c) Transferases (d) Lyases

18. The term 'enzyme' was coined by
(a) Pasteur (b) Buchner
(c) Urey Miller (d) Kuhne

19. From the Lineweaver-Burk plot of Michaelis-Menten equation, K_m and V_{max} can be determined when V is the reaction velocity at substrate concentration S, the X-axis experimental data are expressed as
(a) 1/V (b) V
(c) 1/S (d) S

20. The kinetic effect of purely competitive inhibitor of an enzyme
(a) Increases K_m without affecting V_{max}
(b) Decreases K_m without affecting V_{max}
(c) Increases V_{max} without affecting K_m
(d) Decreases V_{max} without affecting K

21. Coenzymes are
(a) Heat stable, dialyzable, non-protein organic molecules
(b) Soluble, colloidal, protein molecules
(c) Structural analogue of enzymes
(d) Different forms of enzyme

22. A specific inhibitor for succinate dehydrogenase is
(a) Arsenite (b) Malonate
(c) Citrate (d) Fluorid

23. Enzyme inhibition caused by a substance resembling substrate molecule is
(a) Competitive inhibition (b) Non-competitive inhibition
(c) Feedback inhibition (d) Allosteric inhibition

24. Enzymes are required in traces because they
(a) Have high turnover number
(b) Remain unused at the end of reaction and are re-used
(c) Show cascade effect
(d) All are correct

25. In which of the following types of enzymes, water may be added to a C—C double bond without breaking the bond?
(a) Hydrolase (b) Hydratase
(c) Hydroxylase (d) Oxygenase

Answer Key

1. (d)	2. (b)	3. (a)	4. (d)	5. (d)	6. (b)	7. (d)	8. (a)	9. (a)	10. (c)
11. (b)	12. (a)	13. (b)	14. (a)	15. (c)	16. (a)	17. (d)	18. (d)	19. (c)	20. (a)
21. (a)	22. (b)	23. (a)	24. (d)	25. (a)					

Descriptive Questions

1. State the term used to describe an ion that is essential for the enzyme to function.
2. Some enzymes work better in the presence of other molecules or ions. Explain how these molecules or ions increase the activity of enzymes.
3. Vivek Singh is working for a company in a project where immobilized enzyme system is used for production of a high-value biochemical. The Research (R) and Development (D) head of the firm claims that immobilization had improved the stability of the enzymes. As a proof, his experimental data shows that enzymes have half-life of 10 days in free suspended solution, but it increased to 30 days for an immobilized system under identical conditions of pH, temperature and medium composition. The enzymes were immobilized in a polymer matrix. Was the reasoning of *R* and *D* head right? Why or why not?
4. What do you understand by statement "Liposomes are well known for enzyme microencapsulation"? Explain the concept of microencapsulation.
5. Draw the difference between cofactor and prosthetic group? With examples, substantiate.
6. "Covalent modification-attachment of enzymes to matrix" enables their sustained use for prolonged periods of time when compared to physical adsorption. Justify or contradict the statement assigning reasons.
7. "Extremozymes represent the technology of the future". Do you agree? Whether yes/no, assign reasons.

6 Biomass Production in Cell Culture and Growth Kinetics

Microbes or microorganisms are the invisible companions of human beings impacting both positively and negatively on our life and living. Collection of mass of living or non-living microorganisms at a particular ecosystem is called biomass. Since microbes grow in the lap of nature, their cultivation needs recreation of the natural environment in the laboratory. A synthetic culture media must imitate comparable environmental and nutritional conditions that exist in the original natural habitat of the microorganism. During early days, microbial culture broth used to contains ingredients like urine or meat extracts, etc. Potato pieces were used as solid surface for microbial growth. Gradually the requirement of solid media was strongly felt by scientists like Robert Koch, Fanny Eilshemius, Walther Hesse and others. Later on agar was recognized as a solidifying agent for culture.

6.1 CULTURING OF MICROBES

Growing of microorganisms in the laboratory is called culturing. Any species of microorganisms that synthetically grows in the laboratory is referred to as a 'culture' or 'strain'. For effective growth and sustainment of microorganism some nutrients and growth factors are essential. These are carbon, nitrogen, phosphorus, vitamins, proteins, etc. constituting the medium. Thus a medium is a liquid or solid substratum where microorganisms grow. A medium can selectively facilitate growth of a particular microorganism. It can also enrich the growth of certain organisms while suppressing the other. Although carbon and nitrogen are the common nutrients for any type of media, each microorganism has different nutritional and growth requirements. In general there are following three types of media:

(i) *Depending on physical state (medium's normal consistency)*: Solid medium, semisolid medium and liquid medium.

(ii) *Depending on chemical constituents*: Synthetic (Chemically defined) and non-synthetic (non-chemically defined) medium.

(iii) *On the basis of function and use*: Common purpose or supportive media, enrichment media, selective media, differential media.

Table 6.1 summarizes important types of media and their applications.

Table 6.1 Important types of media and their major purposes

Types of media	Specific and general purposes
Chemically defined	Used in the microbial assay, culturing of chemo- and photo-autotrophic microorganism.
Complex	Generally supports culture of heterotrophic bacteria.
Reducing	Growth of obligate anaerobes
Selective	Facilitates the growth of intended microorganisms, while restricting the others.
Differential	It can distinguish colonies of microorganisms of our interest from the rest.
Enrichment	Having similarity with selective media and is able to improve the detectability level of microorganisms.

6.1.1 Solid Medium

Solid medium invariably consists of a solidifying agent such as agar gel (extracted from sea algae and chemically constituted of polysaccharide agarose), gelatin, etc. Microorganisms grow on the surface of the solid media. Unlike broth, solid medium possesses a physical form that allows microbes to grow on its surface (as colonies or streaks). Agar is the most widely used solidifying agent. It is a golden-yellow granular powder prepared from seaweeds. It melts at 98°C and once melted does not solidify, until it reaches 42°C. It is resistant to degradation by most bacteria. Solid medium is usually named as (a) slants, (b) stabs, and (c) petri dishes (Figure 6.1 to Figure 6.3). In slants test-tubes containing a nutrient medium along with solidifying agar are allowed to solidify at an angle in order to get a flat inoculating surface on which microorganisms can grow. The stab tube is made by charging culture material by piercing the inoculation needle deep inside the agar (Figure 6.2). It is used to maintain stock culture for preservation of bacteria. Nutrient media and agar powder are mixed together in appropriate proportion, sterilized and then poured into Petri dishes of dimensions 15 × 100 mm or 15 × 150 mm to solidify. The agar plates act as a solid supporting media for bacterial growth. They remain intact and stable, by resisting.

Figure 6.1 Growth of microorganism on solid media on petri dish.

Figure 6.2 Growth of microorganism on solid media on stab.

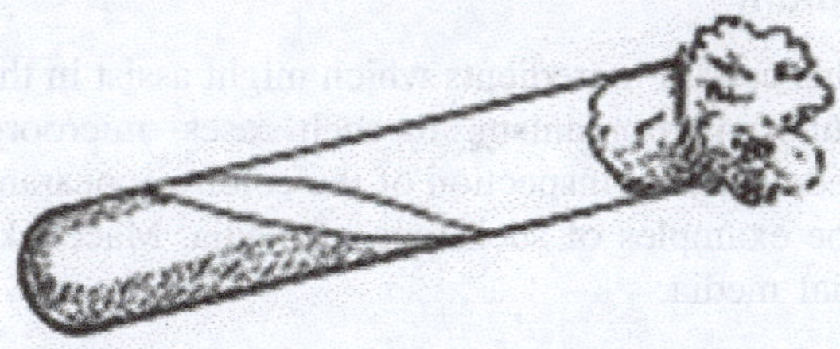

Figure 6.3 A diagram of a slant culture.

Agar media should be dispensed into Petri dishes by keeping uniformly a depth of 3–4 mm. Roughly, fifteen of 100 mm plates can be made by using 20 ml of liquid agar media.

6.1.2 Synthetic and Non-synthetic Medium

A media which is prepared by using pure chemical compounds (either analytical grade (AR) or laboratory reagent (LR) of predetermined quantity is called chemically defined media or synthetic media. This media do not contain any yeast, plant or animal tissue.

To prepare such type of media exact information about the nutritional requirement of microorganisms is required. A media is called non-synthetic (or undefined media), incase at least one of the ingredients is not present in a purified form or vary in quantity from one batch to other. Examples of such ingredients include protein extracts, semi-digested proteins, etc. A simple type of non-synthetic media is useful for growth of non-fastidious microorganisms while a more complex type is used for the growth of fastidious organisms. Czapek-Doxmedium, Dubo's culture medium with Tween 80 (commercial surfactant) is the example of synthetic medium. A non-synthetic media can be composed of yeast, beef extract, peptone and other ingredients.

6.1.3 Reducing Medium

Reducing media is particularly used to grow microorganisms which are anaerobic in nature. Obligate anaerobes need strict exclusion of oxygen to grow. Such media contains chemicals which can selectively react with oxygen and can scavenge it. Examples of such chemicals include sodium thioglyconate, sodium sulphite, ferrous carbonate, cysteine (as antioxidant), etc.

6.1.4 Selective Medium

As the name suggests, selective media facilitates the growth and cultivation of selected microorganisms while disfavoring the growth of undesirable ones. Selective media on solid agar enables us to identify and quantify colonies of intended microorganisms. Such media contain at least one such compound which inhibits or deters the growth of a particular microorganism. For example, growth of gram-positive bacteria can be restricted by using eosin methylene blue dyes in media, or brilliant green agar containing green dye. Sometimes agars having low pH can also retard bacterial growth. For example, malt extract or yeast extract agar.

6.1.5 Differential Medium

Such type of media contains certain ingredients which might assist in the growth and cultivation of more than one particular microorganism. In such cases, microorganisms can be rapidly distinguished and identified by visual inspection of the colony appearances. Mannitol salts agar, blood agar are some of the examples of such type of media. Macconky agar could serve both as selective and differential media.

6.1.6 Enrichment Medium

This kind of media contains additional proportion of nutrients to the base composition. For example, various trace elements, vitamins, egg yolk, serum, etc. are added. This helps the facile growth of fastidious microorganisms requiring more nutrients. Examples are blood agar, chocolate agar, etc. Halophilic bacteria require presence of additional salts in basal medium.

6.2 HOW TO PREPARE CULTURE MEDIA?

A culture medium is the mixture of nutrient substances which support the growth of microorganisms. It essentially contains all the ingredients to maintain microbial life. The constituents are water, carbon, nitrogen, phosphorus, sulphur and many other elements like zinc, iron, molybdenum, etc. These are supplied by inorganic or organic compounds depending upon the organisms to be cultivated. For example, photoautotrophs (organisms which can synthesize their growth materials by photosynthesis) and chemo-autotroph (organisms which can synthesize their growth materials by deriving energy from inorganic chemical oxidation reactions) require the necessary elements in inorganic forms whereas heterotrophs (organisms which derive energy from organic substances only) require the element in complex organic forms.

For the preparation of media distilled water is taken in Erlenmeyer flask. The ingredients are accurately weighed and mixed with water (except the agar in case of solid medium) taken in flask to get a clear solution. If needed, the solution can be filtered and pH needs to be adjusted. The remaining quantity of water is added with agar and is warmed in water bath with stirring to make the agar soluble. The medium is then dispensed to various containers and the mouths of the containers are plugged with non-absorbing cotton swabs and brown paper. Then the containers need to be sterilized by autoclaving at required temperature (generally 120–122°C) and pressure (15 psi) for required time (15 min).

6.3 COMPOSITION OF SOME COMMONLY USED MEDIA

The formulation and composition of nutrient agar is simple and appropriate for the growth of a wide spectrum of microorganisms. It can also be used in the quantification of microorganism in sewer water, dairy wastewater and other materials. Nutrient broth has also similar composition with nutrient agar excepting that it is devoid of agar. A typical composition of nutrient broth and nutrient agar is given in Table 6.2.

Table 6.2 Composition of Nutrient broth and Nutrient agar

Ingredients	Amount (g/lit)
Peptone	10
Beef extract	10
Sodium chloride	5.0
Agar	15
pH after sterilization	7.3 ± 0.1

Fungi and other microbes can easily grow using Czapek-Dox agar medium. It possesses sucrose and nitrate as 'C' and 'N' sources. Amongst the other ingredients, dipotassium hydrogen phosphate acts as buffer. Various other salts of potassium, magnesium and iron are used as sources of cations. The typical composition of Czapek-Dox agar medium is given in Table 6.3.

Table 6.3 Composition of Czapek-Dox agar medium

Ingredients	Amount (g/L)
Sucrose ($C_{12}H_{22}O_{11}$)	30.0
Sodium nitrate ($NaNO_3$)	2.0
Dipotassium hydrogen phosphate (K_2HPO_4)	1.0
Magnesium sulphate ($MgSO_4.7H_2O$	0.5
Potassium chloride (KCl)	0.5
Ferrous sulphate ($FeSO_4$)	0.01
Agar	15.0
Final pH after sterilization	7.3 ± 0.2

6.4 INOCULATION AND SUBCULTURE

Inoculation means the addition of microbes (generally from or to natural environments) to a specific growth medium or to a raw material used as substrate or to a microbial host.

Sub-culturing is the transfer of microbes from pure culture to another fresh growth medium. Inoculation is done in order to isolate microbes from their natural habitats or to see the effect of a particular microbe on their hosts under specific experimental conditions. Sub-culturing is generally done to transfer a microbial pure culture from a depleted medium to a fresh medium.

6.5 PHASES OF MICROBIAL GROWTH

An ideal growth curve of a microorganism contains several distinct phases (Figure 6.4). These are:

(i) Lag phase
(ii) Logarithmic or exponential growth phase
(iii) Deceleration phase
(iv) Stationary phase
(v) Death phase

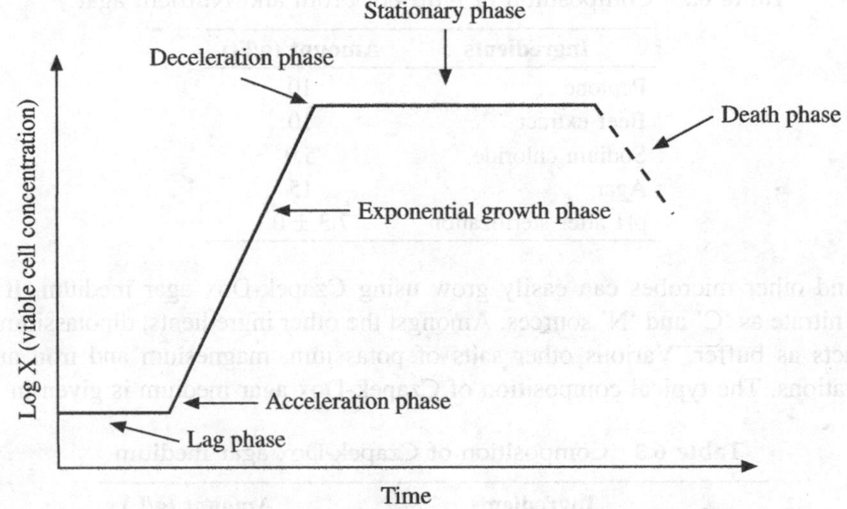

Figure 6.4 Typical phases of microbial growth.

6.5.1 Lag Phase

For any process to start it needs certain time to warm-up. Similar is the case of microbial growth. Indian delay period of the microbial life cycle is known as lag phase. No change in biomass cell concentration is observed during this phase. It is also termed as latent period. The newly inoculated cells into the medium tend to get acclimatized with the culture environment and physico-chemical conditions. Reorganization of molecular constituents, synthesis of new enzymes, adjustment of cellular stress due to water activity like rehydration or dehydration, activation of biochemical pathways, metabolic reprogramming could be some of the mechanisms by which cells attempt to adjust in lag phase prior to the exponential growth. Certain probable reasons for lag phase are as follows:

(i) Change in habitat
(ii) Dearth of key nutrients
(iii) Requisite quantity
(iv) PH, DO and other ambient conditions
(v) Inhibitors
(vi) Cellular stress due to osmosis

Length of lag phase differs from one microorganism to the other. In addition, growth conditions, age of inoculum and quantity have direct impact on the lag phase. The interval of lag phase could be ranging from few hours to even few days.

If the growth medium contains more than one carbon nutrient, microorganisms may exhibit two lag phases in their growth cycles. Such type of phenomenon is known as diauxic growth. Two carbon sources are utilized serially, the simpler one being consumed faster compared to the other one. When one carbon source is used up, microorganisms start using the other. During the cultivation of *E. coli* in a medium containing carbon sources glucose and lactose, it is observed that glucose is used initially as long as it is available. Once glucose is exhausted, lactose is utilized. Similar behavior has also been observed while using carbon sources such as arabinose, maltose, sorbitol, etc.

6.5.2 Exponential Phase

Log phase is another name of exponential phase. This is the most prevalent growth phase in the life cycle of microorganism. Here growth occurs at a very fast rate as cells divide by binary fission resulting in increase of cell count into the millions and billions. As such, the number of cells is calculated using logarithmic functions. In this phase, cells overcome their initial lag and get adequately acclimatized to the culture environment. They start multiplying at a fast rate. Balanced growth occurs during exponential phase. It is actually referred to as periodic doubling of biomass. Asexual cell division takes place by binary fission at a constant rate

In this phase, progressive cell division takes place. For example, from one cell to two cells, from two cells to four cells and so on.

The exponential growth rate is first order:

$$\frac{dX}{dt} = \mu_{net} X, \; X = X_o \text{ at } t = t_o \tag{6.1}$$

Eq.(6.1) can be integrated to get

$$\ln \frac{X}{X_o} = \mu_{net} t , \quad \text{or} \quad X = X_o e^{\mu_{net} \cdot t} \tag{6.2}$$

Where X and X_o are final and initial biomass concentration.
Eq.(6.2) gives the time during which microbial mass doubles.

$$\tau_d = \frac{\ln 2}{\mu_{net}} = \frac{0.693}{\mu_{net}} \tag{6.3}$$

Where τ_d is the doubling time of cell mass or biomass.

Time	Number of generations	Number of organisms
0	0	1
3 hr	9	2^9 (512)
10 hr	30	2^{30} (10^9)

Generation time

$$G = \frac{t}{n} \tag{6.4}$$

where, '*t*' is time duration (min or h)

n = number of generations (no of times the cell population doubles during the time interval

If *B* = initial count of bacteria

b = final count of bacteria

n = number of generations

$$b = B \times 2^n \tag{6.5}$$

$$\log b = \log B + n \log 2$$

$$n = \frac{\log b - \log B}{0.303} \tag{6.6}$$

$$n = 3.3 \log \frac{b}{B} \tag{6.7}$$

$$G = \frac{t}{n}$$

$$= \frac{t}{3.3 \log \frac{b}{B}} \tag{6.8}$$

6.5.3 Deceleration Phase

Towards the end of exponential phase deceleration phase sets in. Growth rate slows down considerably. The reason for slow rate of growth can be attributed to the exhaustion of essential nutrients. However, for many organisms this phase is difficult to notice and quantify.

6.5.4 Stationary Phase

Stationary phase occurs after the end of deceleration phase due to nutrients deficiency. As the nutrients become limiting, cells enter into the stationary phase. Amongst various nutrients, carbon deprivation is believed to be the major cause of stationary phase. In this phase no cell division takes place leading to zero net growth rate. In other words, cell growth rate equals the cell death rate. Despite having zero growth rates, cells can still produce some secondary metabolites. Microorganisms experience a number of physiological and morphological changes during this phase. For example, accumulation of certain chemicals such as glycogen, tetralose, thickening of cell wall, production of superoxide radicals, etc. Scientists have also reported appreciable changes in gene expression and synthesis of protein during this phase. Sometimes, cell lysis or disintegration may occur during stationary phase, whereby surviving cells try to feed on lysed products as these are rich with nutrients. It can be somewhat comparable to predator-prey type behavior. This type of growth is also called cryptic growth.

6.5.5 Death Phase

The death or declining phase marks the ending of microbial life cycle. This phase sets in after stationary phase. Main reasons for death phase include depletion of essential nutrients of the media, accumulation of toxins, contamination and presence of inhibitors. Some spores can withstand harsh nutrient starving conditions of declining phase and even can grow up to certain extent. But overall concentration of microbial population drops drastically in this phase.

6.6 MEASUREMENT OF MICROBIAL GROWTH BY VARIOUS METHODS

Measurement of microbial growth may be done by either of the two processes:

(i) Measurement of cell number density
(ii) Measurement of cell mass concentration

6.6.1 Measurement of Cell Number Density

Following methods are generally used:

Direct microscopic count (DMC)

DMC gives the total cell count of microbial population. It is carried out in Petroff-Hauser counting device or cytometer. The device consists of thick glass slides (0.02 mm depth at the centre) provided with accurately marked chambers and coverslips. Coverslips are laid on the slide. A drop of liquid cell culture or broth is placed on the coverslip by a micropipette. The liquid will be spread under the coverslip by capillary action. All cells in several large squares are counted. The method is quite simple, quick and user-friendly. However, appropriate dilution is required for dense cell suspension. An illustration of Petroff-Hauser counting device is given in Figure 6.5.

Viable plate count

In this method jellified solid agar medium is taken into petri dishes. The solid medium should contain nutrients in required amount. Microbial culture from broth or some other sources are transferred onto the petri dish with the help of inoculation loop or glass spreader after proper dilution. Dense and concentrated broth cannot be used as it leads to formation of colonies with non-discrete cells thereby making the count erroneous. In order to get a fair approximation of the number of cells, serial dilution of broth is done (as shown in the Figure 6.6.). The number of serial dilution is selected based on preliminary estimate of culture density. A fixed

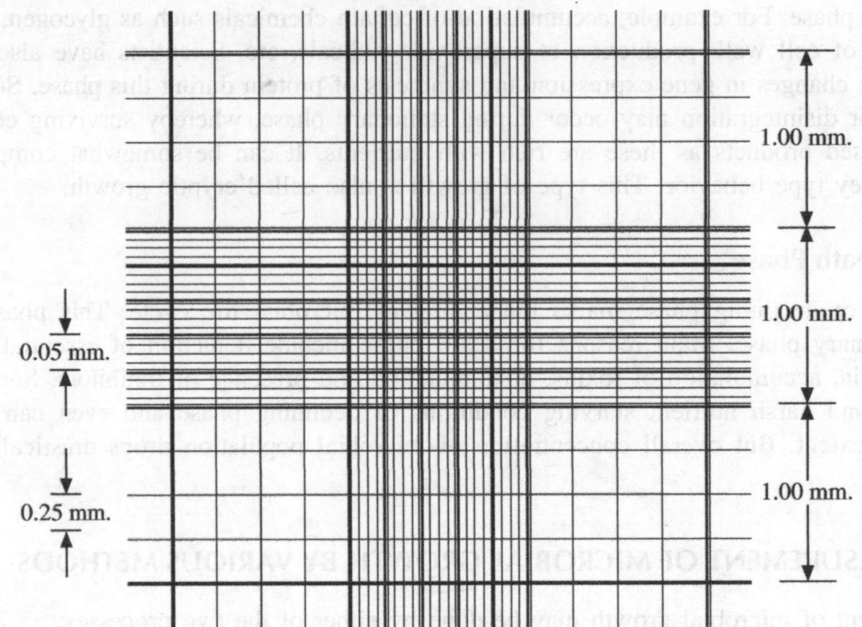

Figure 6.5 An illustration of Petroff-Hauser courting device.

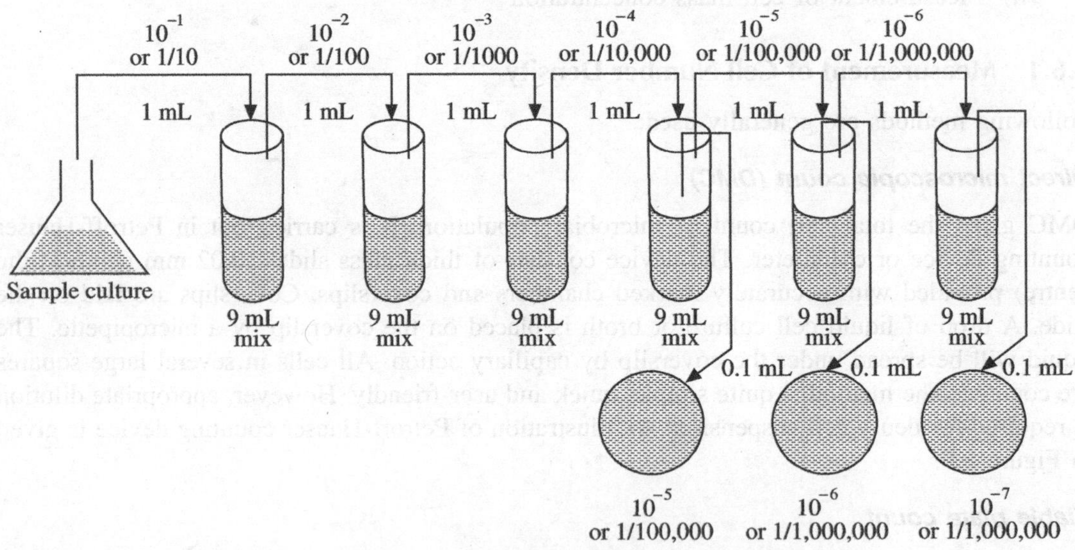

Figure 6.6 Schematic of plate count serial dilution procedure.

volume say 1 mL of original culture is taken in a test tube and is mixed with 9 mL of sterile broth. This represents a dilution factor of 10 or 1:10 or 10^{-1} as compared to original culture. Similarly from this dilution further diluted solutions are made. This process is then repeated.

Petri dishes are then incubated in an incubator maintained at 37°C temperature overnight. After about 24 h colonies of microbial cells are visible on the dishes. Single isolated colonies obtained during the plate count are shown in Figure 6.7. These are counted in terms of number of CFU (colony forming unit). Although this method is simple to perform, only living cells can respond to this method by forming colonies. Therefore, this method can give the count of viable cells only. This is the major difference of viable plate count method with other methods of quantification. Sometimes, clumps or chains of cells are also formed along with colonies.

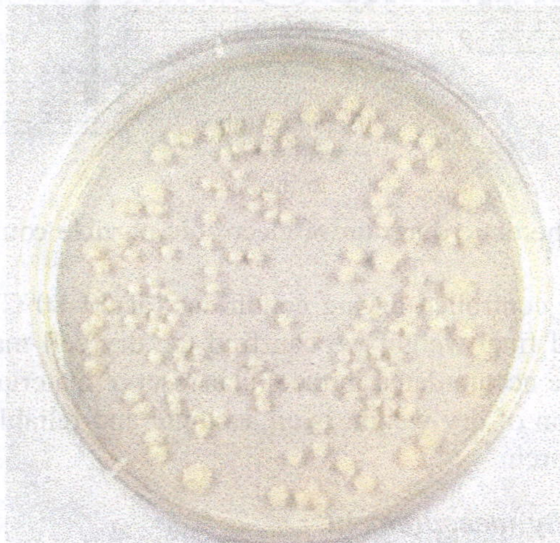

Figure 6.7 Single isolated colonies obtained during the plate count.

Particle counter

In this method a phototube is used to measure the intensity of scattered light from the cell suspension as the light passes through it. The concentration of cells is proportional to the intensity of scattered light.

Electronic particle count

In this method a suspension of microbial cells, whose number are to be determined, is kept in an electronic particle counter device (Figure 6.8). The cells are allowed to pass through an orifice (10-30 μm diameter) connected with two chambers containing electrically conductive solution. If a cell passes through the orifice, there is a momentary increase in electrical resistance between two chambers. As a result electric signals get recorded automatically. However, a disadvantage of this method is its inability to differentiate between dead and living cells.

6.6.2 Measurement of Cell Mass Concentration

Dry cell weight measurement

For dry cell weight (dcw) measurement a sample of broth is collected at a definite time. It centrifuged or filtered followed by washing the residue with water and buffer solution. The

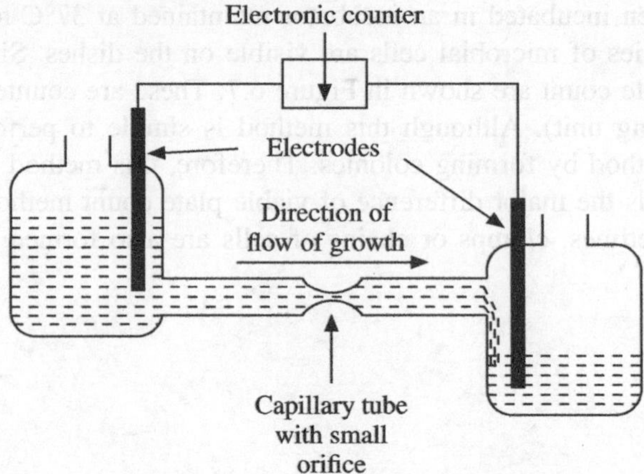

Figure 6.8 Schematic diagram of electronic particle count apparatus.

residue is then dried under controlled drying conditions (at 60–80°C) to avoid charring. From the difference of initial and final weight we can find the dry cell mass. Apart from cell mass dry weight also includes the accumulated intracellular reserve materials. Nonetheless, for many organisms, the determination of dry weight is an accurate and reliable way to measure growth and is widely used in research.

Turbidity or optical density measurement

An easy and quick determination of cell mass concentration is optical density or turbidity measurement. A schematic diagram of turbidity measurement technique is presented in Figure 6.9. Increase in turbidity or optical density is proportional to the cell mass concentration. It is measured in terms of intensity of transmitted light at a wave length 600–660 nm in a spectrophotometer. However, calibration of spectrophotometer needs to be done with known cell mass concentration.

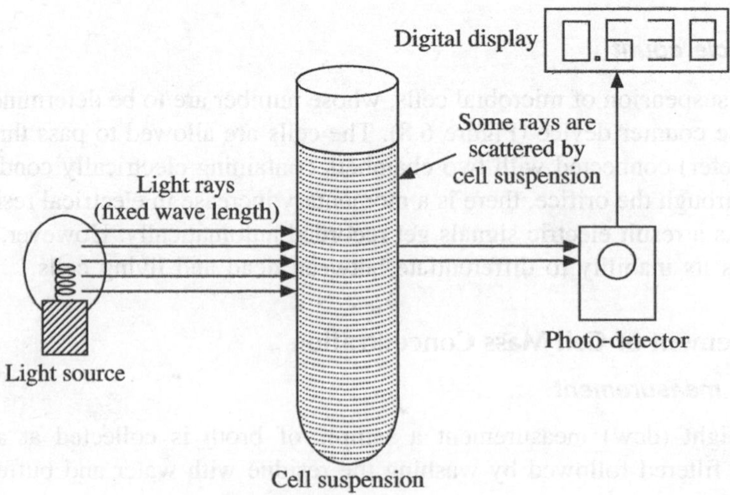

Figure 6.9 Schematic diagram of a turbidity measurement technique.

In addition to the above direct methods, there are some indirect methods of growth measurement. For example, growth can be indirectly measured by determining the substrate consumption or product formation, or the amount of RNA, DNA, and protein produced.

6.7 ENVIRONMENTAL FACTORS AFFECTING MICROBIAL GROWTH

Microbial growth and product formation are functions of several physico-chemical conditions. These are categorized as Extrinsic factors (such as temperature, relative humidity, CO_2 or O_2, number of organisms in the culture) and intrinsic factors (such as pH, moisture content, water activity, redox potential, available nutrients, etc.).

6.7.1 Effect of Temperature

Temperature is a significant parameter of microbial growth and reproduction. For many microorganisms rise in temperature accompanies with the increase the rate of microbial growth, however, up to certain temperature and thereafter it decreases. The temperature at which maximum growth rate occurs is called optimum growth temperature. It is generally 37 ± 1°C for many microbes. At very high temperature deactivation of enzymes responsible for growth takes place. This is due to denaturation of the constituent intracellular proteins at high temperature. High temperature may also lead to evaporative loss of water from intracellular or extracellular environment. In addition, there is an increase in cell maintenance coefficient at elevated temperature resulting in lower yield coefficient. Based on temperature withstanding characteristics of microorganisms, these are classified as psychrophiles, mesophiles and thermophiles. For psychrophilic bacteria optimum growth temperature is less than or equal to 20°C; for mesophillic it ranges between 20–45°C, and for thermophillic it is greater than 50°C. Bacterial species isolated from hot climatic ecosystems like hot springs, deserts, etc. are thermophillic in nature. Temperature dependence of microbial growth rate can be explained using Arrhenius equation.

6.7.2 Effect of Medium pH

Titrable acidity of the media measured in terms of pH has a major bearing on the microbial growth. Each microorganism has an optimum pH at which the growth rate becomes maximal. For most of the cases 6.5 ± 1 is the optimum pH for microbial growth. Acidophillic organisms can withstand low pH (pH: 2) while alkaliphiles withstand high pH conditions (pH: 9–11). Examples of acidophiles are *Lactobacillus*, *Sulfolobus*, etc. and those of alkaliphiles include *Vibrio choleri* (grows at pH: 8). *Helicobacter pylori* can survive at very low pH environment of stomach. This microorganism was identified by Marshall and Warren in 2005 and for which they received Nobel Prize in medicine in 2005.

6.7.3 Effect of DO

Dissolved oxygen (DO) is one of the important parameters for water quality. Strictly aerobic bacteria (obligate aerobes) cannot grow and survive without the presence of molecular oxygen. Aerobic bacteria utilize oxygen as an electron acceptor during respiration. Anaerobic bacteria such as *Streptococcus pyogenes* do not need oxygen for growth. Bacteria which can grow in abundance concentration of CO_2 but with less O_2 concentration are called capnophiles. Another type of microorganisms, microaerophiles can grow even in lower concentration of oxygen (less than 21%).

Oxygen is sparingly soluble in water. The saturation value of DO in water is roughly in the range of 10–15 mg/L depending on temperature. Hence, single serving of oxygen into the broth is not sufficient to sustain the biochemical oxidation reaction to completion. Thus, continuous bubbling of oxygen is provided through aeration. Air spargers are used.

Gas to liquid phase oxygen transfer can be given by basic transport equation (Eq 6.9):

$$N_{O_2} = k_L a (C^* - C_L) = OTR \tag{6.9}$$

where k_L is the oxygen transfer coefficient (cm/h), 'a' is the gas-liquid interfacial area (cm^2/cm^3), k_La is the volumetric oxygen transfer coefficient (h^{-1}), C* is saturated DO concentration (mg/l), C_L is the actual DO concentration in the broth (mg/l) and the NO_2 is the rate of oxygen transfer.

6.7.4 Moisture or Humidity

Available water in the growth media maintains the metabolic activity of cells. For a balanced growth of microorganisms moisture content or the presence of free water molecules is desirable. A relative humidity of 85–90% prevents evaporative loss of water from the media. A drop of moisture content may cause desiccation of microbial cells. Growth of fungal microorganisms is facilitated by high relative humidity and moderate temperature.

6.7.5 Ionic Strength

Ionic strength of the media has some direct and indirect influences on cellular systems. Increased ionic strength results in decreased solubility of proteins. Also a shift of pH optimum takes place with variation of ionic charges. Ionic strength affects surface charge thereby influencing adhesion and transport behavior of bacterial cells. Presence of certain ions such as K^+, Ca^{2+}, Mg^{2+}, Zn^{2+} are beneficial for the synthesis of certain enzymes. Halophilic bacteria require high salt concentration for their growth.

6.7.6 Osmotic Pressure

Osmosis is a natural phenomenon in which solvent flows through a semipermeable membrane from a low concentration region to a higher concentration region. The pressure required for osmotic flow is called osmotic pressure, which is a function of solute particle concentration in the solution. If the osmotic pressure of the external cell environment is reduced, water inflow occurs in the cell. This might result in swelling of cells or even cell lysis. An opposite situation might arise, whereby water from intracellular region may flow out through the cell membrane, causing shrinkage or dehydration. Alteration of water inflow or outflow with change in osmotic pressure can affect many cellular properties.

6.7.7 Light

Light is an important parameter for the growth of photosynthetic microorganism. These microorganisms can convert light energy into biochemically useful energy both under aerobic or anaerobic conditions. Sunlight or even artificial light source can be used for their growth and cultivation. However, intensity of light and duration of exposure are important parameters for the effectiveness of light utilization.

6.8 MICROBIAL GROWTH RATE

Malthus's law of population can be applicable in the exponential growth of microorganisms. This is written as:

$$\frac{dX}{dt} = \mu X \tag{6.10}$$

where X is the mass of cell per unit volume of a batch culture or simply biomass concentration, μ is the specific growth rate of the microorganism and t is the time of growth.

6.8.1 Monod Equation of Growth Kinetics

Jacques Monod was a French biochemist who performed experiments to correlate microbial growth rates with single substrate limiting concentration using bacteria.

Even though microorganisms need several substrates for its growth, but there would be one particular substrate present in excess of requirements and the concentration of which affects the cell growth and changes the growth phase from the exponential to the stationary phase. Such a substrate is termed as 'limiting substrate'.

The Figure 6.10 represents a plot of substrate concentration (S) against specific growth rate (μ). As S increases, specific growth rate also increases but up to a certain extent and ultimately reaches a plateau. The maximum plateau value which μ reaches is called maximum specific growth rate constant (μ_m). This behavior of the substrate in the growth can be explained by Monod equation, which can be written as:

$$\mu = \frac{\mu_m S}{K_S + S} \tag{6.11}$$

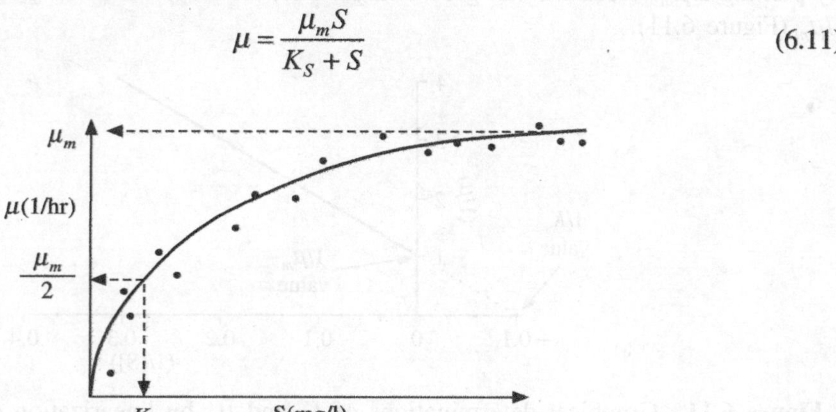

Figure 6.10 Graphical representation of Monod equation.

This equation resembles to the Michaelis–Menten expression for enzyme kinetics. In this equation μ is the specific growth, μ_m is the maximum specific growth rate constant in absence of substrate limitation (which is fixed for a particular microorganism under specified conditions), S is the substrate concentration and K_s is the substrate concentration which allows the organism to grow at 0.5 μ_{max} or half saturation constant, i.e. when:

$$\mu = \frac{\mu_m}{2}$$

which is evident by substituting $\frac{\mu_m}{2}$ for μ in Eq.(6.11) as:

$$\frac{\mu_m}{2} = \frac{\mu_m S}{K_S + S} \tag{6.12}$$

If we cancel μ_m on both the sides and after rearrangement $K_S = S$

For the condition of $S >> K_S$, the denominator of Eq.(6.11) may be equated to simply S, in which case

$$\mu = \frac{\mu_m S}{K_S + S} \approx \frac{\mu_m S}{S} \approx \mu_m \tag{6.13}$$

Therefore, the above equation predicts the percentage increase in μm as compared to μ when S is very large.

6.8.2 Kinetic Parameters of Monod Equation

The kinetic parameters of the Monod equation μ_m and K_S can be determined by linearizing the Monod equation after writing in the reciprocal form.

The Monod equation (Eq 6.11) can be written in a reciprocal form as:

$$\frac{1}{\mu_m} = \frac{K_S}{\mu_m}\left(\frac{1}{S}\right) + \frac{1}{\mu_m} \tag{6.14}$$

By plotting $1/\mu_m$ versus 1/S can get a straight line, the slope of which gives K_S/μ_m and intercept $1/\mu_m$ (Figure 6.11).

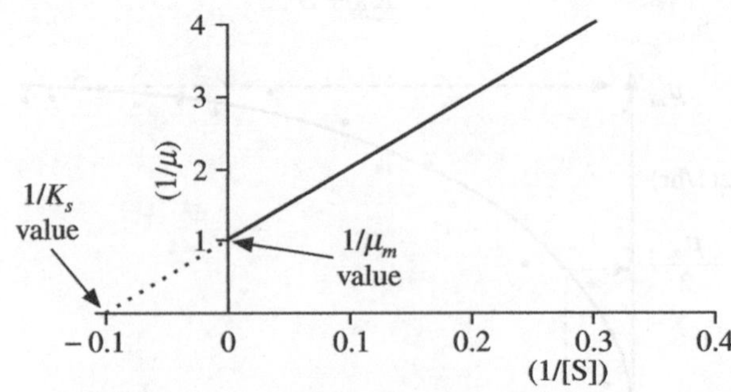

Figure 6.11 Graphical determinations of K_S and μ_m by linearization of Monod equation.

6.8.3 Limitations of Monod Equation

Classical Monod equation in its original version suffers from certain limitation for its wider application to study the growth kinetics of microorganism. These are briefly outlined below:

(i) Monod equation is valid for a single substrate growth media. Many a time when multiple substrates are present in media Monod model does not hold good.
(ii) The equation in its original form does not take into account the maintenance energy requirement of the microbial cells.
(iii) It does not account for substrate inhibition.
(iv) It does not account for product inhibition.
(v) It does not account for death of cells as cell death is a concurrent process along with growth, although its rate is negligible compared to growth rate.
(vi) Monod equation is based on a few assumptions of limiting substrate concentration. For finite value of S, the value of μ will be finite. For the cases where $S > K_S$, $\mu \rightarrow \mu_{max}$ but for the cases of $S < K_S$. Monod equation does not provide any explanation.
(vii) Monod equation does not take care of any other products that might be formed during growth process.

Endogenous and maintenance metabolism

According to Pirt (1965) the energy utilized for carrying out functions other than cellular growth is called maintenance energy. It is also known as endogenous and maintenance metabolism of cells. By this we mean there are reactions in cells which consume cell substance. Examples of cellular functions which are independent of growth rate include shifts in metabolic pathways, cell motility, osmoregulation, change in stored polymeric carbon, synthesis and inter-conversion of molecules such as enzymes and RNA, oxygen stress resistance and so on. Pirt pointed out that the experimental yield coefficient pertaining to cell mass production does not include maintenance energy. Accordingly we can write:

Total consumption of substrate = consumption for growth + consumption for maintenance

$$\frac{ds}{dt} = \frac{dX/dt}{Y_G} + mX \tag{6.15}$$

where, X: biomass concentration, S: substrate concentration, Y_G: true growth yield or yield coefficient.; m: maintenance coefficient.

Dividing both sides of Eq.(6.15) with X, we get

$$\frac{1}{X}\frac{ds}{dt} = \frac{1}{X}\frac{dX}{dt}\frac{1}{Y_G} + m \tag{6.16}$$

But specific growth rate as obtained from Eq.(6.10) can be written as:

$$\mu = \frac{1}{X}\frac{dX}{dt} \tag{6.17}$$

$$\frac{1}{X}\frac{ds}{dt} = \mu\frac{1}{Y_G} + m \quad (6.18)$$

or

$$\frac{1}{X}\frac{ds}{dt}\frac{1}{\mu} = \frac{1}{Y_G} + \frac{m}{\mu}$$

or

$$\frac{1}{X}\frac{ds}{dt}\frac{1}{\frac{1}{X}\frac{dX}{dt}} = \frac{1}{Y_G} + \frac{m}{\mu}$$

or

$$\frac{ds}{dx} = \frac{1}{Y_G} + \frac{m}{\mu} \quad (6.19)$$

Monod equation is modified by the incorporation of additional term to account for the maintenance energy as:

$$\mu = \mu_m \frac{S}{K_S + S} - k_d \quad (6.20)$$

Where k_d is endogenous respiration coefficient or specific maintenance coefficient. k_d is proportional to 'm', the maintenance coefficient.

$$k_d = mY_G \quad (6.21)$$

Substrate inhibition

It has been observed that high concentration of substrate could inhibit growth and multiplication of microorganisms. At low substrate concentrations, the specific growth rate (μ) is nearly zero, indicating a negligible change in microbial concentration. As the substrate concentration is increased beyond a certain threshold, the specific growth rate gradually becomes proportional to the substrate concentration and reaches a maxima. A further increase in the substrate concentration ultimately leads to a decrease in the specific growth rate shown in Figure 6.12. This is called substrate inhibition which is observed in several microbial fermentation processes. The actual mechanism of inhibition has not been known conclusively. At a high substrate concentration, a possible adsorption or complexation type interaction takes place between enzymes, coenzymes, and substrate thereby reducing the enzyme activity. Moreover, an increase in the substrate concentration could also alter the cell metabolism like overproduction of a particular molecule by one pathway and so on.

Edward modified Monod equation by incorporating inhibition term. The modified equation can be written as:

$$\mu = \mu_m \frac{S}{K_S + S + \left(\frac{S^2}{K_I}\right)} \quad (6.22)$$

Differentiating the above equation with respect to Sand setting results equal to zero, we get

$$\frac{d\mu}{dS} = \mu m \left[\frac{\left(K_S + S + \frac{S^2}{K_I}\right) - S\left\{1 + \frac{2S}{K_I}\right\}}{\left(K_S + S + \frac{S^2}{K_I}\right)^2} \right] = 0 \tag{6.23}$$

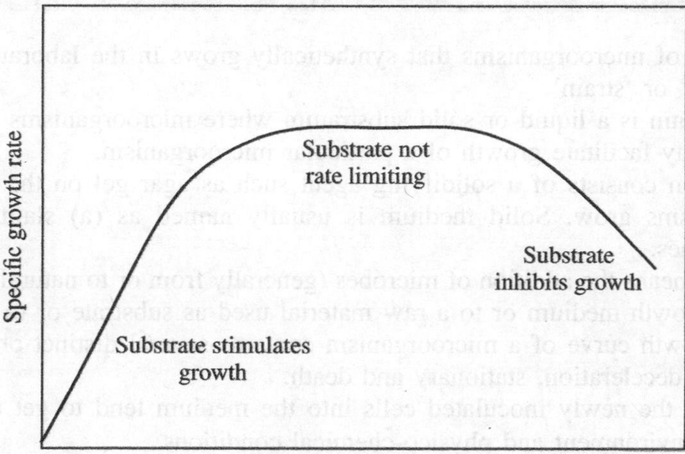

Figure 6.12 Substrate inhibition at higher substrate concentration.

Solving we get $\frac{S^2}{K_I} = K_S$ or $S = \sqrt{K_I K_S}$ (6.24)

Substrate inhibition, as stated earlier, not only results in the decreased specific growth rate but also decreased biomass yield coefficient.

Product inhibition

Another case of inhibition of microbial growth is that of product inhibition. The accumulation of end products of fermentation type reactions such as ethyl, propyl, isopropyl or butyl alcohols, acetone, lactic acid, etc. could lead to decrease in the rate of specific growth and product formation. Accumulation of alkanols may also inhibit transport of sugars and amino acids, adversely affect the cell membrane potential and proton gradient and at times could deactivate certain enzyme functions. Some of the cell metabolic activities cease to take place at very high product concentration.

Aiba et al. modified Monod equation by incorporating product inhibition term. The modified equation can be written as Eq.(6.25):

$$\mu = \frac{\mu_m S}{K_S + S} e^{-K_1 P} \tag{6.25}$$

where P is the product concentration and K_1 is a constant.

An alternative form of above equation was also proposed by Aiba et al.

$$\mu = \frac{\mu_m S}{K_S + S} \cdot \frac{K_P}{K_P + P} \tag{6.26}$$

In both cases expressions reduce to the original (unmodified) Monod equation, when the product concentration is zero.

KEY POINTS TO REMEMBER

- Any species of microorganisms that synthetically grows in the laboratory is referred to as a 'culture' or 'strain'.
- Thus a medium is a liquid or solid substratum where microorganisms grow. A medium can selectively facilitate growth of a particular microorganism.
- Solid medium consists of a solidifying agent such as agar gel on the surface of which microorganisms grow. Solid medium is usually named as (a) slants, (b) stabs, and (c) petri dishes.
- Inoculation means the addition of microbes (generally from or to natural environments) to a specific growth medium or to a raw material used as substrate or to a microbial host.
- An ideal growth curve of a microorganism contains several distinct phases such as lag, exponential, deceleration, stationary and death.
- In lag phase the newly inoculated cells into the medium tend to get acclimatized with the culture environment and physico-chemical conditions.
- The exponential growth rate is first order: $\frac{dX}{dt} = \mu_{net} X$, $X = X_o$ at $t = t_o$.
- Measurement of microbial growth can be done by either measuring cell number density or by cell mass concentration
- Cell numbers can be estimated by direct microscopic count, viable plate count, electronic particle count, etc.
- Cell mass concentration can be estimated by dry cell weight measurement, turbidity or optical density measurement by spectrophotometer.
- Various physico-chemical parameters affecting the cell growth are temperature, pH, dissolved oxygen, humidity, ionic strength, osmotic pressure, light, etc.
- Microbial growth can be explained by Monod equation, which can be written as $\mu = \frac{\mu_m S}{K_S + S}$ where S is the substrate concentration and K_s is the substrate concentration which allows the organism to grow at 0.5 μ_{max} or half saturation constant.
- The kinetic parameters of Monod equation μ_m and K_S can be determined by linearizing Monod equation using Lineweaver-Burk Plot.
- The energy utilized for carrying out functions other than cellular growth is called maintenance energy. It is also known as endogenous and maintenance metabolism of cells.
- As the substrate concentration is increased beyond a certain threshold, the specific growth decreased due to substrate inhibition.
- The accumulation of end products of fermentation type reactions such as ethyl, propyl, isopropyl or butyl alcohols, acetone, lactic acid, etc. could lead to decrease in the rate of specific growth. This is known as product inhibition.

? QUESTIONS FOR PRACTICE

Multiple Choice Questions

1. Which of the following is not the essential equipment used for cell culture?
   (a) Laminar hood (b) Spectrophotometer
   (c) Incubator (d) Vertical autoclave
2. Which of the following piece(s) of equipment is/are required to count cells in cell culture?
   (i) Timer; (ii) Hemocytometer; (iii) Microscope
   (a) i and ii (b) ii only
   (c) ii and iii (d) All of the above
3. Which microorganism(s) among the following perform photosynthesis by utilizing light?
   (a) Cyanobacteria, Fungi and Viruses (b) Viruses
   (c) Cyanobacteria (b) Fungi
4. Bacteria living in salt marshes are most likely which of the following?
   (a) Acidophiles (b) Barophiles
   (c) Halophiles (d) Thermophiles
5. Which of the following environments would harbor psychrophiles?
   (a) Mountain lake with a water temperature of 12°C
   (b) Contaminated plates left in a 35°C incubator
   (c) Yogurt cultured at room temperature
   (d) Salt pond in the desert with a daytime temperature of 34°C
6. Bacteria isolated from a hot tub at 39°C are probably which of the following?
   (a) Thermophiles (b) Psychrotrophs
   (c) Mesophiles (d) Hyperthermophiles
7. In which environment are you most likely to encounter an acidophile?
   (a) Human blood at pH 7.2 (b) A hot vent at pH 1.5
   (c) Human intestine at pH 8.5 (d) Milk at pH 6.5
8. Which of the following is the best definition of generation time in a bacterium?
   (a) The length of time it takes to reach the log phase
   (b) The length of time it takes for a population of cells to double
   (c) The time it takes to reach stationary phase
   (d) The length of time of the exponential phase
9. During batch fermentation, in which phase the microbes in the fermenter are adapting to the new environment?
   (a) Lag phase (b) Log phase
   (c) Stationary phase (d) Death phase
10. A media that contains additional proportion of nutrients to the base composition is known as
   (a) Reducing media (b) Enrichment media
   (c) Selective media (d) Synthetic media

11. Which of the following is considered to be both selective and differential media?
(a) Macconky agar medium (b) Czapek-Dox medium
(c) Blood agar medium (d) Mannitol salt agar medium

12. The rate of generation of new bacterial cell become equal to the death rate during which phase?
(a) Lag phase (b) Log phase
(c) Stationary phase (d) Decline phase

13. During the decline phase
(a) Microorganisms try to settle in the new environment
(b) Shows exponential growth
(c) Death rate is equal to the rate of generation of new cell
(d) Number of live cells decreases

14. Growth of bacteria or microorganisms refer to
(a) An increase in the size of an individual organism
(b) An increase in the mass of an individual organism
(c) Changes in the total population
(d) An increase in number of cells

15. An organic nutrient essential to an organism's metabolism that cannot be synthesized itself is termed a/an:
(a) Trace element (b) Micronutrient
(c) Growth factor (d) Essential nutrient

16. Which of the following statements is true for all organisms?
(a) They require organic nutrients (b) They require inorganic nutrients
(c) They require growth factors (d) They require oxygen gas

17. Active transport of a substance across a membrane requires:
(a) A gradient (b) The expenditure of ATP
(c) Water (d) Diffusion

18. A cell exposed to a hypertonic environment will ________ by osmosis:
(a) Gain water (b) Lose water
(c) Neither gain nor lose water (d) Burst

19. The temperature-pressure combination for an autoclave is
(a) 100°C and 4 psi (b) 121°C and 15 psi
(c) 131°C and 9 psi (d) 115°C and 3 psi

20. Which of the following is used to grow bacterial cultures continuously?
(a) Chemostat (b) Thermostat
(c) Haemostat (d) Coulter counter

21. Which phase has the condition of specific growth rate $\mu = 0$
(a) Lag phase (b) Log phase
(c) Stationary phase (d) Death phase

22. Which of the following is a limitation of the basic form of Monod equation for microbial growth?

(a) It is applicable for single substrate system only
(b) It does not account for substrate inhibition of growth
(c) It does not consider product inhibition of growth
(d) All of the above

23. Which growth phase is usually longer in continuous culture?

(a) Lag (b) Exponential
(c) Stationary (d) Death

24. Presence of two lag phases are generally encountered in which type of growth?

(a) Diauxic growth (b) Cryptic growth
(c) Exponential growth (d) Chronic growth

25. A higher K_s value in Monod equation of microbial growth means

(a) Lower affinities to substrates
(b) Greater affinities to substrates
(c) Lower Dissociation constant value
(d) Unaffected with the substrate concentration

26. The Monod model relates

(a) Specific growth rate to the substrate availability
(b) Substrate utilized with the biomass consumption
(c) The biomass concentration with specific growth rate
(d) Yield with the biomass utilization

27. Population doubling time τ_d can be expressed as (where μ is the specific growth rate)

(a) $\ln 2/\mu$ (b) $\log 2/\mu$
(c) $\mu/\log 2$ (d) $\mu/\ln 2$

Answer Key

1. (b) 2. (c) 3. (c) 4. (c) 5. (a) 6. (c) 7. (b) 8. (b) 9. (a) 10. (b)
11. (a) 12. (c) 13. (d) 14. (d) 15. (c) 16. (b) 17. (b) 18. (b) 19. (b) 20. (a)
21. (d) 22. (d) 23. (b) 24. (a) 25. (b) 26. (a) 27. (a)

Descriptive Questions

1. Name a solidifying agent used for the preparation of solid media.

2. Name carbon and nitrogen sources (02 each) for media preparation.

3. In connection to microbial growth state the following:

(i) Reasons for lag phase and measures to reduce it.
(ii) Serial utilization of carbon substrate.
(iii) Environmental factors affecting growth.

4. Write down an approximate synthetic medium composition for yeast with all ingredients and amounts. Take a volume of 2 litre.

5. Cite an example of the following in microbial systems:
   (i) Symbiotic relationship
   (ii) Parasitic relationship
   (iii) Predator-Prey relationship
6. Calculate the hydrogen ion concentration for blood plasma and gastric juice having pH 7.4 and 1.9 respectively.
7. A biokinetic study of dephenolation of petrochemical wastewater is being carried out with *Pseudomonas putida*. Outline the approximate concentration profiles of substrate (carbon substrate, take phenol in this case) and product, assuming substrate inhibition takes place after 75% of reaction time is elapsed in a batch of 40h.
8. State various methods of cell growth measurements. Frame out equations to correlate the simulated values of substrate concentration (S) as function of time (t) in terms of biomass concentration (X), yield coefficients ($Y_{X/S}$) and constants of Monod equation.
9. Write down Monod equation of microbial growth kinetics. Explain various terms in the same equation. How do you determine the kinetic parameters of the above equation graphically?
10. A strain of *Aspergillus niger* was grown in a batch reactor using fructose as the carbon source. The following experimental data were recorded.

Time (h)	Biomass concentration (g/L)	Substrate concentration (g/L)
0.0	1.20	98
9.0	2.35	94
16.0	5.1	89.0
23	10.5	80.1
30	20.5	75.2
36	37.5	48.1
40	41	20.6

Calculate the maximum net specific growth rate from the above data. What would be the maximum cell concentration if 140 g of fructose were used with same inoculum size?

11. A batch fermentation generated the following data.

Time (h)	Biomass (g/L)	Substrate (g/L)
0	0.1	10.0
1	0.15	9.9
2	0.2	9.8
3	0.25	9.7
4	0.68	8.8
5	1.8	6.5
5.5	3.0	4.1
6	5.0	0.2
7	5.3	0
8	5.3	0

Determine the length of the lag phase, specific growth rate, and yield coefficient based on biomass.

11. State with a diagram different phases of microbial growth. What are diauxic growth and cryptic growth?

12. The specific growth rate for inhibited growth in a chemostat is given by the following equation:

$$\mu_g = \frac{\mu_m S}{K_s + S + \dfrac{IK_s}{K_I}}$$

where S_o = 10 g/L, K_s = 1 g/L, I = 0.05 g/L, $Y_{x/s}$ = 0.1 g cells/g subs; μ_m = 0.5 h^{-1}, K_I = 0.01 g/L

(i) Determine X and S as a function of D when I = 0.

(ii) With inhibitor added to a chemostat, determine the effluent substrate concentration and X as a function of D.

13. Consider a fermentation reaction is being carried out in an Erlenmeyer flask of 1L capacity. The medium is constituted of glucose as principal carbon source together with some inorganic nitrogen source vitamins and trace elements. It is inoculated with 10% (v/v) culture of *Saccharomyces cerevisiae* (yeast) grown overnight. The temperature is 37°C and initial pH of the medium is 6.5. You are required to study the growth kinetics of the process. State the following:

(i) Which basic equation would you use? Explain different parameters of the equation and how would you find out those parameters?

(ii) Formulate an approximate composition of the medium (take 500 ml as liquid volume).

(iii) Explain how do you find out experimentally substrate and biomass concentration? (State only the name of protocol).

(iv) Is there any limitation of the above equation you have used? Explain.

(v) It is observed that pH of the medium drops in course of reaction. Would it affect the fermentation process? If so, how?

7 Transport in Bioprocess Systems

7.1 INTRODUCTION

Various transport phenomena are observed during growth and cultivation of microorganisms in bioreactor systems. Transport of nutrients and oxygen in aerobic fermentation processes involves various modes of gas-liquid mass transfer depending on broth characteristics, type of gas liquid contact and other physico-chemical environment. There is also solid-liquid mass transfer mostly in case of immobilized enzyme systems. Many of the bioreactions being mildly exothermic generate heat energy. This heat transfer is also an important consideration. In addition cultivation of a particular microorganism in the laboratory essentially requires the strict exclusion of any other interfering organisms. The presence of other undesirable microbes results in contamination of the media and reactor. The contaminating microorganisms compete for available nutrients thereby hinder its growth. As a result, the growth of intended microorganism is taken over by some other species with accumulation of toxic products. To avoid this, sterilization becomes an extremely essential operation prior to any microbial culture.

7.2 MASS TRANSFER

In the following section different types of mass transfer that are encountered in microbial systems are briefly presented along with the important governing equations, underlying theories and correlations.

7.2.1 Molecular and Convective Mass Transfer

Mass transfer is the transport of any component in a mixture from a region having high concentration to the one with low concentration. The process is fundamentally driven by the difference in concentration. Mass can be transferred by random molecular motion in quiescent fluids. This mode of transport is called molecular mass transfer. When mass is transferred from a surface into a moving fluid, flow behavior, the mode of mass transfer is called convective mass transfer. Various types of molecular and convective mass transfer are presented diagrammatically in Figure 7.1. This classification is based on diversity of driving forces responsible for mass

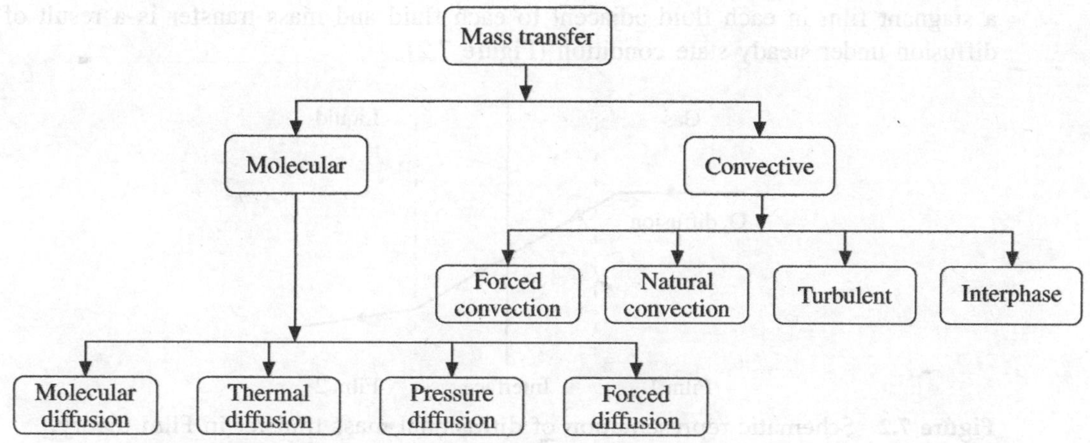

Figure 7.1 Various types of molecular and convective mass transfer.

transfer. For example molecular diffusion occurs due to concentration, whereas thermal diffusion is an outcome of temperature gradient. Similarly pressure diffusion arises from a pressure gradient while forced diffusion results from external forces other than gravity.

For convective mass transfer apart from forced and natural convections, turbulent mass transfer results from eddy currents in a fluid, while interface mass transfer results from non-equilibrium at an interface.

7.2.2 Fick's Law and other Theories

Aerobic fermentation process requires continuous supply of oxygen into the broth for growth and maintenance of microbial cells. The movement of oxygen from the source to the cells via the liquid medium experiences a series of mass transfer resistances. The major mechanism of oxygen transfer in microbial cells is considered to diffusion.

Mass transfer by diffusion

The rate equation for mass diffusion is represented by the classical equation as denoted by Flick's law which is an empirical relationship that defines the diffusion of species *A* in system where temperature and pressure are constant. Considering diffusion only in *x*-direction, the Fick's rate equation is given by Eq.(7.1):

$$J_{A,x} = -D_{AB}\frac{dC_A}{dx} \tag{7.1}$$

In the above equation *J* stands for flux of solute '*A*' in *x*-direction; D_{AB} stands for diffusivity of solute *A* through *B* (m^2/s); $\frac{dC_A}{dx}$ is the concentration gradient (considering *x*-direction). Use of minus sign indicates a trend of decreasing solute concentration C_A with increasing *x*. D_{AB} is a constant for a particular system and is affected by temperature, pressure and composition.

There are several theories of diffusional mass transfer. These are briefly outlined here.

(i) *Film theory*: It is the earliest and the simplest theory of interphase mass transport proposed by the Lewis and Whitman (1924). According to this theory there exists a stagnant film in each fluid adjacent to each fluid and mass transfer is a result of diffusion under steady state condition (Figure 7.2).

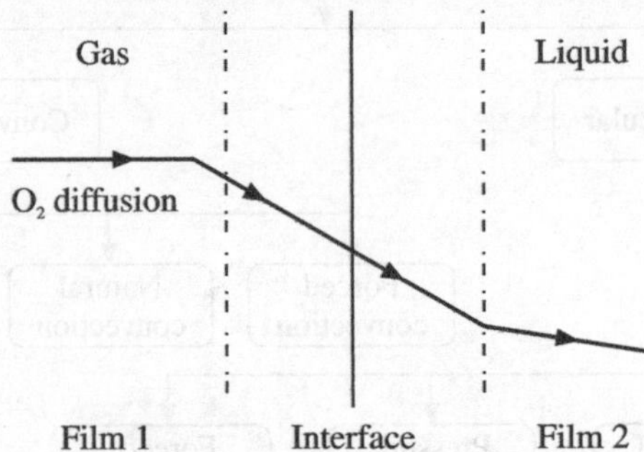

Figure 7.2 Schematic representation of diffusional mass transfer in Film theory.

Fick's law equation can be integrated considering boundary conditions at the two end of the film. A linear plot is obtained. Accordingly following equation for mass transfer coefficient (k_L) can be represented as Eq.(7.2):

$$k_L = \frac{D_{AB}}{\delta} \tag{7.2}$$

(ii) *Penetration theory*: Higbie's penetration theory (proposed in 1935) is based on the transient mass transfer. It considers short contact time which is quite likely in many industrial gas-liquid operations. This theory predicts $k_L \propto D^{05}$. The exponent is much lower than film theory's prediction of diffusivity dependence on k_L. The model involves only one unknown parameter θ, the contact time between gas and liquid element. Hence, the simple mathematical expression can be presented as Eq.(7.3):

$$k_L = 2\sqrt{\frac{D_{AB}}{\pi\theta}} \tag{7.3}$$

(iii) *Surface renewal theory*: An important modification of penetration theory was put forward by Danckwerts in 1951 in his surface renewal theory. This was based on the assumption that there are wide spectra of contact time causing a continual replacement of surface element. Surface agae of an eddy element may be anything from zero to infinity. The probability of replacement of an eddy with a newer one is independent of its residence time at the interface. Mathematical expression of this theory can be given by Eq.(7.4):

$$k_L = \sqrt{D_{AB}s} \tag{7.4}$$

where's' is the fraction of surface area that is replaced with fresh liquid in a unit time and will have the units of s^{-1}.

7.2.3 Correlations for Mass Transfer Coefficients

For estimation of k_L and a several mathematical correlations are found in literature. Calderbank and Moo-Young (1961) proposed a correlation for the determination of k_L for mold pellets and other suspensions. This is stated in Eq.(7.5):

$$k_L = \frac{0.13\{(P/V)\mu_L/\rho_L{}^2\}^{1/4}}{(N_{Sc})^{2/3}} \tag{7.5}$$

where P/V is the power input per unit volume supplied by the agitator. N_{Sc} is the Schmidt number defined as $\frac{\mu_L}{\rho_L D_{AB}}$ (μ_L is the liquid viscosity and ρ is the liquid density. In bubble column and air lift bioreactors freely rising bubbles can be observed. The gas-liquid mass transfer for swarms of rising bubbles when bubble diameter is less than 2.5 mm can be expressed by the correlation as per Eq.(7.6):

$$N_{Sh} = k_L dp/D_{O_2} = 0.31 N_{Sc}{}^{1/3} N_{Gr}{}^{1/3} \tag{7.6}$$

Where, D_{O_2} is the diffusion coefficient of oxygen in the fluid. N_{Gr} is the Grashof number.

7.3 GAS LIQUID TRANSPORT IN CELLS

For aerobic bioprocess oxygen is introduced into the bioreactor through air sparger. There are several governing parameters which determine the rate of oxygen transfer into the broth. These include hydrodynamics of gas bubbles, temperature, medium composition, cellular activity and so on. There exists a series of mass transfer resistance in the whole process of oxygen transfer right from source up to cellular bodies.

Mass transfer of any solute A (say oxygen) through the liquid film is given by Eq.(7.7):

$$N_{A_L} = k_L a(C_{A_2}{}^* - C_{A_2}) \tag{7.7}$$

Similarly mass transfer through the gas film is given by Eq.(7.8):

$$N_{A_G} = k_G a(C_{A_1} - C_{A_1}{}^*) \tag{7.8}$$

Here $C_{A_1}{}^*$ and $C_{A_2}{}^*$ are the equilibrium solute concentrations in the gas phase and liquid phase respectively. The ratio of $C_{A_1}{}^*$ to $C_{A_2}{}^*$ is called distribution coefficient. Specifically in terms of oxygen transfer in a bioreactor, the commonly used equation is given by Eq.(7.9):

$$N_A = k_L a(C_{AL}{}^* - C_{AL}) \tag{7.9}$$

$k_L a$ is popularly termed as volumetric oxygen transfer coefficient. Its unit is s^{-1}.

The dissolution and absorption of oxygen in fermentation broth is crucial. It not only depends on physical and chemical parameters but also on the nature of substrate, morphology of the culture and its growth characteristics. In the above equation N_A is the oxygen transfer to the broth and this is also called specific oxygen uptake rate (OUR). If OUR needs to be improved, either $k_L a$ should be high or the concentration gradient ($C_{AL}{}^* - C_{AL}$) should be high. But in practical situation it becomes difficult to get a high value of N_A. By increasing rate of agitation, gas flow rate, and by selecting $k_L a$ can be improved.

7.4 ESTIMATION OF VOLUMETRIC MASS TRANSFER COEFFICIENT

The knowledge of volumetric mass transfer coefficient ($k_L a$) is required during the designing of an aerated bioreactor and further scale-up. There are various traditional methods of determination of k_La whereby oxygen balance in gas-liquid contactor and basic mass transfer correlations are used. Static method and dynamic gassing out methods are widely used in fermenter to determine k_La from a set of experimental data points. Accurate determination of oxygen gas concentration in the entry and exit streams of a bioreactor is important. Oxygen electrodes are generally used for the same. Sulphite oxidation method is a chemical method where chemical oxidation of sulphite to sulphate is carried out by dissolved oxygen in presence of selective catalysts (Cu^{2+} or Co^{2+}). In this method the amount of oxygen consumed is determined to obtain the reaction rate. In peroxide method hydrogen peroxide is used for the onsite generation of oxygen in presence of an enzyme catalase. The oxygen, thus produced, is transferred to the gas phase and water is produced as a byproduct. The order of enzymatic dissociation reaction of hydrogen peroxide is one with respect to enzyme and peroxide is given by Eq.(7.10):

$$r_1 = K_1[\mathrm{H_2O}][\mathrm{Catalase}] \tag{7.10}$$

At steady state

$$\frac{Q_i[\mathrm{H_2O}]}{V_1} = r_1 \tag{7.11}$$

where r_1 is the rate of reaction; Q_i is the hydrogen peroxide flow rate to the reactor (m^3s^{-1}) V_1 is the liquid volume (l). On the assumption that the rate of oxygen production and that of transfer are equal, we can get oxygen transfer rate (OTR) according to Eq.(7.12):

$$(\mathrm{OTR}) = \frac{r_1}{2} = k_L a\Delta c \tag{7.12}$$

where Δc is the concentration driving force (Gogate and Pandit 1999).

7.5 FACTORS AFFECTING OXYGEN TRANSFER RATE

Important factors affecting oxygen transfer rate and k_La are briefly stated in the following section:

(i) *Gas hold-up*: Gas hold-up (ε) is defined as the ratio of volume fraction of gas held-up to the total volume of gas and liquid held-up together according to Eq.(7.13):

$$\varepsilon = \frac{V_G}{V_G + V_L} = \frac{V_G}{V} \tag{7.13}$$

Higher value of gas hold-up results in higher interfacial area and thereby facilitates high rate of mass transfer.

(ii) *Superficial gas velocity*: As the superficial gas velocity increases rate of mass transfer increases as well. However, gas flow rate is to be optimized with stirrer speed to avoid impeller flooding.

(iii) *Bubble size*: If the bubble size is small (<2 mm), mass transfer rate should increase as the interfacial area increases. But small bubbles have rigid surfaces and also their rise velocities are less. As a result mass transfer is retarded. On the contrary larger bubbles (>3 mm) have higher rise velocity resulting it less residence time in the broth for gas liquid contacting. Hence an optimum bubble size needs to be selected.

(iv) *Temperature*: Gas diffusivity increases with increase in temperature and hence k_L increases. But with rise in temperature solubility of gas decreases, thereby minimizing the driving force for mass transfer. The temperature 40°C is found to be suitable for better mass transfer rate and beyond 40°C it falls.

(v) *Pressure*: Pressure affects the mass transfer by increasing the solubility of the gas in the liquid phase, as shown by Henry's law (Eq 7.14):

$$p_{AG} = HC_{AL}* \tag{7.14}$$

The partial pressure and total pressure of the system are related by Eq.(7.15):

$$p_{AG} = p_T \times y_A \tag{7.15}$$

Thus, with increase in the total pressure p_T, p_{AG} and C_{AL}^* also increases. This again increases the driving force for mass transfer.

(vi) *Type of agitators*: Various types of agitators such as propeller, turbine, paddle, vaned disc, anchor, etc. are widely used in mixing operations in fomenters providing different mixing patterns. Generally turbine and vane disc type impellers present very good gas-liquid contacting. Helical, ribbon, and anchor agitators find their uses in case of viscous solutions and slurries.

7.6 SOLID LIQUID MASS TRANSFER

`Solid liquid mass transfer is encountered in heterogeneous reaction systems involving immobilized enzymes, cells in clump or pellet form and flocks of cell. In such systems solid catalyst particles are composed of numerous interconnected tortuous micro-channels and micro-pores.

7.6.1 Various Mass Transfer Resistances

Ideally the solid particle is surrounded by the thin film of substrate fluid or media through which mass transfer occurs. It experiences both internal and external mass transfer resistances (Figure 7.3). Following are the major steps involved in the transport processes across the solid particle:

(i) Diffusion from bulk gas phase to gas-liquid interface.
(ii) Movement through gas-liquid interface.
(iii) Diffusion through relatively stagnant liquid film surrounding the bubble.
(iv) Diffusion through the bulk of the liquid.
(v) Movement across liquid cell interface.
(vi) Diffusion through cell aggregates.

Following correlation (Eq. 7.16) holds good for liquid-solid mass transfer coefficients:

$$\frac{k_s d_p}{D_m} = 1 + cN_{Sc}^n N_{Re}^m \tag{7.16}$$

Particle Reynold's number is given by Eq.(7.17):

$$N_{Re} = c\left[\frac{\varepsilon d_p^4}{\upsilon^3}\right]^p \tag{7.17}$$

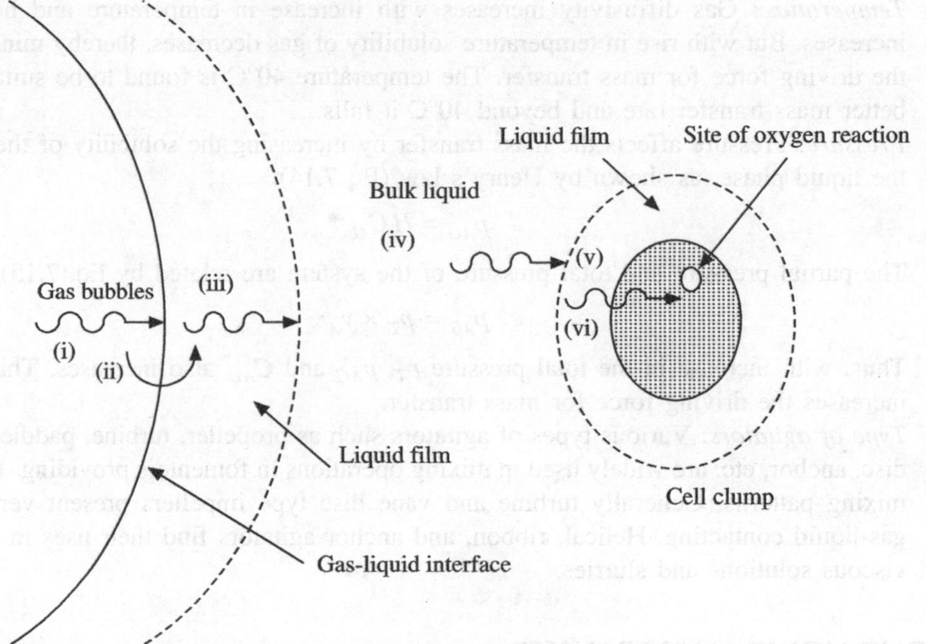

Figure 7.3 Various resistances during gas-liquid mass transfer in cellular system.

For industrial fermenters correlation of Sherwood number is given by Eq.(7.18):

$$N_{Sh} = 2 + 0.545 N_{Sc}^{1/3} \left[\frac{\varepsilon d_p^4}{\upsilon^3} \right]^{0.264} \tag{7.18}$$

Considering steady state, reaction rate equals the quantity of substrate diffusing from the bulk liquid to the microbes. A Monod type equation can be developed (Eq 7.19):

$$k_s (S_0 - S_s) = \frac{1}{Y_{x/s}} \frac{\mu_{\max} S_s}{K_s + S_s} \tag{7.19}$$

where S_0 and S_s are the substrate concentration at the bulk liquid and on the surface respectively and $Y_{x/s}$ is the yield coefficient.

7.6.2 Effectiveness Factor, Thiele Modulus and Damkohler's Number

Effectiveness factor: It is the ratio of the observed rate of enzymatic reaction with diffusion limitation to that without any diffusion limitation. This factor is an indication of intra-particle diffusion in an immobilized bioreactor system.

Thiele modulus: It is another parameter which gives an indication of the inter-particle diffusion. It is defined as Eq.(7.20):

$$\varphi = R \sqrt{\frac{\mu_{\max} / Y_{x/s} K_s}{D_s}} \tag{7.20}$$

Damkohler's number: It is the ratio of the maximum rate of reaction to maximum rate of mass transfer (Eq. 7.21).

$$Da = \frac{\text{maximum rate of reaction}}{\text{maximum rate of mass transfer}} \tag{7.21}$$

7.7 RHEOLOGY OF FERMENTATION BROTH

Rheology is a branch of study that includes viscosity, flow behavior and deformation of fluid (or even solid).The term was first used by Eugene C. Bingham (1878–1945), professor at Lafayette College. Rheological properties of fermentation broth involves the specific cases of fermentation where growing microbial cells are affected by fluid shear or viscosities and which in turn affects the overall rate of mass and heat transfer. Control of viscosity of fermentation broth is an important consideration to improve process performance. For many fermentation broths appropriate dilution with water have found to improve its yield and performance. In addition, information of broth rheology helps in better understanding of transport processes.

Equation for Newtonian fluid

Behavior of non-Newtonian fluids can be given by Eq.(7.22):

$$F = \mu A \frac{dv}{dx} \tag{7.22}$$

$$\mu = \frac{F/A}{dv/dx} = \frac{\tau}{\upsilon} = \frac{\text{Shear stess}}{\text{Shear rate (velocity gradient)}} \tag{7.23}$$

The plot of shear stress τ vs. shear rate dv/dx for the Newtonian and the non-Newtonian fluids are depicted in Figure 7.4.

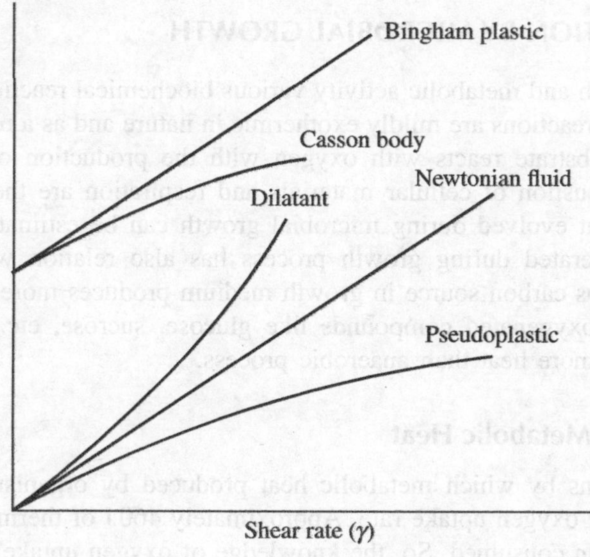

Figure 7.4 Shear stress as a function of shear rate in various non-Newtonian fluids.

Bingham plastic

Typical examples of this type of fluid are *Streptomyces kanamyceticus* and *Penicilium chrysogenum* broths.

The characteristic equation is given by Eq.(7.24):

$$\tau = \tau_0 + \eta v \tag{7.24}$$

Since fermentation broth of kanamycin exhibits Bingham plastic behavior, dispersing nutrients and oxygen transfer become difficult due to increased viscosity.

Pseudoplastic and dilatant

The characteristic equation is given by Eq.(7.25):

$$\tau = K\left(\frac{dv}{dx}\right)^n \tag{7.25}$$

Taking log on both the sides, we get

$$\log \tau = \log K + n \log v \tag{7.26}$$

For pseudoplastic $n < 1$.

Examples of this type of fluid include the fermentation broth of *Streptomyces* sp. and *Endomyces* sp. Casson in 1959 reported a typical type of non-Newtonian fluid which is named as Casson body fluid. It is closely related to pseudoplastic and to some extent the Bingham plastic fluid. The flow characteristic of Casson body fluid can be described by Eq.(7.27):

$$\sqrt{\tau} = \sqrt{\tau_o} + K_c\sqrt{\gamma} \tag{7.27}$$

K_c is the casson viscosity.

7.8 HEAT GENERATION BY MICROBIAL GROWTH

During microbial growth and metabolic activity various biochemical reactions take place inside the cell. Many of these reactions are mildly exothermic in nature and as a result heat is evolved. For aerobic process substrate reacts with oxygen with the production of CO_2, water vapor and heat energy. Combustion of cellular materials and respiration are the two major sources of heat generation. Heat evolved during microbial growth can be estimated by using various calorimeters. Heat generated during growth process has also relation with ΔG, Gibbs free energy. Hydrocarbons as carbon source in growth medium produces more heat on combustion compared to partially oxygenated compounds like glucose, sucrose, etc. In general, aerobic fermentation produces more heat than anaerobic process.

7.8.1 Evaluation of Metabolic Heat

There are various means by which metabolic heat produced by organism can be evaluated. One such method is the oxygen uptake rate. Approximately 460 J of thermal energy is evolved for each mole of oxygen consumed. So, the knowledge of oxygen uptake rate of fermentation helps in the determination of heat evaluation in aerobic processes. Similarly, the maximum rate of metabolic heat evolution can be calculated from oxygen transfer rate.

Metabolic heat produced during microbial growth process can be related to Gibbs free energy G. For aerobic processes, where considerable heat generation occurs, a part of free energy of the substrate is dissipated to the surroundings. From the basic thermodynamics we can write

$$G = G^o + RT \ln c \tag{7.28}$$

Where G^o is the standard Gibbs free energy and c is the concentration of the reacting species in moles per liter.

For a chemical reaction to occur spontaneously, the change in free energy, ΔG (i.e., free energy of the products minus the free energy of the reactants), must be negative. This requirement is independent of the path of the reaction. Although ΔG indicates whether a given reaction is possible, ΔG describes the amount of energy needed to 'drive' the reaction.

Using this equation the change of Gibbs free energy for intracellular reactions, ΔG_c can be calculated (Eq. 7.29);

$$\Delta G_c = \alpha_{substrate}\ \Delta G_{substrate} + \beta_{biomass}\ \Delta G_{biomass} + \gamma_{meta\ apdts}\ \Delta G_{meta\ pdts} \tag{7.29}$$

α, β, and γ represents the stoichiometric coefficients of substrate, biomass and metabolic products of a particular growth equation.

Further we know, Gibbs energy (G) combines the heat related enthalpic (H) and entropic (S) contributions according to Eq.(7.30):

$$\Delta G = \Delta H - T\Delta S \tag{7.30}$$

where ΔH is the change in enthalpy which is equal to the generation heat for the reaction and and ΔS is the change in entropy of the particular system. In line with Eq.(7.28) the enthalpy balance equation can be set up as Eq.(7.31):

$$\Delta H_c = -Q_S = \alpha_{substrate}\ \Delta H_{substrate} + \beta_{biomass}\ \Delta H_{biomass} + \gamma_{meta\ pdts}\ \Delta H_{meta\ pdts} \tag{7.31}$$

where Q_S representing the amount of heat generated by a particular biochemical reaction.

EXAMPLE 7.1 Calculate the heat generated during the growth of a yeast *Candida tropicalis* under aerobic and anaerobic conditions as shown below:

(a) Aerobic growth with stoichiometric equation (without the formation of ethanol) in a chemically defined medium

$0{:}600\ CH_{1:62}\ O_{0:53}\ N_{0:15} + 0{:}400\ CO_2 + 0{:}649\ H_2O \rightarrow CH_2O + 0{:}090\ NH_3 + 0{:}384\ O_2$

(b) Anaerobic growth with stoichiometric equation;

$0{:}122\ CH_{1:62}\ O_{0:53}\ N_{0:15} + 0{:}582\ CH_3O_{0.5} + 0{:}297CO_2 + 0{:}05H_2O \rightarrow CH_2O + 0{:}018\ NH_3$

[Given that: Heat of combustion of *Candida tropicalis*, $\Delta H^o_{biomass} = 545$ kJ/mole; $\Delta H^o_{Glucose}$ = 467 kJ/mole; $\Delta H^o_{ammonia\ (g)} = 383$ kJ/mole; $\Delta H^o_{ethanol} = 683$ kJ/mole]

***Solution*:**

Using Eq.(7.31) the heat generated when *Candida tropicalis*is grown aerobically:

$$-Q_{aerobic} = (0.600)(545) - 467 - (0.090)(383) = -165.5 \text{ kJ/mole}$$

Similarly for the anaerobic growth of yeast:

$$-Q_{anaerobic} = (0.122)(545) + (0.582)(683) - 467 - 0.018(383) = -8.1 \text{ kJ/mole}$$

The results clearly indicate that the amount of heat generated in an aerobic process is greater than that in an anaerobic process. This is because of that fact that anaerobic cell respiration consisting of glycolysis and fermentation can produce 2 ATP per mole of glucose consumed. On the contrary, aerobic respiration consisting of glycolysis, Krebs cycle and the respiratory electron transport is able to produce 36 ATP per mole of glucose consumed. Aerobic respiration is roughly 18 times more efficient than anaerobic cell respiration. On an average aerobic respiration converts about 40% of the available energy of glucose into ATP. The remaining 60% is dissipated. Due to high exothermicity the cooling requirement in aerobic processes is much higher compared to anaerobic ones.

7.8.2 Evaluation of Heat Exchange with Surrounding Environment

This term is actually responsible for the balances of heat sources and thus allows us to control the temperature. The magnitude of the exchange must be carefully regulated in existing equipment and accurately estimated for effective equipment design. The capacity to remove heat must be more than the magnitude of the potential heat sources. Usually temperature is controlled by flowing coolant through a jacket installed around the fermentation vessel, or through coils installed inside the vessel or both. Heat generated can be estimated by a general equation (Eq. 7.31):

$$Q_{ex} = UA\,\Delta T \tag{7.32}$$

where, U = overall heat transfer coefficient ($Wm^{-2}°C^{-1}$)
A = heat transfer area (m^2)
ΔT = temperature difference between bulk fermentation fluid and temperature control fluid (°C)
ΔT can also be the log mean temperature difference between two fluids.

7.9 THERMAL DEATH RATE KINETICS

Death rate of microorganisms by heat treatment follows a first order kinetics and similar relationships of chemical kinetics can be applied. Mathematically it is represented as Eq.(7.33):

$$\frac{dN}{dt} = -kN \tag{7.33}$$

where N = Number of living microorganisms at a definite time 't' (min)
k = reaction rate constant (min^{-1})

The reaction rate constant is a direct function of temperature and indirect function of the nature of microorganism.

$$\int_{N_0}^{N} \frac{dN}{N} = -k\int_{0}^{t} dt \tag{7.34}$$

$$\ln \frac{N}{N_0} = kt \tag{7.35}$$

where,

N_0 = starting number of microbes at zero time.

N = Number of organisms at time $t = t$

N_0 is also known as 'Contamination level' (initial number of microbes responsible for contamination prior to sterilization)

N is the sterility level (expressed as Eq.(7.35):

$$N = N_0 e^{-kt} \tag{7.36}$$

In terms of decimal reduction time D (time needed to reduce the initial number of microorganism to 1/10th)

$$\frac{N}{N_0} = \frac{1}{10} = e^{-kD} \tag{7.37}$$

Putting log on both sides of Eq.(7.37) and solving for D:

$$D = \frac{2.303}{k} \tag{7.38}$$

$$t = D \log_{10} \frac{N_0}{N} \tag{7.39}$$

Death rate of microorganisms can be determined experimentally. Samples of cell suspension or broth (for various time intervals of growth) are taken in sealed test tube under sterile conditions. The tubes are immersed into hot bath for fixed period of time. They are then removed and cooled immediately. The number of microorganism before and after exposure can be counted by several techniques (plate counts or colony counts, etc.).

The death kinetic rate constant k is similar to chemical reaction rate constant for a first order reaction and its dependence on temperature can be conveniently represented by Arrhenius equation (Eq. 7.40):

$$\frac{d(\ln k)}{dT} = \frac{E}{RT^2} \tag{7.40}$$

Where E is the activation energy (in J/mol) which is a specific constant for the population, R Universal gas constant, T (temperature in K).

Integration of Eq.(7.40) yields Eq.(7.41):

$$k = k_0 e^{-E/RT} \tag{7.41}$$

where, k_0 is the Arrhenius constant. The constant k_0 and E/R can be calculated by the Arrhenius plot drawn between ln k and $1/T$.

For this purpose, a set of experiments is conducted at one temperature and the values of k are evaluated by using equation ($\ln N_t/N_0 = -kt$) by measuring N_0 and N_t at different time intervals. The procedure can be repeated for other temperatures. The slope of the line gives ($-E/R$) and intercept is equal to ln (k_0):

$$\ln \frac{N_0}{N_t} = k_0 t e^{-E/RT} \tag{7.42}$$

This equation sets the design criteria for sterilizer and this is called "Del factor" (Δ) or Nabla factor or Sterilization criterion represented by Δ (Deindoerfer and Humphery, 1959).

'Δ' indicates the fractional reduction in count of living microorganisms from the initial numbers.

Thus,

$$\Delta = \ln \frac{N_0}{N_t} = k_0 t e^{-E/RT} \tag{7.43}$$

Deindoerfer and Humphrey suggested that a risk factor of one in thousand being contaminated could be considered as reasonable for all practical purposes. However, to have a complete profile of all the contaminating spores is seldom possible. Hence, it is assumed that all the contaminants can be clubbed into a single category of spores of *Bacillus streothermophilus* which is one of the most heat-resistant microbial types.

Deindoerfer and Humphrey reported for *B. streothermophilus* spores:

Activation energy E = 283 kJ/mol; Arrhenius constant $k_O = 1 \times 10^{36.2}\ s^{-1}$

Risk factor of 1 in 1000 means $N_t = 10^{-3}$

Although, sterilization temperature is set at 121°C (corresponding steam pressure 15 psig) destruction of microorganisms starts even during heating (from ambient temperature to 121°C) and cooling (from 121°C to ambient temperature) periods. Therefore, the total time sterilization is the summation of heating, holding and cooling time (Eq. 7.44).

$$\Delta_{overall} = \Delta_{heating} + \Delta_{holding} + \Delta_{cooling} \tag{7.44}$$

$$\Delta_{overall} = \ln \frac{N_0}{N} = \Delta_{heating} + \Delta_{holding} + \Delta_{cooling} \tag{7.45}$$

$$\Delta_{heating} = \ln \frac{N_0}{N_1} = \int_0^{t_1} k\,dt = \alpha' \int_0^{t_1} e^{-E/RT}\,dt \tag{7.46}$$

$$\Delta_{holding} = \ln \frac{N_1}{N_2} = \int_{t_1}^{t_2} k\,dt = \alpha' \int_{t_1}^{t_2} e^{-E/RT}\,dt \tag{7.47}$$

$$\Delta_{cooling} = \ln \frac{N_2}{N} = \int_{t_2}^{t_3} k\,dt = \alpha' \int_{t_2}^{t_3} e^{-E/RT}\,dt \tag{7.48}$$

where N_1 is the number of contaminating microbes after heating period t_1, and N_2 is the number of contaminating microbes after holding period t_2. E is the activation energy and α' is the pre-exponential constant for the death of microorganism. As the temperature during holding period is constant, by choosing a holding temperature, k is fixed and Δ_{hold} can be calculated according to Eq.(7.49):

$$\Delta_{hold} = k(t_2 - t_1) \tag{7.49}$$

Compared to heating or cooling time, holding time is normally smaller; but during this period temperature is maximum. As a result maximum death of microbes occurs during this period. Sterilization time in a fermenter can be estimated from the nature and volume of the medium. In addition nutrient media, other accessories of the bioreactor (fermenter) such as pH and DO electrodes, fittings, tubing, valves, etc. need to be sterilized. Following heat sources are generally used in batch sterilization:

- Steam sparging (bubbling of live steam through the medium)
- Electrical heating
- Heating or cooling with a two-fluid heat exchanger.

Batch sterilization is a relatively simple process and can be done *in situ*. But chances of thermal damage of heat-sensitive materials cannot be always avoided. Moreover, heating and cooling takes long time. To avoid this sterilization should operate at the highest optimum feasible temperature and for the shortest time of exposure (HTST: High Temperature Short Time).

7.10 STERILIZATION

Sterilization is the treatment by which all forms of microorganisms, sporous and non-sporous, other living entities, fungi and protozoa from surface, medium and instruments are destructed or inactivated by various physical and chemical methods. The objectives of sterilization are:

(i) To ensure that process is carried out only with the desired organism
(ii) To permit the safe use of product
(iii) To prevent deterioration of the product
(iv) To avoid environmental contamination

The prime purpose of sterilization is to maintain an aseptic environment for microbial culture—free from contamination by undesired microorganisms. As the attainment of highest level of sterile environment is next to impossible, it is always quantified in terms of probability of existing microorganism.

Contamination can be avoided by several ways, such as:

- Medium containing nutrients must be sterilized.
- Inoculum should be fresh and pure.
- Reactor vessel needs sufficient sterilization.
- Auxiliary equipments such as pipes, valves, bends, etc. should be sterilized.
- Any materials added into the fermenter must be sterilized.
- Air should introduced after proper sterilization.
- Aseptic conditions should be consistently maintained during the course of fermentation.
- Maintaining an appropriate pH during fermentation helps mitigation of contamination and undesired product formation.

7.11 STERILIZATION VS PASTEURIZATION

In the process of sterilization, microorganisms are killed or get inactivated by various means such as by denaturation of constituent proteins, interruption of DNA synthesis/repair or disruption of cell membranes. Pasteurization: Another almost similar term is pasteurization, widely used in dairy industries. Pasteurization is a mild thermal process used for the destruction of only pathogenic microorganisms by applying a cycle of successive heating and cooling. In contrast to sterilization pasteurization does not destroy all microorganisms in the media. Rather, it tends to reduce the number of viable microorganisms thereby reducing their propensity to cause contamination or infection. Difference between sterilization and pasteurization are given in Table 7.1.

Table 7.1 Differences between sterilization and pasteurization

	Sterilization	Pasteurization
Function	Destruction of all forms of microorganisms, including pathogenic and non-pathogenic	Destruction of only pathogenic microorganisms
Temperature	Above 100°C	Below 100°C
Advantage	Storage life is more, no other preservatives are required	Flavour, texture and nutrients value remain unaffected
Disadvantage	Chances of overheating, damage of flavor and nutrients	Storage life is less, needs secondary preservative methods, refrigeration
Application	Wider applications across the biotechnology, food processing, packaging, microbiology, and surgery	Mainly applied in food industry

7.12 STERILIZATION METHODS

Sterilization methods are broadly classified in two types: physical methods and chemical methods. The choice of a method of sterilization depends on several factors like degree of sterility, reliability, effect on medium quality and cost (both operating and capital).

7.12.1 Physical Methods

Physical methods of sterilization involve application of heat, radiation or employing filtration.

Heat treatment: Pressurized steam sterilization is the most economical and convenient way of sterilization. Use of elevated temperature to get the desired level of sterilization may lead to thermal degradation, damage of medium. A balance between destruction of unwanted microorganisms and retention of useful nutrients, proteins, vitamins, etc. which are thermo-labile, is considered to be the most important criterion for the selection of temperature in sterilization or pasteurization process. High pressure saturated steam in autoclave destroys microorganisms by hydrolysis followed by coagulation of cell proteins. This is accomplished by using pressurized steam having high heat content.

High temperature short time (HTST) is the conditions that are most suitable for general purpose sterilization. According to this, materials and media that undergo sterilization are exposed to a high temperature but with minimum holding time. This allows the destruction of all the unwanted microorganisms but at the same time ensures no degradation of any constituents of the media. Under these conditions, temperature is kept at 121°C pressure at 15 psi pressure and holding time 15 min. Total time of steam sterilization is the summation of heating time, holding time and cooling time (Figure 7.5).

7.12.2 Radiation

Both ionizing and non-ionizing types of radiations are used for sterilization. Non-ionizing radiations (like UV, sonic irradiations, etc.) have mild penetration power and hence it causes minimal damage to the cells. Ionizing radiation (like X-rays, Gamma rays, etc.) has more penetration power and unless it is precisely-controlled it could damage the nutrients or the microbial cells on exposure. Exposure of ionizing radiation could also be dangerous to

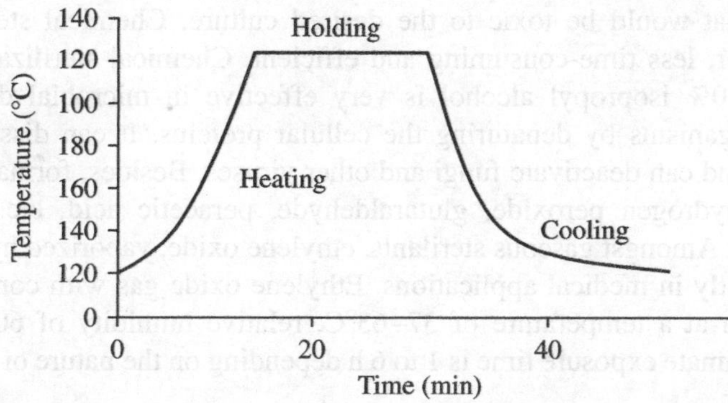

Figure 7.5 Heating, holding and cooling time during steam sterilization.

the working personnel in the laboratory. Many drugs, pharmaceutical derivatives, active pharmaceutical ingredients (API) and medical apparatus undergo ionizing radiations during the final processing steps; for example gamma radiation. Suitable radioisotope sources like cobalt 60 or energized electron beams are used in ionization radiation. Such type of radiation can also lead to mutation of strain. Ultraviolet radiation is one type of electro-magnetic radiation in the wave length of 100–400 nm. Exposure to UV rays leads to inactivation of microorganisms. It damages the genetic materials of the nucleus of cell and nucleic acids in virus.

7.12.3 Filtration

Use of elevated temperature to get the desired level of sterilization may lead to thermal degradation and damage to medium. A common alternative approach applicable equally to gases and liquids is the use of appropriate filters to remove undesirable microorganisms. Filtration sterilization differs from other sterilization techniques because; organisms are not killed, but physically removed. Solutions containing thermally unstable compounds like vitamins, hormones, pharmaceutical ingredients, eye drops, antibiotics, vitamins and other thermosensitive solutions are sterilized by using filtration techniques.

Sterilization by filtration is accomplished through microporous membrane filter with 0.2 micron or smaller pore sizes to arrest microbial contaminants. Typical membrane materials are organic polymeric compounds such as cellulose acetate, polycarbonate, polyether sulfone, polyvinyl fluoride, nylon, etc. Inorganic compounds like sintered glass, porous ceramic are also used. Materials which are less prone to protein binding are generally selected. Membrane fouling is another problem which occurs as a result of adsorption of protein or other components on to the membrane surface. This eventually leads to membrane pore blocking. High flow rate, high throughput, chemical compatibility, availability in different pore sizes, cost, and mechanical stability are some of the important selection criteria of membranes.

7.12.4 Chemical Methods

Sterilants and high-level disinfectants are used as chemical scavengers to deactivate the microorganisms. The chemicals agents should be selected in such way that they should not leave any residue that would be toxic to the desired culture. Chemical sterilization methods are relatively cheaper, less time-consuming and efficient. Chemical sterilization involves both liquid and gasses. 70% isopropyl alcohol is very effective in microbial decontamination. It can destroy microorganisms by denaturing the cellular proteins. It can dissolve lipid layer of bacterial cell walls and can deactivate fungi and other viruses. Besides, formalin (dilute solution of formaldehyde), hydrogen peroxide, glutaraldehyde, peracetic acid, etc. are also used as chemical scavengers. Amongst gaseous sterilants, ethylene oxide, vaporized hydrogen peroxides, ozones are used mostly in medical applications. Ethylene oxide gas with concentration ranging from 450–1200 ppm at a temperature of 37–63°C, relative humidity of 60–80% is used for sterilization. Approximate exposure time is 1 to 6 h depending on the nature of the contamination.

7.13 METHODS OF AIR STERILIZATION

Aerobic fermenter requires a continuous supply of sterile air to assist in the growth of microorganisms. According to Aiba (1973) the permissible limit of air borne microbes should be approximately 10^4 particle/m^3 of air. Various methods such as heating, passage of UV rays, other electromagnetic radiations, corona discharge, use of chemical scavengers, membrane filtration, etc. are used for air sterilization. Although air can be sterilized by heating, it is not economical as it is poor conductor of heat thereby resulting in low heat transfer coefficient. Exposure time is also long. Ideally, sonic waves, high energy cathode rays and gamma rays can be applied for sterilizing air. But these are not always economical for large-scale use. Cottrell precipitator applies corona discharge for the removal of air-borne microbes which adhere to the fine dust particles. Spraying of germicides such as vapors or aerosols of hexyl resorcinol, glycols, formaldehyde, ethylene oxide, phenols, etc. can significantly reduce the air-borne bacteria. These chemical scavengers are sprayed in to the rooms along with recirculation ducts like air-conditioners or pedestal fans, etc.

Common filter materials used in air filtration equipment include porous alumina, silica, granular activated charcoal, glass-wool, asbestos fibre, slag wool, etc. There are various factors responsible for efficient filter operation. These are particle size, fibre diameter, depth of filter bed, filter packing density and air flow rate. Either one or combination of more than one mechanism as stated below is responsible for filtration of microbes by air filter:

- Direct interception of particles
- Inertial impaction
- Diffusion
- Turbulence
- Brownian movement
- Electrostatic attraction
- Settling by gravitational force.

While selection of air filter low pressure drop, long service life and efficiency of removing microorganisms are considered. HEPA (High efficiency particulate air or High efficiency particulate absorbing) filter is a popular commercial air widely used by several industries or laboratories (ASME, U.S. DOE). It has capacity to remove more than 99% of particles having diameter close to 0.3 µm. It can retain droplets of moisture, pollen grains, dirt particles, bacterial cells (0.2–2.0 µm), virus (0.02–0.3 µm), and submicron liquid aerosol (0.02–0.5 µm). HEPA is composed of a matrix of polymeric fibres such as polypropylene or fibre-glass (diameters between 0.5 and 2.0 micrometers).

7.14 CONTINUOUS STERILIZATION

Continuous sterilization is of two types (Figure 7.6):

1. Continuous steam injection type sterilizer
2. Continuous plate or spiral heat exchanger

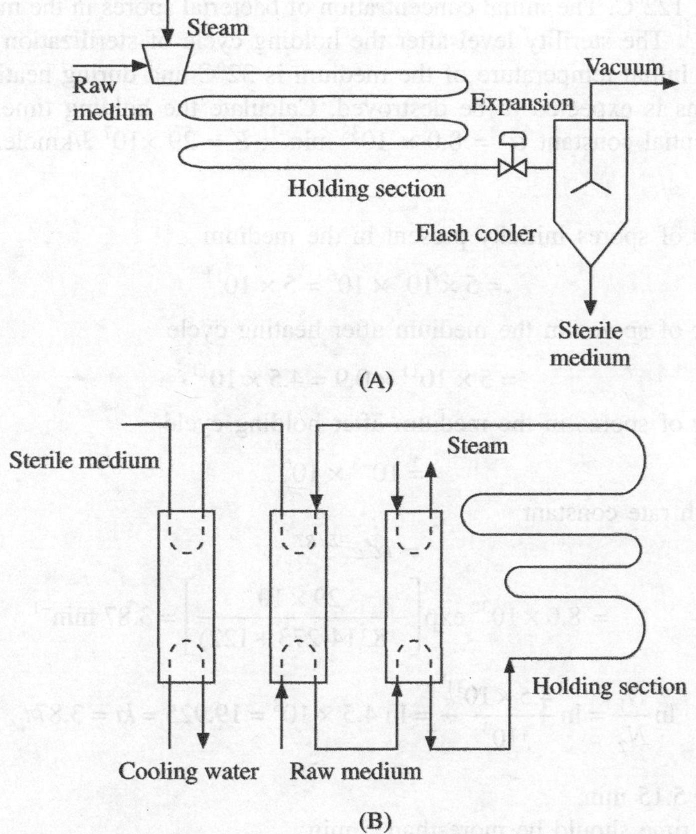

Figure 7.6 Schematic of Continuous sterilization (A) Steam injection type (B) Plate exchanger type.

In steam injection type continuous sterilizer raw medium comes in direct contact with injected steam. This results in almost instantaneous rise of set-point temperature for sterilization. The length of pipe in the holding section determines the holding time, during which most of the sterilization occurs. It is followed by immediate cooling of medium via an expansion valve connected to a vacuum chamber.

In plate heat exchanger steam-heated plates help attaining the temperature of the raw medium. The medium is then maintained at the elevated temperature for fixed time interval and is finally cooled in another section of the plate heat exchanger. Although the time required to heat and cool the raw medium is much longer than with the steam injection type of continuous sterilization, the contribution of these, periods to the sterilizing cycle will be much smaller (1 to 2%) than in the case of batch sterilization. Advantages of continuous sterilization include better steam utilization, ease of scale-up and ease of interfacing with continuous fermenter. However, sometimes foaming problems occur during operation.

NUMERICAL PROBLEM

EXAMPLE 7.1 A liquid medium in a batch fermenter having working volume 1000 L is sterilized *in situ* at 122°C. The initial concentration of bacterial spores in the medium was found to be 5×10^5 ml^{-1}. The sterility level after the holding cycle of sterilization is desired not to exceed 10^{-3} ml^{-1}. Initial temperature of the medium is 32°C and during heating cycle 10% of the microorganisms is expected to be destroyed. Calculate the holding time. Additional data given: Pre-exponential constant $\alpha' = 8.0 \times 10^{38}$ min^{-1}; $E = 29 \times 10^7$ J/kmole.

Solution:

N_0 = total number of spores initially present in the medium

$$= 5 \times 10^5 \times 10^6 = 5 \times 10^{11}$$

N_1 = total number of spores in the medium after heating cycle

$$= 5 \times 10^{11} \times 0.9 = 4.5 \times 10^{11}$$

N_2 = total number of spores in the medium after holding cycle

$$= 10^{-3} \times 10^6$$

k = specific death rate constant

$$= \alpha' e^{-E/RT}$$

$$= 8.0 \times 10^{38} \exp\left[-\frac{29 \times 10^7}{8314(273+122)}\right] = 3.87 \text{ min}^{-1}$$

$$\ln\frac{N_1}{N_2} = \ln\frac{4.5 \times 10^{11}}{10^3} = \ln 4.5 \times 10^8 = 19.925 = kt = 3.87t$$

Therefore $t = 5.15$ min

Thus holding time should be more than 5 min.

PROBLEMS FOR PRACTICE

QUESTION 7.1 A liquid medium initially contains bacterial spores at a concentration of 9.2×10^{11} m^{-3}. The medium is sterilized thermally at 125°C and spores density is recorded. The data are as below:

Time (min)	0	5	10	20	30	35
Spore density (m^{-3})	9.2×10^{11}	2×10^{10}	3.32×10^9	2.3×10^5	115	3

Find out the thermal death kinetic rate constant. Find out the inactivation factor after 40 min.

QUESTION 7.2 An unsterile bath contains initially 1.0×10^{10} viable cells. Find the del factor of sterilization ($N_t = 10^{-3}$).

KEY POINTS TO REMEMBER

- Transport of nutrients and oxygen in aerobic fermentation processes involves various modes of gas-liquid mass transfer depending on broth characteristics, type of gas liquid contact and other physico-chemical environment.
- The rate equation for mass diffusion is represented by the classical equation as denoted by Fick's law $J_{A,x} = -D_{AB}\frac{dC_A}{dx}$
- Important hypotheses of diffusional mass transfer are film, penetration and surface renewal theory.
- A series of mass transfer resistances exist during the whole process of oxygen transfer right from source up to cellular bodies.
- For oxygen transfer in a bioreactor, the commonly used equation is given by $N_A = k_L a(C_{AL}{*} - C_{AL})$.
- $k_L a$ is popularly termed as volumetric oxygen transfer coefficient. Its unit is s^{-1}.
- Static method and dynamic gassing out methods are widely used in fermenter to determine $k_L a$ from a set of experimental data points.
- Important factors affecting oxygen transfer rate and $k_L a$ are gas hold-up, superficial gas velocity, bubble size, temperature, pressure, type of agitators and so on.
- The ratio of the observed rate of enzymatic reaction with diffusion limitation to that without any diffusion limitation is called effectiveness factor. This factor is an indication of intra-particle diffusion in an immobilized bioreactor system.
- The ratio of the maximum rate of reaction to maximum rate of mass transfer is known as Damkohler's number.
- Fermentation broth often exhibits non-Newtonian behavior. These are called pseudoplastic, dilatant, Casson body fluids, etc.
- Approximately 460 J of thermal energy is evolved for each mole of oxygen consumed. So, the knowledge of oxygen uptake rate of fermentation helps in the determination of heat evaluation in aerobic processes.
- Death rate of microorganisms (number of microorganisms dying per unit time) by heat treatment is directly proportion to the number of organisms.
- Pasteurization is a mild thermal process used for the destruction of only pathogenic microorganisms by applying a cycle of successive heating and cooling.
- Sterilization is the treatment by which all forms of microorganisms and other living entities are destructed or inactivated by various physical and chemical methods.
- High temperature short time (HTST) is the conditions that are most suitable for general purpose sterilization. Under these conditions, temperature is kept at 121°C pressure at 15 psi and holding time 15 min.

- Methods, such as heating, passage of UV rays, other electromagnetic radiations, corona discharge, use of chemical scavengers, membrane filtration, etc. are used for air sterilization.

QUESTIONS FOR PRACTICE

Multiple Choice Questions

1. According to Film theory mass transfer co-efficient (K) and diffusivity (D) are related as
   (a) $K \propto D$ (b) $K \propto \sqrt{D}$
   (c) $K \propto D^{1.5}$ (d) $K \propto D^2$
2. Mass transfer co-efficient is directly proportional to $D_{AB}^{0.5}$, according to ____________ theory.
   (a) Penetration (b) Surface renewal
   (c) Film (d) None of these
3. Mass transfer rate between two fluid phases does not necessarily depend on the ____________ of the two phases.
   (a) Physical properties (b) Chemical properties
   (c) Degree of turbulence (d) Interfacial area
4. For which type of mass transfer does oxygen balance method is used?
   (a) Gas-gas (b) Gas-solid
   (c) Gas-liquid (d) Liquid-liquid
5. What is the unit of volumetric mass transfer coefficient (K_{La})?
   (a) h^{-1} (b) ml/h
   (c) mmol/h (d) ml/s
6. Which of the following is the efficient value for oxygen concentration?
   (a) Equal to C_{crit} (b) Below C_{crit}
   (c) Above C_{crit} (d) None of the above
7. Which of the following methods of K_{La} determination does not require the measurement of dissolved oxygen concentration?
   (a) Dynamic gassing out methods (b) Static method
   (c) Oxygen balance method (d) Sulphite oxidation method
8. Which of the following is not a chemical method to measure K_{La}?
   (a) Dynamic gassing out method (b) Sodium sulphite oxidation method
   (c) Carbon dioxide absorption method (d) Glucose oxidase method
9. Which of the following factors does not affect the K_{La} value?
   (a) Air flow rate (b) Presence of enzymes
   (c) Presence of antifoam agents (d) Degree of agitation
10. Which of the following physical property cannot be associated with rheological property?
    (a) Body and slip (b) Spreadability
    (c) Surface tension (d) Viscosity

11. According to Newton's law of viscosity, the shear stress in flowing fluid is ____________ to the rate of shear.
(a) Inversely proportional (b) Directly proportional
(c) Square root (d) Perpendicular

12. The unit of viscosity is
(a) Ns m^{-2} (b) Ns^2 m^{-2}
(c) Ns^{-1} m^{-1} (d) Ns

13. Kinematic viscosity is the ratio of dynamic viscosity to the
(a) Density of the fluid (b) Plastic viscosity
(c) Volume of liquid to flow (d) Specific gravity of fluid

14. Pseudoplastic flow is also known as
(a) Shear thinning system (b) Shear thickening system
(c) Elastic deformation (d) Plastic deformation

15. Dilatant flow is also known as
(a) Shear thinning system (b) Shear thickening system
(c) Elastic deformation (d) Plastic deformation

16. Rheology is the study of
(a) Flow of a material (b) Particle size
(c) Formation of colloids (d) Formation of complex

17. The resistance of a liquid to flow is termed as
(a) Viscosity (b) Density
(c) Interfacial tension (d) Surface tension

18. As the temperature increases the viscosity of liquid
(a) Decreases (b) Increases
(c) First decreases then increases (d) Remains constant

19. In pseudoplastic flow as shear stress increases, shear rate
(a) Increases (b) Increases but linear
(c) Increases but not linear (d) Decreases

20. Which of the following is target microbe in commercial sterilization?
(a) *Pseudomonas aeruginosa* (b) *Bacillus anthracis*
(c) *Salmonella typhi* (d) *Clostridium botulinum*

21. Which of the following statements about sterilization are correct?
Statement 1: Sterilization is achieved by heat and chemicals alone
Statement 2: Ultraviolet irradiation is not used for sterilization
(a) True, False (b) True, True
(c) False, True (d) False, False

22. ____________ is the quickest way to sterilize solutions without heating.
(a) Filtration (b) Chemical addition
(c) Steam (d) U V irradiation

23. ____________ is used for sterilization of metallic devices like needles, scalpels, etc.
(a) Flaming (b) Incineration
(c) Hot air oven (d) Steam

24. Which of the following processing methods involves testing food at high temperatures for short periods of time in order to reduce the risk of food poisoning?
(a) Blanching (b) Ohmic heating
(c) Sterilization (d) Pasteurization

25. Which of the following is called cold Pasteurization?
(a) Freeze drying (b) Heating under low pressure
(c) Irradiation (d) Heating at low temperature

Answer Key

1. (a)	2. (a)	3. (b)	4. (c)	5. (a)	6. (c)	7. (d)	8. (a)	9. (b)	10. (c)
11. (b)	12. (a)	13. (a)	14. (a)	15. (b)	16. (a)	17. (a)	18. (a)	19. (c)	20. (d)
21. (d)	22. (a)	23. (a)	24. (d)	25. (c)					

Descriptive Questions

1. Explain the significance of volumetric mass transfer coefficient in aerobic fermenter.
2. State the advantages and disadvantages of both bigger and smaller size bubbles in the gas liquid mass transfer.
3. Why is it important to provide continuous supply of oxygen in aerobic fermentation?
4. Explain the significance of volumetric mass transfer coefficient in aerobic fermenter.
5. Write the equation for oxygen transfer rate in an aerated fermenter and explain various terms involved therein along with their units.
6. What are the temperature, pressure and time conditions for steam sterilization? Name a few methods for sterilization of medium other than using steam.

8 Bioreactors and Scale-up

8.1 CLASSIFICATION OF BIOREACTORS: BATCH AND CONTINUOUS TYPES

Bioreactor is a vessel where any sort of biochemical reaction or conversion takes place in presence of a biocatalyst. The biocatalyst may be either living cells or isolated enzymes. If microorganisms are used, it may be either native cells or recombinant cells. Bioreactors employed in industry may be broadly classified into two groups as (a) Batch reactor, and (b) flow or continuous reactor. The latter may be again classified in to two groups (i) mixed flow reactor (or chemostat or continuous stirred tank reactor), and (ii) plug flow reactor. Packed-bed reactor and tubular reactor come under plug flow reactor provided their departure from ideality is not appreciable. On the other hand, recycle reactor with high recycle ratio is considered as a mixed flow reactor. Fluidized-bed reactors and fed batch reactor are also used in biochemical industries.

8.2 BATCH BIOREACTOR: BASIC FEATURES

In its simplest form it is a cylindrical vessel slightly conical at the bottom and fitted with a mechanical agitator. Reactants along with the biocatalyst are charged inside the reactor all at a time and the products are withdrawn all at a time. The reactor contents are well stirred. The reactor may be fitted with different attachments to measure temperature, pH, dissolved oxygen, etc. At a particular instant concentration of any component is uniform throughout the reactor. However, it changes with progress of time. Batch reactor is thus an unsteady state reactor. In laboratory what we usually refer as a fermenter is nothing but a batch reactor.

8.2.1 Culturing in Batch Reactor

For batch culture, spores of microorganisms from a source stock are inoculated into the specific media containing nutrients. Air is bubbled through the media if microbes of interest are aerobic in nature. Unsteady state material balance can be constructed on substrate and biomass:

[Flow of material in] + [Formation by biochemical reaction] – [Flow of material out] = [Accumulation]

Since it is a batch process, the terms for the flow of material in and flow of material out would be zero. Thus the balance equation will take the form

[Formation by biochemical reaction] = [Accumulation]

If it is written in the form of equation, we may get Eq.(8.1):

$$R_x V = \mu V X \tag{8.1}$$

where, R_x is the rate of conversion of biomass per unit volume of reactor, V is the working volume of the reactor, μ is the specific growth rate and X is the biomass or cell mass concentration.

As we know specific growth rate μ can be denoted following Malthus's law (Eq. 8.2) as:

$$\mu = \frac{1}{X}\frac{dx}{dt} \tag{8.2}$$

By incorporating Eq.(8.1) into Eq.(8.2), we get

$$R_x V = \mu V X = \frac{1}{X}\frac{dx}{dt} X V = \frac{dX}{dt} V \tag{8.3}$$

Equivalent expression for substrate can be written as $R_s V = \frac{ds}{dt} V$ (8.4)

where, R_s is the rate of conversion of substrate per unit volume of reactor.

If Y is the overall yield coefficient for the formation of biomass and S is the limiting substrate concentration, the yield coefficient of growth process can be written as Eq.(8.5):

$$Y = -\frac{\Delta X}{\Delta S} \quad \text{or} \quad Y\frac{ds}{dt} = -\frac{dX}{dt} \quad \text{or} \quad YR_s = -\mu X \tag{8.5}$$

Y can also be represented differently according to Eq.(8.6):

$$Y = \frac{X - X_o}{S_o - S} \quad \text{or} \quad S = S_o - \frac{X - X_o}{Y} \tag{8.6}$$

If the growth rate follows Monod kinetic equation $\mu = \frac{\mu_m S}{K_S + S}$

We may incorporate into Eq.(8.3) to obtain Eq.(8.7):

$$\mu V X = \frac{dX}{dt} V \quad \text{or}$$

$$\frac{\mu_m S}{K_S + S} X = \frac{dX}{dt} \tag{8.7}$$

Integrating Eq.(8.7) within the limits X and X_o and time between initial and the final we can obtain Eq.(8.8):

$$\int_{X_o}^{X} \frac{K_S + S}{\mu_m S}\frac{dX}{X} = \int_0^t dt \tag{8.8}$$

But S in Eq.(8.8) is also a function of X, therefore by incorporating Eq.(8.6) into Eq.(8.8) and rearranging, we get Eq.(8.9):

$$\int_{X_o}^{X} \frac{K_S + S_o - \dfrac{X - X_o}{Y}}{\mu_m \left(S_o - \dfrac{X - X_o}{Y}\right)} \frac{dX}{X} = \int_0^t dt \tag{8.9}$$

By solving Eq.(8.9), we get

$$t = \frac{K_S Y + S_o Y + X_o}{\mu_m (Y S_o + X_o)} \ln\left(\frac{X}{X_o}\right) + \frac{K S_Y}{\mu_m (Y S_o + X_o)} \ln\left[\frac{Y S_o}{Y S_o + X_o + X}\right] \tag{8.10}$$

A similar type of expression can be obtained for substrate concentration as Eq.(8.11):

$$t = \frac{K_S + Y + S_o Y + X_o}{\mu_m (Y S_o + X_o)} \ln\left[1 + \frac{Y(S_o - S)}{X_o}\right] - \frac{K_S Y}{\mu_m (X_o + Y S_o)} \ln\frac{S}{S_o} \tag{8.11}$$

8.2.2 An Alternative Method of Representation of Batch Reactor Design Equation

For a batch bioreactor microbial growth can be represented by the two equations for cell mass and substrate, using Monod Model:

$$\frac{dX}{dt} = \mu X = \frac{\mu_m S X}{K_S + S} \tag{8.12}$$

$$\frac{ds}{dt} = -\frac{1}{Y_{X/S}} \frac{\mu_m S X}{K_S + S} \tag{8.13}$$

With initial conditions $X(0) = X_o$ and $S(0) = S_o$

These equations can be added after multiplication of Eq.(8.13) by $Y_{X/S}$ to yield Eq.(8.14):

$$\frac{d(X + Y_{X/S} S)}{dt} = 0 \tag{8.14}$$

and thus

$$X + Y_{X/S}.S = X_o + Y_{X/S} S_o \tag{8.15}$$

Expressing X in terms of S from Eq.(8.15) and substituting into Eq.(8.13) gives Eq.(8.16):

$$\frac{dS}{dt} = -\frac{\mu_m}{Y_{X/S}} \frac{(X_o + Y_{X/S}\{S_o - S\})S}{(K_S + S)} \tag{8.16}$$

Eq.(8.16) can be analytically integrated to give:

$$[X_o + Y_{X/S}(S_o + K_S)] \ln\left(\frac{X_o + Y_{X/S}\{S_o - S\}}{X_o}\right) - K_{S.} Y_{X/S} \ln\frac{S}{S_o} = \mu_m (X_o + Y_{X/S}.S_o) t \tag{8.17}$$

The Eq.(8.17) is implicit in 'S' and the time course of substrate consumption can be best found by calculating 't' values which correspond to a specified value of substrate concentration. Substitution of values of 'S' into Eq.(8.17) provides the time course of biomass concentration.

8.2.3 Advantages and Disadvantages of Batch Culture

Advantages:

- Batch cultivation of microorganisms can be replicated for various other reactions.
- Safety and aseptic conditions are adequately maintained.
- Mutation or contamination of growing culture is minimal.
- Very high substrate to cell mass transformation.

Disadvantages:

- High labour cost
- Longer time for pre-treatment and post-treatment

8.3 CONTINUOUS (MIXED FLOW) BIOREACTOR

This is featured by continuous adding of reactants as well as continuous retrieving of products. The reactor contents are well stirred with the help of an agitator. Concentration is uniform throughout the reactor and it does not change with time. This reactor, is therefore, a steady state reactor.

8.3.1 Culturing in Continuous Reactor

Both tubular reactor and plug flow fermenter and mixed reactor or back mix fermenter are used for continuous culture.

As the reactor contents are homogeneously mixed it can be well assumed that concentration throughout should be uniform and this should also be same as that in the exit stream.

Since the fermenter is well mixed, samples taken from any location in the fermenter will be identical and exit stream composition will be the same as that of the liquid inside. CSTF can be operated as a 'Turbidostat' where the feed is metered in such a way as to maintain a constant biomass concentration (as measured by its turbidity) in the fermenter. It is also known as a basic chemostat system. Here, the mixing is done by an impeller, rising gas bubbles or both. We assume that mixing is so vigorous that each phase of the vessel contents is of uniform composition. Chemostat cultures have a number of advantages over other culture system (e.g. batch system) for the study of cell growth and metabolism.

Let us consider volume of reactor V and feed flow rate F as shown in Figure 8.1. S_o and X_o are the influent substrate and biomass concentration and S and X are the exit substrate and biomass concentration. A general material balance equation over the fermenter may be written as:

[Cells in – Cells out + growth of cells by biochemical reaction – Death of cells] = [Accumulation]

The similar balance equation can be applicable both for the cell mass and the substrate. For the cell mass equation can be written as Eq.(8.18):

$$FX_o - FX + \mu VX - \alpha X = V\frac{dX}{dt} \tag{8.18}$$

where α is the specific death rate. However, it is negligible compared to very high growth rate of the biomass. Also, the feed to the fermenter is sterile initially, it means that it does not contain any microorganisms; hence X_o can be taken as zero.

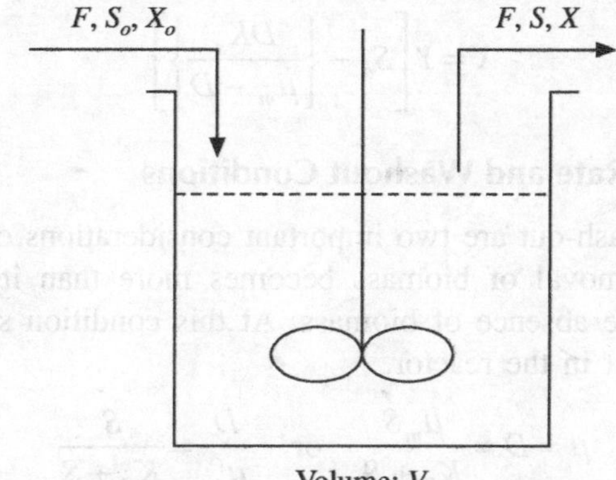

Figure 8.1 Schematic diagram of a continuous stirred tank bioreactor.

Using above conditions the Eq.(8.18) can be now written as:

$$-FX + \mu VX = V\frac{dX}{dt} \tag{8.19}$$

But assuming steady state approximation, the term *dX/dt* will be zero. The Eq.(8.19) can be modified as Eq.(8.20):

$$FX = \mu VX \quad \text{or} \quad F = \mu V \quad \text{or} \quad \mu = \frac{F}{V} = D \tag{8.20}$$

where *D* is dilution rate and which is the ratio of feed flow rate to the working volume of the reactor.

For substrate balance, the equation can be written as Eq.(8.21):

$$FS_o + R_s V - FS = \frac{ds}{dt}V; \tag{8.21}$$

where R_s is the rate of formation of substrate (*−ve*)

Incorporating Monod equation we can write:

$$\mu = D = \frac{\mu_m S}{K_s + S} \tag{8.22}$$

Thus, steady state substrate concentration can thus be obtained as Eq.(8.23):

$$S = \frac{DK_s}{\mu_m - D} \tag{8.23}$$

From the expression of yield coefficient, we get,

$$Y = \frac{X - X_o}{S_o - S} \quad \text{or} \quad X = X_o + Y(S_o - S) \tag{8.24}$$

After incorporating Eq.(8.23) in Eq.(8.22) the biomass concentration at steady state from the continuous stirred tank fermenter can be derived as Eq.(8.25):

$$X = Y\left[S_o - \left\{\frac{DK_s}{\mu_m - D}\right\}\right] \tag{8.25}$$

8.3.2 Critical Dilution Rate and Washout Conditions

Critical dilution rate and wash-out are two important considerations of continuous bioreactors. At critical dilution rate removal of biomass becomes more than its generation. It leads to washout which indicates the absence of biomass. At this condition substrate concentration in the feed solution equals that in the reactor.

$$\mu = D = \frac{\mu_m S}{K_S + S} \quad \text{or} \quad \frac{D}{\mu_m} = \frac{S}{K_S + S} \tag{8.26}$$

Inserting the conditions, $S \to S_o$; as $D \to D_{crit}$

$$\frac{D_{crit}}{\mu_m} = \frac{S_o}{K_S + S_o} \tag{8.27}$$

The quantity $\frac{S_o}{K_S + S_o}$ will be always less than unity and D_{crit} for a CSTF will always be less than μ_m by that factor.

8.3.3 Merits and Demerits of Continuous Culture

Relative merits and demerits of continuous culture of microorganisms are discussed below:

Merits:

- Growth can be controlled and maintained indefinitely.
- The effect of changes in physical and chemical parameters on growth and product formation at constant growth rate can be examined.
- Biomass concentration may be maintained constant by varying dilution rate (D).
- Various analyses of metabolic activity and cell composition as function of limiting nutrient concentration are possible.
- Using the Pulso-and-Shift-Technique medium composition can be optimized for biomass and product formation. This technique involves injecting (pulsing) nutrients directly into the chemostat and observing changes in X (biomass concentration) and S (substrate concentration). If X increases the nutrient is added to the medium supply reservoir and a new steady state is established. The procedure is then repeated.
- Products of secondary metabolism can be obtained and monitored along with cell mass production.
- Enables accurate determination of kinetic parameters, maintenance energy and actual yield coefficients.
- Results obtained in continuous culture are frequently more reliable and reproducible.
- Higher productivity per unit volume and less time lost in 'down-time' involving cleaning, preparation, medium sterilization, etc. Continuous culture is frequently more economic due to these factors and because it is low labor-intensive and involves smaller initial capital costs.

- Chemostat culture can be used to maintain mixed cultures under steady state condition. In batch culture one organism would usually outgrow the other.

Demerits:

- Production of some non-growth-related products cannot always be achieved and often requires fed-batch culture techniques.
- Wall growth and cell aggregation can cause wash-out or prevent true steady state growth.
- As the process continues for longer period there are chances of loss of the original strain due to take over by a faster growing strain, not of our interest.
- Growth of filamentous organism is often difficult due to the viscosity and heterogeneous nature of the culture preventing steady state growth.
- Growth over long periods may result in problems of contamination.
- For extended growth periods the plant and machinery must be reliable and of best quality to prevent mechanical break, closure, etc.

8.4 CONTINUOUS STIRRED TANK FERMENTERS IN SERIES

Continuous stirred tank fermenters are widely used in biochemical processes, primarily in homogenous liquid phase flow reactions in which continuous stirring is required. Instead of a stand-alone CSTF, sometimes these are used in series or in a battery, thereby enabling better conversion of the substrate. In a series of stirred tank fermenter in cascade, effluent from one tank constitutes the feed for the next.

Thus, excepting the very first CSTF in the series, no other will have a sterile feed. In case of two CSTF in series, as shown in Figure 8.2, the first one will behave as a simple chemostat

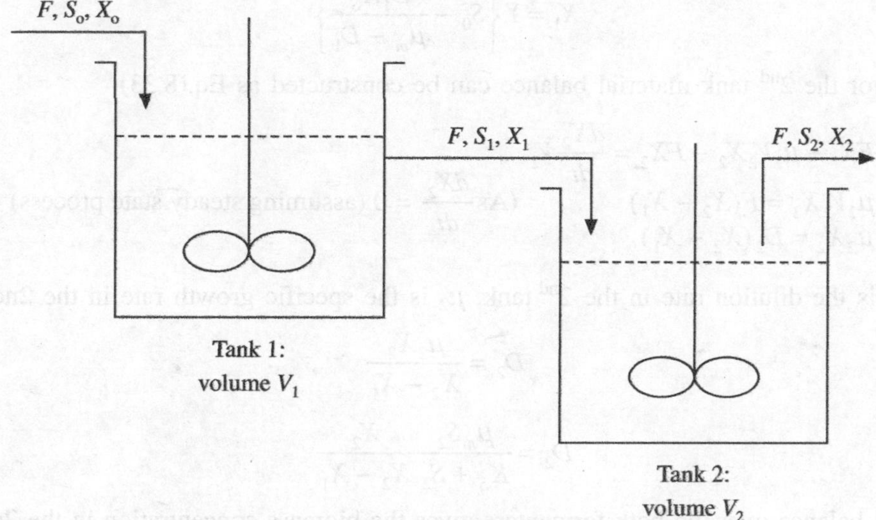

Figure 8.2 Schematic diagram of CSTF in series.

and all the design equations of simple chemostat will be applicable. However, this will not be the case for the rest of the CSTF in the series.

For the first tank Biomass concentration X_0, substrate concentration S_0, for the 2nd tank Biomass concentration X_1, substrate concentration S_1

For the 1st Tank:

$$FX_0 + R_x V_1 - FX_1 = \frac{dX_1}{dt} V_1 \tag{8.28}$$

Since $X_0 = 0$ (sterile feed), and $\frac{dX_1}{dt} = 0$ (assuming steady state process):

$$\mu_1 X_1 V_1 = FX_1 \quad \text{or} \quad \mu_1 = \frac{F}{V_1} = D_1 \tag{8.29}$$

where μ_1 = specific growth rate for the 1st tank, D_1 = Dilution rate for the 1st tank

$$\mu_1 = \frac{\mu_m S_1}{K_S + S_1} = D_1 \tag{8.30}$$

By rearrangement, we can get Eq.(8.31)

$$\begin{aligned} \mu_m S_1 &= D_1 K_S + D_1 S_1 \\ S_1(\mu_m - D_1) &= D_1 K_S \\ S_1 &= \frac{D_1 K_S}{\mu_m - D_1} \end{aligned} \tag{8.31}$$

From definition of yield coefficient (Y) (Eq. 8.6) and by inserting Eq.(8.31) in the yield coefficient equation (Eq. 8.6), we can get Eq. 8.32

$$X_1 = Y\left\{S_0 - \frac{D_1 K_S}{\mu_m - D_1}\right\} \tag{8.32}$$

Similarly for the 2nd tank material balance can be constructed as Eq.(8.33):

$$\begin{aligned} FX_1 + \mu_2 V_2 X_2 - FX_2 &= \frac{dX_2}{dt} V_2 \\ \mu_2 V_2 X_2 &= F(X_2 - X_1) \qquad \left(\text{As } \frac{dX_2}{dt} = 0 \text{ (assuming steady state process)}\right) \\ \mu_2 X_2 &= D_2 (X_2 - X_1) \end{aligned} \tag{8.33}$$

where D_2 is the dilution rate in the 2nd tank, μ_2 is the specific growth rate in the 2nd tank:

$$D_2 = \frac{\mu_2 X_2}{X_2 - X_1} \tag{8.34}$$

$$D_2 = \frac{\mu_m S_2}{K_S + S_2} \frac{X_2}{X_2 - X_1} \tag{8.35}$$

A material balance over the both fermenters gives the biomass concentration in the 2nd vessel can be written as Eq.(8.36):

$$X_2 = Y(S_0 - S_2) \tag{8.36}$$

Incorporating Eq.(8.36) into Eq.(8.35) and rearranging, we can get Eq.(8.37):

$$
\begin{aligned}
D_2 &= \frac{\mu_m S_2}{K_S + S_2} \frac{Y(S_0 - S_2)}{Y(S_0 - S_2) - X_1} \\
&= \frac{\mu_m S_2}{K_S + S_2} \frac{(S_0 - S_2)}{(S_0 - S_2 - \dfrac{X_1}{X_1}\left\{S_0 + \dfrac{D_1 K_S}{\mu_m - D_1}\right\}}
\end{aligned}
\tag{8.37}
$$

Eq.(8.37) can be solved for S_2 to obtain a quadratic equation Eq.(8.38):

$$
(\mu_m - D_2)S_2^{\,2} + \left\{\frac{D_1 D_2 K_S}{(\mu_m - D_1)} - D_2 K_S - \mu_m S_0\right\} S_2 + \frac{D_1 D_2 K_S^{\,2}}{(\mu_m - D_1)} = 0 \tag{8.38}
$$

Only one of the roots of Eq.(8.37) will give a positive value of the substrate concentration S_2 in the 2nd fermenter in the train.

8.5 CONTINUOUS STIRRED TANK FERMENTER WITH RECYCLE OF BIOMASS

Operation of continuous stirred tank bioreactor (fermenter) at high dilution rate sometimes proves necessary in wastewater treatment plants. For simple CSTF under such conditions substrate concentration in the effluent may rise considerably and the system tends to be unstable. One solution to this problem is to use a fermenter vessel with large working volume. However, there could be an alternative strategy to devise a method to retain the biomass in the fermenter while allowing the spent feed to flow out. Alternatively the effluent from the fermenter vessel can be passed to a settler-thickener and concentrated biomass or a portion of it is returned or recycled to the growth vessel. Different arrangements such as use of centrifuge (Figure 8.3), submerged internal filter (Figure 8.4) or external cross flow filter (Figure 8.5) can be installed with the CSTF.

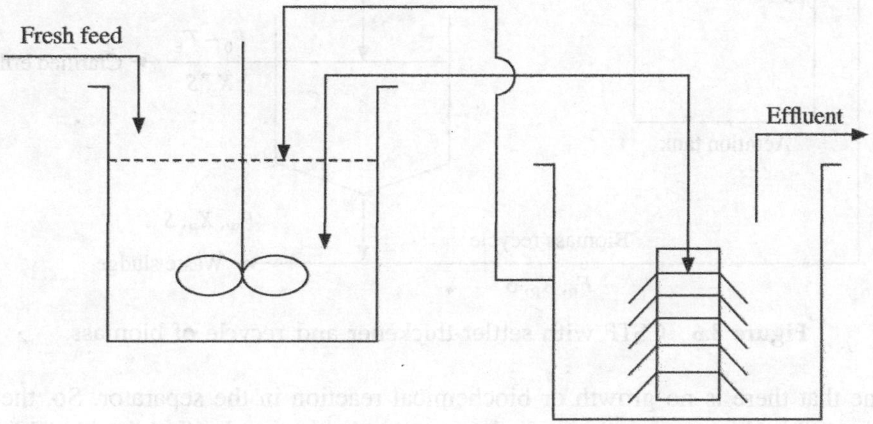

Figure 8.3 Use of centrifuge for cell recycle.

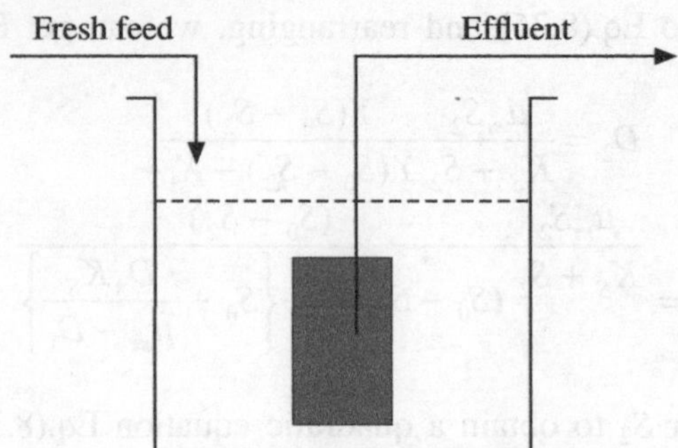

Figure 8.4 Use of submerged internal filter for cell recycle.

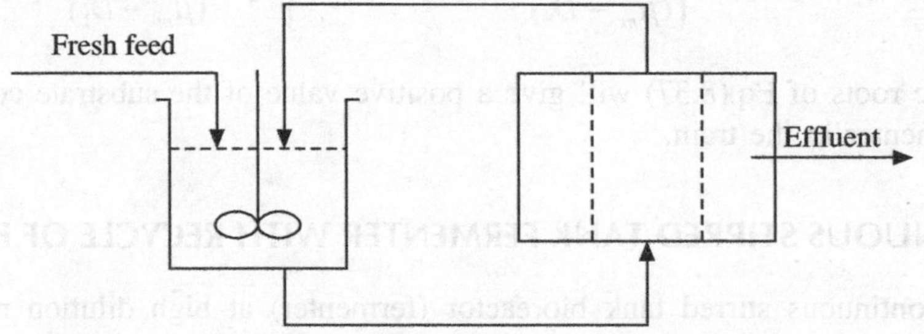

Figure 8.5 Use of external cross flow filter for cell recycle.

Let us consider a case of CSTF with settler-thickener and recycle of biomass as outlined in Figure 8.6.

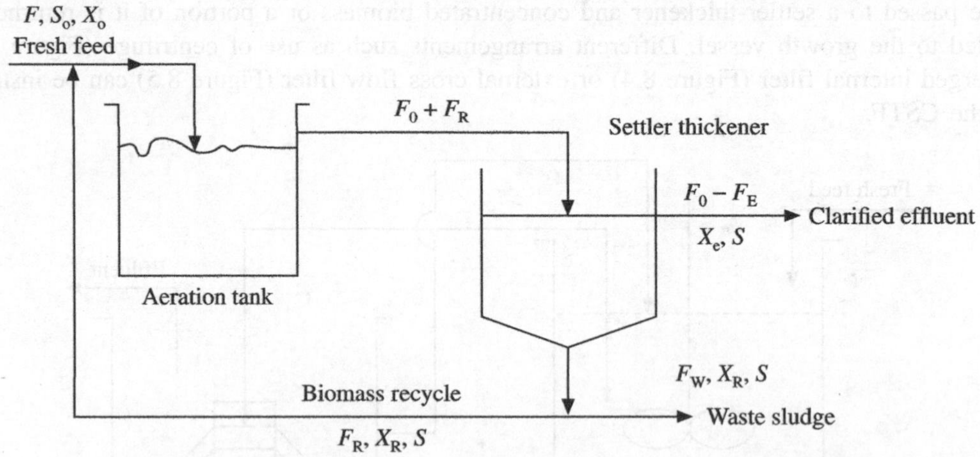

Figure 8.6 CSTF with settler-thickener and recycle of biomass.

Assume that there is no growth or biochemical reaction in the separator. So, the substrate concentration 'S' in the entering stream is the same as that in the clarified liquid effluent stream, in the recycle stream and in the exit biomass-rich stream. Constructing a material balance over the fermenter vessel we get:

[Materials in with fresh stream] + [Materials with recycle stream] + [formation by biochemical reaction] – {Output to separator} = [Accumulation]

The above relation can be written in the form of notations for biomass and substrate in the system.

For biomass:

$$F_0X_0 + F_RX_R + R_xV - (F_0 + F_R)X = V\frac{dX}{dt} \tag{8.39}$$

For substrate:

$$F_0S_0 + F_RS_R + R_sV - (F_0 + F_R)S = V\cdot\frac{dS}{dt} \tag{8.40}$$

where rate of formation of biomass R_x, $R_x = \mu X$.

Assuming steady state approximation $\frac{dx}{dt} = 0$; $X_0 = 0$ (assuming sterile feed) Eq.(8.39) can be written as:

$$F_RX_R + \mu XV - F_0X - F_RX = 0 \tag{8.41}$$

Dividing both sides of Eq.(8.41) by F_0:

$$\frac{F_R}{F_0}X_R + \mu X\frac{V}{F_0} - X - \frac{F_R}{F_0}X = 0 \tag{8.42}$$

But $\frac{F_R}{F_0}$ is the recycle ratio R (based on the fresh feed) and the dilution rate D, also based on the fresh feed, is $\frac{F_0}{V}$ so that substitution in Eq.(8.42) results in:

$$RX_R + \mu X\frac{1}{D} - X - R.X = 0 \tag{8.43}$$

or

$$\frac{\mu X}{D} = R.X + X - RX_R = X(1+R) - RX_R \tag{8.44}$$

8.6 FED-BATCH REACTOR

These are also called semi-batch or semi-continuous reactors and are very commonly used in biochemical processes. For example many gas-liquid two phase reactions occur in semi-batch reactors, where gas is continuously bubbled into the liquid reactant. As the name suggests, their operation is in between batch and continuous. In such reactors, any one or more streams are in continuous mode and the rests are in batch mode. It also an unsteady state reactor since concentration changes with time. Octave Levenspiel indicated three options of feeding of reactants into fed-batch or semi-batch reactor. These are as follows:

- One reactant is kept in the reactor as a batch while the other is fed continuously. As a result, volume as well as composition varies.
- Both the reactants are added continuously but the products are not withdrawn continuously. This would lead to the continuous change in volume, but keeping the composition constant.
- One of the reactants is fed initially in batch mode, while the other reactant is fed continuously. Product is withdrawn continuously. Consequently volume remains constant, but composition varies.

A general mass balance equation can be written for a fed batch reactor as:

$$V\frac{dS}{dt} + S\frac{dV}{dt} = FS_o - R_s V \tag{8.45}$$

Where F and S_o are the inlet flow-rate and influent feed concentration.

For a fed-batch reactor where feed in maintained constant, it leads to constant substrate concentration (i.e. $dS/dt = 0$). Under such conditions, the mass balance equation can be written as Eq.(8.46):

$$S\frac{dV}{dt} = FS_o - R_s V \tag{8.46}$$

Fed-batch reaction system is also useful in mitigating the substrate inhibition by intermittent feeding of substrate into the system. In repeated fed-batch culture, a portion of broth volume is removed at a definite interval of time to overcome the limitation of reactor volume. A good example of intermittent fed batch operation is antibiotic fermentation where glucose is added at a define tie interval into the fermentation broth

8.7 PLUG FLOW REACTOR

In this type of reactor the reactants are fed continuously and the products are withdrawn continuously. No mixing in the longitudinal direction takes place although slight mixing in the radial direction may occur. The velocity profile is flat. Concentration changes along the length of the reactor. However, at a particular position concentration does not change with time. This type of reactor is, therefore, a steady state reactor. While using the above reactors as bioreactors, the reactor system must be a closed one, i.e. the catalyst must be retained or recovered from the product system. If the support materials themselves become an integral part of the reactor, no separate retention system is required. In general, where the catalyst particles are in suspension, an outlet filter may be used. Particulate immobilized catalysts may be separated from the product stream by filtration or centrifugation. In special cases like magnetized immobilized enzyme particles a magnetic field is to be applied to recover the catalyst. For an ideal plug flow reactor the residence time (τ_{PFR}) can be expressed by Eq.(8.47):

$$\tau_{PFR} = \frac{V}{F} = \int_{S_o}^{S} \frac{dS}{-R_s} \tag{8.47}$$

8.8 CHOICE OF BIOREACTOR

Design and selection of bioreactor proceed through a series of interconnected stages of laboratory experiments, engineering analysis, economic and regulating considerations. Although design and selection of each type of bioreactor have their own unique characteristics, some fundamental principles are uniform for all. It starts with identification of product based on commercial objective and market need. Then the biochemical process to produce the particular product is to be established. It consists of biological system identification (selection of microorganisms, enzymes, animals or plant cells, etc.), raw materials and medium formulation, stoichiometric analysis, study of kinetic and transport processes. Stoichiometric analysis and reaction kinetics are the key information that are needed for the selection of type and capacity of bioreactors. Other considerations include whether the reaction will be in suspended cells or in immobilized bed, requirement of heat removal, adequate aeration, agitation, power requirement, need of recycling, low shear stress and product separation. The basic facts that are needed to establish the size of reactor are:

- Rate equation, yield data
- Maximum biomass concentration required
- Flow patterns of reactor

Apart from the bioreactors discussed earlier, there are several other types available. A few of these are mentioned below:

- Tower Fermenter
- Bubble column Bioreactors
- Gas/Airlift Bioreactors
- Packed-bed Bioreactors
- Fluidized-bed Bioreactor
- Membrane Bioreactors
- Photo-Bioreactors
- Wave Bioreactors
- Rotary Drum Bioreactors
- Deep-Jet Fermenter
- Mist Bioreactors
- Cyclone Column Fermenter

8.9 SCALE-UP OF BIOREACTOR

The term 'scale-up' indicates augmenting the scale of operation from laboratory scale to the pilot plant or from pilot plant to the production scale. By increase scale we mean the increase in volume which may affect the process parameters (Figure 8.7).

Since the biochemical processes are generally complex, we have to depend on specific solution rather than general solution. In general, for laboratory scale bioreactor large Erlenmeyer flasks of jar reactors of volume 500 mL to 100 mL are used while for pilot scale large vessels with volume ranging from 3000 litre–4000 litre are used.

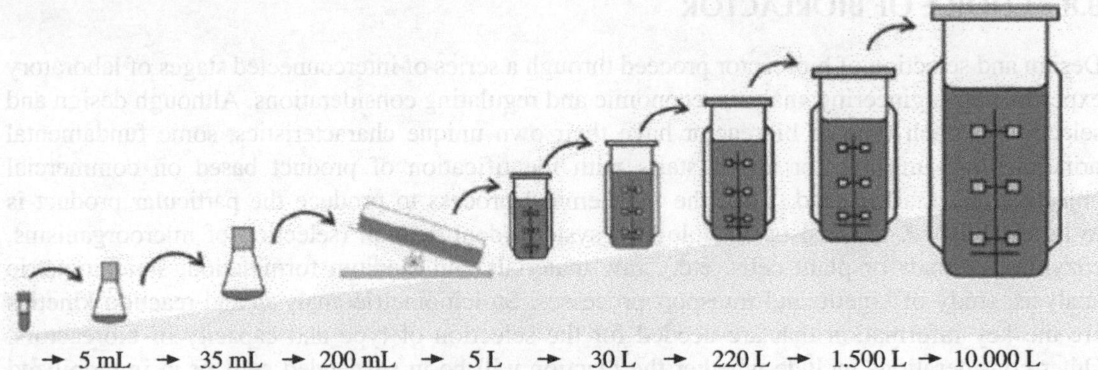

Figure 8.7 Conceptual diagram of a scale-up train of bioreactors (By permission from PharSol, Pharsol Ltd., Stegne 33, 1000 Ljubljana, Slovenia).

Potential benefits of scale-up of bioreactor are:

- Higher yield of biomass and associated products
- Availability of real-time bioprocess information
- Increased adaptability as compared to shake flask
- Ensuring better parametric control

If one could develop the solution of complete equations for mass, momentum and energy balance for a biochemical system under consideration the scale-up would have been easier. But since the above is not possible, we need to use a simplified approach following some major kinetic, transport and energy utilization processes.

In the past, dimensional analysis method as applied to ordinary chemical engineering or technology may also be adopted for biochemical engineering processes. While we use the above analysis, we look for some dimensionless quantities and based on similarity principle, some criteria for scale-up are found out. The consistency or invariance of all dimensionless quantities relevant to the particular problem provide key for scale-up.

While we adopt the technique of partial similarity, we are to maintain identical operating condition assaying the existence of geometrical similarity for extrapolation in fermentation. Some simple rule of thumb is followed with the following process parameters for scale-up:

- Identical Reynold's number
- Identical mass transfer coefficient
- Identical tip speed of the agitator if it is stirred reactor
- Identical volumetric power input
- Identical mixing time, etc.

Constant volumetric mass transfer coefficient is considered to be one of the convenient ways to consider a basis for scale of a fermenter. A general correlation based on geometrical similarity between a large and small fermenter is given by Eq.(8.48) (Aiba et al.):

$$\frac{(k_{La})_1}{(k_{La})_2} = \frac{(F/V)_1}{(F/V)_2}\left(\frac{H_{L_1}}{H_{L_2}}\right)\left(\frac{d_{B_2}}{d_{B_1}}\right)^{3/2}\left(\frac{v_{B_2}}{v_{B_1}}\right)^{1/2} \tag{8.48}$$

In the Eq.(8.48) subscripts 1 and 2 indicates small-scale and large-scale reactor (fermenter) respectively. In the above equation k_{La} is the volumetric mass transfer coefficient, F/V is the aeration rate, d_B is the bubble diameter, H_L is the depth of liquid, V_B is the ascending terminal velocity of bubble.

Another approach to scale-up is to consider constant power per unit volume of liquid in case of aerobic fermentation. Accordingly:

$$\frac{P}{V} \propto n^3 d_i^3 \tag{8.49}$$

where n is the rotational speed of impeller, d_i is then impeller diameter. Physical parameters in large fermenter can never be exactly replicated from those obtained from a lab scale unit. According to Oldshue (1996), increase in diameter of a stirred tank vessel by a factor of 5, can result in 125 times increase in volume, if height to diameter ratio is kept constant. Conceptual steps of a scale-up process (from shake flask to large bioreactor) is shown in Figure 8.8. In addition to physical concepts, biological concepts also need to be well understood for successful scale-up operation. These are various rate equations, kinetic constants, yield coefficients and even metabolic pathways.

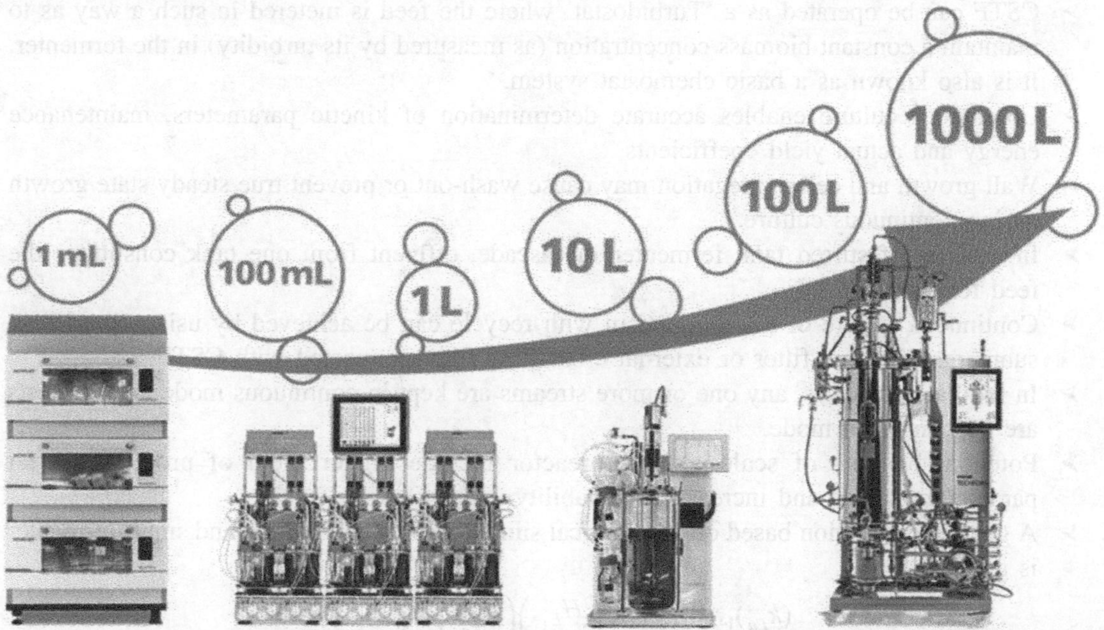

Figure 8.8 Steps of a scale-up process (from shake flask to large bioreactor) ((By permission from INFORS HT, Wuhrmattstrasse, Bottmingen, Switzerland www.infors-ht.com).

8.10 SCALE-DOWN OF BIOREACTOR

The consideration of scale-up of bioreactor arises at the time of commissioning of new large scale reactors. But there could be possibility of replicating some already operational bioreactors in the laboratory or pilot scale for the purpose of process improvement, parametric investigation or trouble shooting. Here comes the question of scaling down. For a full-scale plant various reaction conditions such as agitation, aeration, temperature, etc. must be well-established. Thus, for the process improvement it must be scaled down with identical conditions. Scale-down reactor is used for rapid validation of data and this should exactly mimic the full-scale reactor. It can monitor growth rate, various patterns of product formation or can assess any level of contamination, by products formation and so on. Mostly it is also done by assuming geometric similarity based on various constant parameters such as constant *P/V*, constant impeller speed and constant Reynold's number. It has been observed that *P/V* parameter is the most preferred for scale-up and scale-down problem.

KEY POINTS TO REMEMBER

- Bioreactors employed in industry may be broadly classified into two groups as (a) Batch reactor, and (b) flow or continuous reactor. The latter may be again classified in to two groups (i) mixed flow reactor (or chemostat or continuous stirred tank reactor), and (ii) plug flow reactor.
- CSTF can be operated as a 'Turbidostat' where the feed is metered in such a way as to maintain a constant biomass concentration (as measured by its turbidity) in the fermenter. It is also known as a basic chemostat system.
- Continuous culture enables accurate determination of kinetic parameters, maintenance energy and actual yield coefficients.
- Wall growth and cell aggregation may cause wash-out or prevent true steady state growth during continuous culture.
- In a series of stirred tank fermenter in cascade, effluent from one tank constitutes the feed for the next.
- Continuous culture of microorganism with recycle can be achieved by using centrifuge, submerged internal filter or external cross-flow filter connected with CSTF.
- In fed-batch reactors, any one or more streams are kept in continuous mode and the rests are kept in batch mode.
- Potential benefits of scale-up of bioreactor include higher yield of products, better parametric control and increased adaptability.
- A general correlation based on geometrical similarity between a large-and small-fermenter is given as
- $$\frac{(k_{La})_1}{(k_{La})_2} = \frac{(F/V)_1}{(F/V)_2}\left(\frac{H_{L_1}}{H_{L_2}}\right)\left(\frac{d_{B_2}}{d_{B_1}}\right)^{3/2}\left(\frac{v_{B_2}}{v_{B_1}}\right)^{1/2}$$

  The subscripts 1 and 2 indicate small-scale and large-scale reactor (fermenter) respectively.
- *P/V* (constant power per unit volume) parameter is the most preferred for scale-up and scale-down problem.

QUESTIONS FOR PRACTICE

Multiple Choice Questions

1. A chemostat has a liquid volume of 2 litres and is being fed at a rate of 4 litres per hour. Dilution rate for this reactor will be
   (a) 2 litre (b) 2 litre per hour
   (c) $2h^{-1}$ (d) 4 litre per h
2. What are the main industries that use bioreactors?
   (a) Food industry (b) Pharmaceutical industry
   (c) Biofuels industry (d) All of the above
3. Which type of bioreactor is used to produce microalgae, photosynthetic organisms and secondary metabolite?
   (a) Photobioreactor (b) Airlift bioreactor
   (c) Bubble column bioreactor (d) Fluidized-bed bioreactor
4. What are the limitations of bioreactors?
   (a) High cost (b) Technical complexity
   (c) Contamination risk (d) All of the above
5. The small-scale bioreactors have volume of
   (a) 5–10 litres (b) 10–20 litres
   (c) 1–10 litres (d) 1–20 litres
6. Which of the following fermenters are characterized by height to diameter ratio?
   (a) Tower fermenter (b) Airlift fermenter
   (c) Hollow fibre (d) Perfusion bioreactor
7. In World War II, the fermentation was used for the production of
   (a) Alcohol (b) Antibiotics
   (c) Wine (d) Beer
8. In which of the following fermenters the impellers are replaced by the constant flow of gas?
   (a) Airlift fermenter (b) Tower fermenter
   (c) Hollow fibre (d) Perfusion bioreactor
9. The reverse of hydraulic retention time is
   (a) Sedimentation rate (b) Dilution rate
   (c) Filtration rate (d) Hydrolysis rate
10. In which of the following reactors mechanical agitation is required?
   (a) Packed-bed (b) Airlift reactor
   (c) Stirred tank (d) Bubble column
11. Microcarrier beads can be used in
   (a) Fluidized-bed (b) Stirred vessel
   (c) Packed-bed (d) Trickle bed

12. Which material is disengaged in the disengagement zone?
(a) Culture broth (b) Culture media
(c) Bubbles (d) Microbes

13. The number of baffles in standard stirred tank bioreactor
(a) 6 (b) 8
(c) 2 (d) 4

14. In airlift bioreactor, the air sparging region is called
(a) Disengagement zone (b) Down comer
(c) Air riser (d) None of the above

15. Turbidostat is recommended when continuous fermentation needs to be carried out at
(a) Low dilution rate near the washout point
(b) High dilution rate near the washout point
(c) Any intensity of dilution rate near the washout point
(d) Moderate dilution rate at the washout point

16. The height to diameter ratio for a tank fermenter is
(a) > 3 (b) < 3
(c) > 1.5 (d) < 1.5

17. Chemostats work on the principle of
(a) Maintaining continuous flow of nutrients
(b) Maintaining constant volume of culture medium
(c) Operating at high pressure
(d) Maintaining uniform nutrients concentration

18. Which bioreactor is generally used in wastewater treatment?
(a) Trickle bed reactor (b) Plug flow reactor
(c) CSTR (d) Tower fermenter

19. The relatively large volume (10 to 40% of the total fermeneter volume) of the seed culture is employed to
(a) Increase the overall yield of biomass
(b) Increase the overall fermentation time
(c) Decrease the overall yield of products
(d) Decrease the overall fermentation time

20. The contents in stirred tank reactor is mixed through
(a) A mechanical agitator (b) Compressed air
(c) External pumping (d) All of the above

21. Which of the following is not a scale-up process?
(a) Laboratory to Pilot scale (b) Pilot scale to Industrial scale
(c) Industrial to Pilot scale (d) Laboratory to Industrial scale

22. Which of the following factors are not involved in the scale-up process?
(a) Inoculum development (b) Sterilization
(c) Temperature (d) Medium design

23. Which of the following is not a rule while doing scale-up studies?

(a) The similarity in the geometry of the fermenter used
(b) The similarity in the configuration of the fermenter used.
(c) There should be a minimum of three or four stages of increment
(d) Increasing power is a jump in scale

24. Which of the following is not an aspect in the context of scale-down

(a) Number of generations (b) Oxygen transfer rate
(c) Mixing (d) Temperature

25. In which of the following bioreactors, the particles are not immersed in liquid?

(a) Airlift reactor (b) Trickle bed reactor
(c) Plug flow reactor (d) Stirred tank reactor

26. What do you mean by Quasi steady state?

(a) Cell concentration remains virtually constant
(b) Cell concentration is virtually variable
(c) Total biomass remains constant with time
(d) Total biomass decreases with time

27. In fed-batch culture, the feed solution is

(a) Less concentrated (b) Highly concentrated
(c) Highly diluted (d) Diluted

28. Which of the following Fed-batch system is described as a fixed volume?

(a) The same medium used to establish the batch culture
(b) A solution of the limiting substrate
(c) A concentrated solution of the limiting substrate
(d) A very concentrated solution of the limiting substrate

29. Which of the following is not a disadvantage of continuous culture of microorganism?

(a) Wall growth and cell aggregation
(b) Loss of the original strain due to take over by a faster growing strain
(c) Higher productivity per unit volume and less 'down-time'
(d) Wash-out may prevent true steady state growth

30. The valve used to prevent accidental reversal of flow of liquid/gas in a pipe is

(a) Pressure control (b) Check
(c) Safety (d) Gate

31. Which amongst the following represent aerobic process?

(a) Trickling filter (b) Rotating disc contactor
(c) Rotating drums (d) All of the above

Answer Key

1. (c) 2. (d) 3. (a) 4. (d) 5. (d) 6. (a) 7. (b) 8. (a) 9. (b) 10. (c)
11. (a) 12. (c) 13. (d) 14. (c) 15. (b) 16. (b) 17. (d) 18. (a) 19. (d) 20. (a)
21. (c) 22. (d) 23. (d) 24. (d) 25. (b) 26. (a) 27. (b) 28. (d) 29. (c) 30. (b)
31. (d)

Descriptive Questions

1. The steady state substrate and biomass concentration for a continuous stirred tank fermenter operated at various dilution rates are given below. Given that the fresh feed concentration is 700 mg/L, calculate the values of Monod constants μ_m and K_S, the yield coefficient Y and the endogenous respiration coefficient k_d.

Dilution rate (h^{-1})	Substrate concentration (mg/L)	Biomass concentration (mg/L)
0.30	4	326
0.25	41	328
0.20	16	340
0.12	8	342
0.08	3.8	344

2. Using substrate and biomass balance derive the design equation for a continuous stirred tank system for biochemical process. State the assumptions. Comment on critical dilution rate and washout factor.
3. Calculate the steady state substrate and biomass concentration in a continuous fermenter which has an operating volume of 25 L. When the sterile feed stream contains limiting substrate at 2000 mg/L and enters the vessel at 8 L/h. The values of K_s and μ_m are 10.5 mg/L and 0.45 h^{-1} respectively and the yield coefficient may be taken as 0.48.
   (Ans: 25.8 mg L^{-1}; 948 mg L^{-1}**)**
4. The Monod rate model is valid for a CSTR bioreactor with maximum specific growth rate of 0.5 h^{-1} and K_s 2 gL^{-1}. What would be a suitable dilution rate at steady-state condition, where there is no cell death if initial substrate concentration is 50 gL^{-1} and yield of biomass on substrate is 100%? **(Ans:** 0.4 h^{-1}**)**
5. Illustrate various probable configurations of fed-batch bioreactors with schematic diagrams.
6. What are the various factors which were followed while scaling up process in the bioreactor? Comment on its significance.
7. Briefly explain the mass balance reactions in an Ideal plug-flow reactor.
8. Explain in detail about the working of CSTR reactors with cell recycle and wall growth conveniences.
9. What is the relationship between the dilution rate, D and growth rate, μ, at steady state? What happens if D is very large?
10. The design of a bioreactor can be different from that of a chemical reactor in many ways. What are some of the issues that must be considered to assure the health and productivity of microorganisms? Based on these, make a list of features, that make a bioreactor different from a standard chemical reactor.
11. A bacterial fermentation was carried out in a reactor containing broth with average density 1200 kg/m^3 and viscosity 0.02 N s/m^2. The broth was agitated at 90 rpm and air was introduced through the sparger at a flow rate of 0.4 vvm. The fermenter was equipped with two sets of flat blade turbine impellers and four baffles. Given that tank diameter, 4 m; impeller diameter, 2 m; baffle width, 0.4 m; also the liquid depth was 6.5 m. Determine ungassed power required. (Refer Power number versus Reynolds number standard plot for various impellers (flat blades, turbine, vaned disk and marine propeller)
   (Ans: 180 hp**)**

9 Fermentation Technology

9.1 FERMENTATION AS A UNIT PROCESS

The term fermentation has evolved from a Latin word 'fervere' which means 'to leaven' or cause to rise. It also means 'to boil or bubble or effervesce'. The boiling appearance results from effervescence of CO_2 which is produced as a byproduct of fermentation reaction. Fermentation is a very ancient art or technique associated with the life and living of the people from time immemorial. History of early civilizations abounds with the instances of such techniques applied in making tastier food, stimulating drinks and other applications. However, the technique of earlier fermentation was developed sans the knowledge of microorganisms and enzymes effecting biochemical reactions.

The modern conceptual understanding of fermentation as a microbial transformation process was put forth by a famous French Chemist Louis Pasteur (1822–1895). Pasteur was the first to report the action of yeast in the conversion of sugars into ethyl alcohol in the year 1857. Later on his famous research on eliminating the contamination of milk and other foodstuffs led to the development of milk pasteurization process. Today, fermentation is acknowledged as the single most important biochemical unit process, responsible of driving the production of a wide spectrum of bioproducts.

Chemistry of fermentation

Fermentation is a biochemical process where carbonaceous organic compounds are enzymatically converted into useful products in presence of suitable microorganisms. It takes place both in presence or absence of oxygen. Most of the microbial fermentation is anaerobic process while fermentation taking place in higher organisms are aerobic in nature. In anaerobic fermentation sugars are metabolized to volatile fatty acids with the additional production of other compounds and gaseous products. Generally fermentation leads to the production of many useful products of commercial value. These could be:

(i) Biomass as product
(ii) Microbial enzymes
(iii) Microbial metabolites

(iv) Recombinant products
(v) Modified compounds

Table 9.1 gives a glimpse of five stage development of fermentation process over the centuries.

Table 9.1 Stage-wise development of major biochemical products

Stage	Period	Products
Stage 1	Pre–1900	Major products were alcohol, vinegar
Stage II	1900–1940	Baker's yeast, glycerol, citric acid, lactic acid, acetone, butanol
Stage III	1940–1960	Major products were penicillin, other antibiotics, vitamins, gibberellin, amino acids, enzymes and steroids.
Stage IV	1960–1980	Several development of fermentation industries, e.g. continuous system, process control through computer, etc., SCP using hydrocarbons, and other feedstock.
Stage V	1980-onward	Production of very high value, low-volume products, development is mainly based on genetic manipulation e.g. production of heterogeneous proteins such as insulin, human growth hormone, monoclonal antibodies, etc.

9.2 BASIC FUNCTIONS OF A FERMENTER

Fermenter provides appropriate, aseptic and pre-defined physico-chemical factors, conducive for the growth of microorganism and synthesis of high-value useful products from low-value substrates. It can perform different functions, such as:

1. Production of valuable chemicals like alcohol, organic acids, enzymes, antibiotics vaccines, therapeutic proteins, monoclonal antibodies and so on.
2. To enable optimal growth and multiplication of microorganisms and help bulk production of specific products,
3. To provide a reaction environment where microorganisms can breakdown complex molecules into simpler and more useful products.
4. To increase the efficiency and yield of bioproducts
5. To control factors such as temperature, pH and levels of oxygen, carbon dioxide and nutrients to ensure ideal fermentation conditions.
6. To monitor and control the progress of the bioprocess through the use of sensors and other laboratory devices.
7. Sample withdrawal facility from time to time needs to be provided.
8. Should ensure less energy consumption and prevent evaporation loss.
9. To recycle fermentation byproducts for the benefit of the environment.
10. Affordable raw material as the precursors.
11. To provide reproducible result of yield and other factors so that scale-up will be facilitated.

9.3 GENERAL REQUIREMENTS OF FERMENTATION PROCESS

In the following section various mechanical parts of fermenter vessel and their functions are briefly discussed. In addition, strategies for sterilization of fermenter and air supply systems to maintain aseptic conditions, media preparation, inoculum development and methods of inoculation are covered.

9.3.1 Construction and Mechanical Parts of a Fermenter

Basic parts of a fermenter

A fermenter is a bioreactor which facilitates the growth of microorganism under precisely controlled physico-chemical conditions for the production of useful end-products along with cell mass. It is a cylindrical vessel of suitable working volume provided with various accessories and auxiliary instruments for aseptic growth and production. A schematic diagram of a laboratory fermenter with different accessories is presented in Figure 9.1.

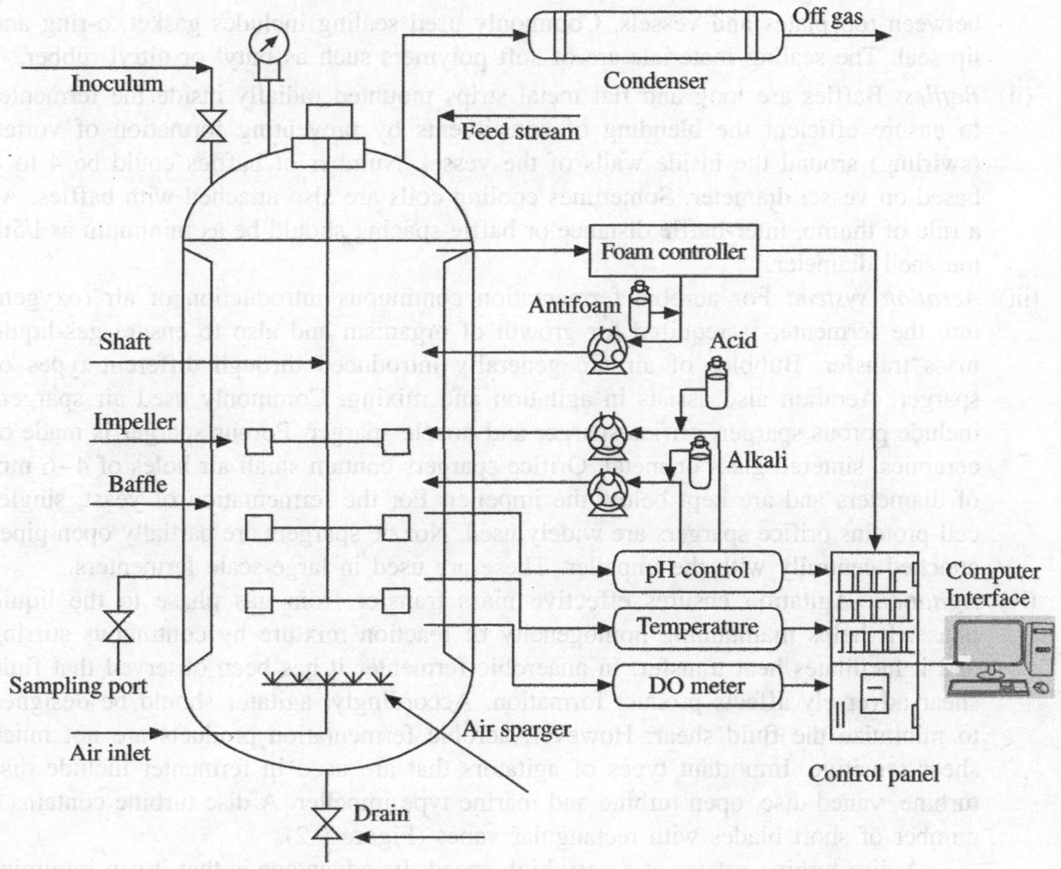

Figure 9.1 Schematic diagram of a typical laboratory fermenter.

The important parts of an ideal fermenter are discussed below:

(i) *Materials of construction*: Depending on the scale of operation fermenter vessels can be made of high quality stainless steel, glass, glass-lined mild steel or even some polymeric materials. The vessels are having smooth internal surface and non-corrosive nature. Borosilicate glasses are preferred for small laboratory-scale fermenter. Stainless steel (10–13% chromium) is the most preferred materials of construction. It is non-corrosive due to the presence of chromium. Glass vessels could be of either jars or cylindrical with flat bottoms and flanged tops. Generally cylindrical vessels are fitted with head plates at both ends. Vessel size is dependent on several factors such as required fermentation capacity, cost and ease of fabrication. Sometimes stainless steel clad vessels are also used. Even if for cost reduction outside jacket be made of ordinary steel, the vessel must be made of good quality stainless steel to avoid problems of corrosion and contamination. Flange connections should be minimum to avoid chances of contamination

For leakage free aseptic operation there must be some provisions of sealing between top plates and vessels. Commonly used sealing includes gasket, o-ring and lip seal. The sealing materials are of soft polymers such as butyl or nitryl rubber.

(ii) *Baffles*: Baffles are long and flat metal strips mounted radially inside the fermenter to ensure efficient the blending of constituents by preventing formation of vortex (swirling) around the inside walls of the vessel. Number of baffles could be 4 to 8 based on vessel diameter. Sometimes cooling coils are also attached with baffles. As a rule of thumb, inter-baffle distance or baffle spacing should be as minimum as l/5th the shell diameter.

(iii) *Aeration system*: For aerobic fermentation continuous introduction of air (oxygen) into the fermenter is required for growth of organism and also to ensure gas-liquid mass transfer. Bubbles of air are generally introduced through different types of sparger. Aeration also assists in agitation and mixing. Commonly used air spargers include porous sparger, orifice sparger and nozzle sparger. Porous sparger is made of ceramics, sintered glass or metal. Orifice spargers contain small air holes of 4–6 mm of diameters and are kept below the impeller. For the fermentation of yeast, single-cell-proteins orifice spargers are widely used. Nozzle spargers are partially open pipes attached centrally with the impeller. These are used in large-scale fermenters.

(iv) *Agitation*: Agitation ensures effective mass transfer from gas phase to the liquid phase. It helps maintaining homogeneity of reaction mixture by continuous stirring and it facilitates heat transfer. In anaerobic fermenter it has been observed that fluid shear adversely affects product formation. Accordingly, agitator should be designed to minimize the fluid shear. However, aerobic fermentation products are not much shear sensitive. Important types of agitators that are used in fermenter include disc turbine, vaned disc, open turbine and marine type impeller. A disc turbine contains a number of short blades with rectangular vanes (Figure 9.2).

A disc turbine rotates at a very high speed. Its advantage is that it can minimize flooding by air bubbles. However, it is subjected to rotational stress and fatigue cracking. Vaned disc has several rectangular vanes attached to it. Variable pitch open turbine and marine type impeller contain vanes attached directly on the agitator shaft.

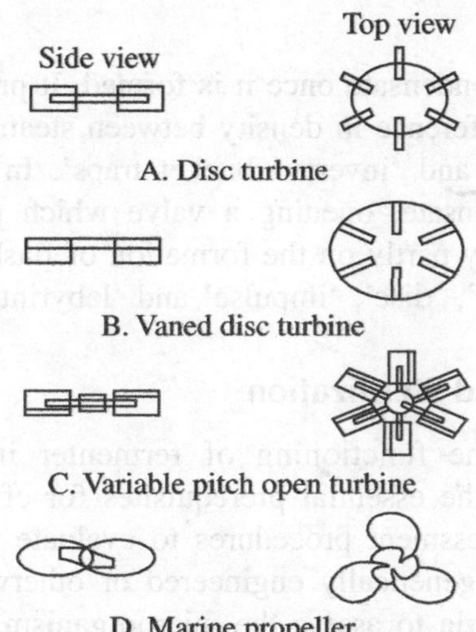

Figure 9.2 Common types of impellers.

(v) *Stirrer glands and bearings*: Maintaining aseptic conditions in the fermenter is of utmost importance for efficient fermentation. Stirrer gland and bearings ensure leak proof functioning of the fermenter vessel.

- **The Stuffing Box:** This is an assembly which is used to house a gland seal with packing rings of asbestos or compressed cotton. High pressure is needed to make the packing tight-fit. As it is quite problematic to properly sterilize the packing, it is better to be replaced at regular intervals.
- **The Mechanical Seal:** It has three parts attached together. These are a stationary seat, a rotating face on shaft, and a spring for attachment.
- **Magnetic Drives:** It has two magnets, one is called driven and the other is driving plate. The driven magnet is joined with a drive-shaft and is held in bearings. This can be used for stirring the vessel at speed of 300–2000 rpm.

(vi) *Valves and steam traps*: There are four types of commonly used addition valve, namely simple ON and OFF, rough control, precise adjustment and safety valve. Gate valve are used for fully open or fully shut-off operation, globe valve is used for fine regulation of flow. Piston and needle valve are also similar in function to globe valve excepting that the flow is controlled by piston in the former while in the latter the disc is substituted with a cone-shaped plug or needle. Butterfly, diaphragm and pinch valves are also used. Pinch and diaphragm valves are particularly used for highly sterile operation. Check valves are useful in preventing sudden reversal of flow due to some error. Some important types of check valves include swing check, lift check, etc. In addition pressure control valves are used for either reduction or maintaining pressure. Safety valves are used as a protection against sudden shoot-up of pressure. These can be spindle type or rupture disc. Ball valves are generally preferred in fermentation industry as these are easy to maintain and can be operated aseptically. They are particularly useful in inoculation and sample lines.

Steam Traps

Steam traps help in releasing condensate once it is formed. It prevents steam to escape. Steam traps operate by sensing the difference in density between steam and condensate. These steam traps include 'ball float traps' and 'inverted bucket traps'. In the 'ball float trap', the ball rises in the presence of condensate, opening a valve which passes the denser condensate. Thermodynamic steam traps rely partly on the formation of flash steam from condensate. This group includes 'thermodynamic', 'disc', 'impulse' and 'labyrinth' steam traps.

9.3.2 Aseptic Operation and Sterilization

Aseptic operation indicates the functioning of fermenter in an environment free from contamination. This is one of the essential prerequisites for efficient operation of fermenter. There are various standard assessment procedures to evaluate the pathogenicity or virulence level of microorganisms, both genetically engineered or otherwise, used in the fermentation process. There are several criteria to assess the microorganism. The amount of inoculums to be developed, presence of any other local vectors or potential reserves within the system, past history of contamination and literature information are some of the important considerations. A few important considerations are:

- Flange connections should be minimum to avoid chances of contamination.
- Welded constructions should be used and welded portions need to be polished.
- Avoid dead space, crevices, and idle pockets.
- Various parts of the vessel should be independently sterilizable.
- Valves should be so selected that those are easily cleaned, maintained, and sterilized.
- All vessel connections should be steam sealed.
- A slightly positive pressure is maintained inside the fermenter so that leakage is always out (for pathogens it is not applicable).

An aseptic operation of a fermenter requires that pipings for gas (air/oxygen supply), inoculation line, and other materials or accessories be sterilized appropriately by steam.

Sterilization of fermenter and air supply systems

Aseptic operation is of prime importance in any fermentation process. This is achieved by appropriate and adequate sterilization of the reactor vessel (fermenter), media to be introduced, air supply or any other associated devices or accessories. For a smaller size batch fermenter steam sterilization can be done by placing it in an autoclave. In case of large fermenter *in situ* sterilization is preferred. Steam is introduced into the reactor through jacket or coil. Care should be taken to minimize the generation of condensate. Steam should reach to all bends and corners of the reactor vessel as well as associated piping network. This ensures prevention of potential contamination.

Uninterrupted supply of sterile air is essential for aerobic fermentation. Traditionally, air is filtered through glass wool, fibres, mineral slag or even cartridge filters. Nowadays membrane filters are also used. Figure 9.3 presents a simple schematic representation of air sterilization process in a fermenter. There are three valves designated as A, B, and C in air and steam inlet line. A is non-sterile air inlet valve, while B is sterile air inlet valve.

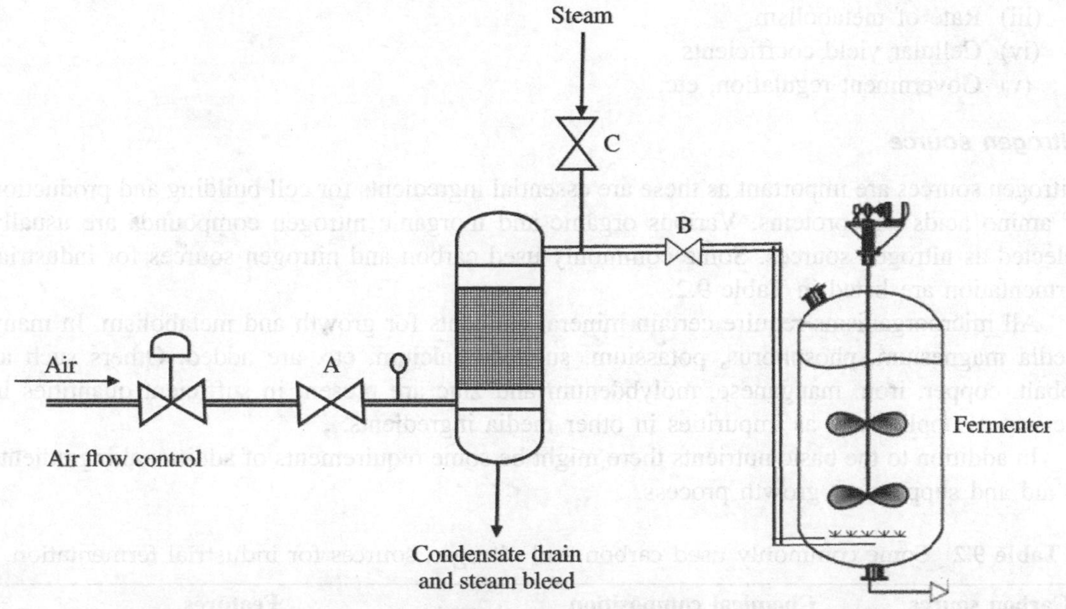

Figure 9.3 A schematic of the working of packed air filter in a fermenter.

The valve C is used for steam purging. At the beginning of operation valves A and B are kept closed. Steam is introduced at valve C. Air is purged downwards through the packed section to a bleed valve at the bottom. After sometime, when steam is freely issued through the bleed valve, B is opened to ensure steam to enter into the fermenter. Bleed valve needs to be regulated to maintain correct sterilization pressure in the fermenter and filter bed. There is an alternative method of air sterilization by using steam jacketed air filter.

Exhaust gases that are generated during fermentation should also be sterilized before they are vented out to atmosphere. These gases are composed largely of carbon dioxide, water vapor, nitrogen and small amount of other gases like hydrogen, methane, hydrogen sulphide, etc. There are also chances of aerosol formation. These may be sterilized using 0.2 µm filters on the exhaust pipe. Sometimes, cyclone separator and coalescer are also used to arrest particulate matter and liquid droplets respectively.

9.3.3 Media Design for Fermentation

Medium formulation is one of the essential stages of fermentation process. Media contains required quantity of carbon source, nitrogen source, minerals and vitamins.

A carbon source supplies energy for the microorganisms to grow and reproduce. It also helps in the biosynthesis of various primary and secondary metabolites. Popular carbon sources include molasses, malted barley, starch and dextrin, sulphite waste liquor, alkanes and so on.

Selection criteria for carbon source are:

(i) Cost of product
(ii) Ease of availability

(iii) Rate of metabolism
(iv) Cellular yield coefficients
(v) Government regulation, etc.

Nitrogen source

Nitrogen sources are important as these are essential ingredients for cell building and production of amino acids and proteins. Various organic and inorganic nitrogen compounds are usually selected as nitrogen sources. Some commonly used carbon and nitrogen sources for industrial fermentation are listed in Table 9.2.

All microorganisms require certain mineral elements for growth and metabolism. In many media magnesium, phosphorus, potassium, sulphur, calcium, etc. are added. Others such as cobalt, copper, iron, manganese, molybdenum and zinc are present in sufficient quantities in the water supplies and as impurities in other media ingredients.

In addition to the basic nutrients there might be some requirements of additional ingredients to aid and supplement growth process.

Table 9.2 Some commonly used carbon and nitrogen sources for industrial fermentation.

Carbon source	Chemical composition	Features
Molasses	35–37% sucrose, 15–20% (fructose + glucose), 10-12% ash, rest solid and some smaller quantities of polysaccharides.	By-product from sugar industry, cheap and abundantly available. Examples beet molasses and cane molasses.
Malt extract	More than 50% maltose, small amounts of fructose and sucrose, vitamin B (niacins), amino acids, slats of potassium and magnesium.	Brown coloured viscous liquid, available in both liquid and dried extract form. Prepared by crushing and mashing malted grains, followed by concentration and drying.
Whey	Mostly water, with 7–8 % of lactose, lactic acids, whey proteins (bovine serum albumin, bovine lactoferrin) and rest solids.	It is a by-product from dairy processing units and is used for the fermentative production of vitamin B12, alcohol, lactic acid, single-cell-proteins, etc.
Starch and starch hydrolysates	A polysaccharide consisting of long chain of glucose.	soft, white, tasteless powder that is insoluble in cold water,
Dextrin	Dextrins are a group of carbohydrates obtained from starch or glycogen by hydrolysis or roasting.	Prepared from corn starch by roasting followed by enzymatic hydrolysis (amylase)
Spent liquor of pulping	Mainly consists of lignin residues, hemicellulose, and other organic and inorganic materials.	It is generated as a wastewater from the paper and pulp industry employing sulphite (or sulphate) pulping. Used as a carbon source in various fermentation processes.
Nitrogen source	**Chemical composition**	**Features**
Corn steep liquor	Contains about 50% of water and the rest containing crude proteins (about 40%), nitrogen free extracts, lactic acid, minerals and vitamins.	It is cost effective and rich source of nitrogen produced during starch production from corn by wet milling.

(Contd.)

Nitrogen source	Chemical composition	Features
Urea	It is diamide of carbamic acid, $CO(NH_2)_2$.	It is a cheap and abundant source of nitrogen, with available nitrogen being 46%. Can also act as carbon source.
Yeast extract	It is rich in amino acids, peptides, vitamin B groups and minerals like magnesium, zinc.	Produced by hydrolysis or autolysis of waste yeast obtained from breweries.
Soy meal	It is rich in nitrogen in the form of proteins with 47–49% protein and 3% crude fibre.	It is obtained as a by-product from the extraction of soybean oil.
Peptone	Water soluble protein hydrolysate containing various amino acids such as lysine, arginine, histidine.	Obtained from enzymatic digest of animal tissue.
Ammonium salts	Ammonium sulphate, ammonium phosphate.	Rich source of ammonia, soluble in water. Ammonium sulphate 21%; ammonium chloride 26%; mono-ammonium phosphate 10-11%; di-ammonium phosphate 18%; ammonium thiosulphates: 12%.

Chelators

These are added to form metal ion complexes which survive heat degradation and could be taken up by microorganism for a long time. Example: EDTA (Ethylene diamine tetraacetic acid), citric acid, polyphosphate, etc.

Vitamins and growth factors

Some bacteria, filamentous fungi and yeasts, are unable to synthesize vitamins. So, for growth of such microorganism medium must be supplemented by vitamins. Sometimes, natural carbon and nitrogen sources could also supply vitamins to the media.

Precursors

They are growth stimulator which tends to augment the yield and product quality. For example, corn steep liquor, phenyl acetic acid and similar other compounds in the fermentation of penicillin, cyanides in the production vitamin B_{12} (cyanocobalamin) using *Streptomyces* sp., and so on.

Inducers and elicitors

Enzymes can act as inducers. Generally they are often substrates such as starches and dextrins for amylase. Synthesis of secondary metabolites like flavanoids is induced by addition of elicitors.

Inhibitors

These are specific chemical compounds used for selective reduction of metabolic pathways for a targeted product or to suppress the formation of other undesired products. An example of an inhibitor specifically employed to redirect metabolism is sodium bisulphate.

9.3.4 Inoculum Development

Inoculation or seeding the fermenter with the intended microorganism is an important step towards efficient fermentation. The inoculums should be collected from the exponential growth phase so that it remains in a healthy and active condition and does not contribute to a lengthy lag phase in the subsequent fermentation. It should be available in sufficiently large volumes to provide an inoculum of optimum size and free from contamination. Normally inoculums volume is kept 3% to 10% of the medium volume. A relatively large inoculums volume can minimize the length of the lag phase and increase productivity of biomass as well as other products.

9.3.5 Inoculation Method

There are various methods for inoculating during fermentations. The most essential requirement is to prevent the contamination during transfer of inoculums as much as possible. To prevent contamination during the transfer process it is essential that both vessels be maintained under a positive pressure and the inoculation port be equipped with a steam supply. A sterile spore-suspension vessel can be used for inoculation. The sterile spore-suspension vessel is batched with the spore suspension in the sterile room. Fermenter vessel may also be inoculated by aseptic piercing of a membrane. In this system the inoculum vessel is connected to an inoculating needle assembly and the sterile needle pierces a rubber septum set.

9.4 CONTROL OF VARIOUS PARAMETERS

Temperature

Since fermentation is mildly exothermic reaction, there is a possibility of increase in temperature during the course of reaction. Thus control of temperature is important. Generally water is used for providing cooling through either jacket or cooling coils. The fermenter may also be placed in thermostatically controlled bath. In case of large fermenter vessels internal coils are used to circulate cold water through them for removing the excess process heat. Temperature is generally measured using various ranges of thermometer, thermocouples, thermist or a metal resistant thermometer. Nowadays thermistor coupled to a cascade control system appears to be a preferred device for temp regulation.

Pressure

Pressure can be measured by diaphragm gauges. They also should be sterilized. Pressure of the fermenter can be regulated by a simple back-pressure regulator valve in the exit gas line.

Agitator shaft power

Two major systems used for the measurement of agitator shaft power in bench-scale fermentor are a torsion dynamometer and strain gauge.

Foam

Foaming is an unavoidable problem during fermentation. Presence of surface active substances such as, proteins, lipids, cellular components, etc. in the fermentation broth may cause foaming or frothing. Media containing corn steep liquor, soybean meals or other ingredients are prone to foam formation. These substances get denatured at the air-broth interface with the formation of a film that cannot be ruptured. Foaming causes several physical and biological problems. These are:

1. Reduction of effective working volume of fermenter because of circulation of oxygen-lean gas bubbles in the system.
2. Affects the gas bubble size and hinders proper mass transfer.
3. Cause interference in the working of sensing electrodes such as pH and DO probe.
4. Deposition of cell fragments inside the fermenter.
5. Product loss due to siphoning of culture broth.

Foam detection and control can be achieved by either by mechanical foam breaking or by adding antifoaming agents. Sometimes a combination of the both can work. Silicone emulsion antifoams are widely used and it is sterilizable. Oil-based antifoams are used for most industrial fermentations.

Table 9.3 Major parts of a fermenter, its salient features and functions

Parts of fermenter	Salient features	Major functions
Top plate	Standard torispherical heads made of good quality stainless steel (operating pressure: 15 bar, with various ports and openings for insertion of accessories.	As a closure/top cover.
Clamp	Made of Stainless steel, different types are available pinch type, rocket type, ring type, tri-clamp, etc.	Top plate is compressed onto the vessel flange by an easily released clamping system.
Baffle	Metal strips with approximately 01–0.5 vessel diameter, 4 to 6 baffles are used based on the volume of the reactor.	To prevent radial swirl (vortex) and increases aeration efficiency.
Impeller	Blade-shaped vanes are arranged around a central shaft. Pitch blade/marine impellers, helical ribbon, Rushton turbine types are commonly used.	Ensures uniform mixing.
Driver motor	Usually induction 3-phase AC motor, 208 V, Motor torque rating 34 N-m, with standard programmable VFD.	To drive the impeller shaft.
Air Sparger	Orifice sparger with diameter 75% of impellers diameter. It has circular rings with perforation of various diameters as per requirement.	To distribute air/oxygen into the vessel.
Inlet air filter	High efficiency particulate air (HEPA) filters are used, 02. It is a pleated mechanical air filter, can remove 99.9% of air borne microbes with 0.3 μm diameter.	Sterilization of air for aeration.
Sampling port	Either sterilizable hooded samplers (vacuum driven) or bottom-harvesting valve (*in situ* diaphragm or piston valve)	To withdraw samples of biomass and substrate aseptically for analysis.

(*Contd.*)

Parts of fermenter	Salient features	Major functions
Exhaust air filter	Sterilizable membrane filters are used, consisting of PTFE, PES membranes.	To prevent escape of microorganism to the environment.
Rotameter	It is a tapered, vertical tube having small diameter at bottom, provided with a calibrated scale and float.	To measure the volumetric flow rate of air or liquid feed.
Pressure gauge	Different types are available to measure gauge, absolute or vacuum pressure. Dial type, size 6 inch/150 mm.; Analog display, SS materials.	To measure and visual monitoring of gas/air pressure.
pH probe	Consists of one measuring electrode (glass electrode) with electrolyte (KCl of pH 7) and a reference electrode. This is sterilizable.	To measure and monitor the pH during the course of fermentation.
DO probe	Consists of a PTFE membrane, a cathode, and an anode soaked in an electrolyte. Oxygen diffuses through the DO probe membrane and as they are reduced at the cathode, a small voltage generates.	To measure and monitor the dissolved oxygen.
Temperature probe	Pt-100 temperature sensor, MOC: high quality SS, sterilizable, range: –40 ... + 85°C, Standard length: 46–55 mm.	To measure and monitor the temperature of fermenter.
Foam control	Done by measuring impedance, small magnitude of alternating current is applied through the foam or the liquid under test. Guided wave radar is used to measure the foam level.	To detect the presence of foams.
Sight glass	15 cm window glass surrounded by a SS 304 Collar which is less expensive compare a switch for LED light for better viewing.	To view the fermenter inside while in operation.
Valves		
Cooling coils /jackets	Usually jacket size is 50 mm for small vessels and 300 mm for large vessels. Made of SS 304 with thickness same as fermenter. Fermenter wall thickness 6 mm with 2 mm allowance for corrosion.	To provide cooling of the system in case of temperature shoot up.
Drive shaft	It consists of a shaft, an impeller. Bearings Casing/volute, shaft sleeves and a sealing arrangement.	Mixes the medium evenly with its impeller.
Feed pump	It is used in feeding nutrients, acid-base and antifoaming agents. Generally peristaltic pumps are used.	Regulates the flow-rates of additives (medium, nutrients) variable speed.
Support	Skirt support, with angles of the base to be 80°, thickness should be greater than 5 mm.	To support the fermenter vessel.

Flow rate (Gas and liquid)

Among various devices that are used to measure and control gas flow rates variable area flow meters such as rotameters are important. The position of the float in the meter is converted to an electric signal through either a capacitance or resistance principle. This signal is used with the aid of a controller, to regulate a flow rate in the gas line.

Turbidity

Turbidity is an indirect measure of cell mass concentration. The more the turbidity of the medium, the more will be the biomass concentration. Turbidity meter (Nephelometric Turbidity Unit) is used to quantify the turbidity.

Viscosity

As a broth exhibits distinct non-Newtonian characteristics, it becomes more difficult to disperse nutrients uniformly and to supply oxygen to the microbial population. Thus, control of viscosity is required to improve performance of fermentation. Viscosity can be controlled by appropriate dilution of the broth using sterile water or other diluents.

Chemical parameters

pH

It is measured by a glass reference electrode. It is sterilizable. The device is inserted inside the fermenter. It contains half-cells of glass and reference electrodes. Both of these electrodes are composed of Ag/AgCl saturated with solid KCl. An asbestos or porcelain cylinder is used as the junction material.

Redox

Redox control has been achieved by gaseous sparging with nitrogen or oxygen or by chemical control with cystine, ascorbic acid or sodium thioglycolate. Redox environments are almost universally made with a combined platinum reference electrode system. However, the reading of the redox potential may subject to some errors as dissolved oxygen interact with the redox system in case of aerobic fermentation.

DO

Dissolved oxygen electrode is covered by a pol an electrolyte sandwiched between the cathode and the membrane.

Exit gas composition

Exit gas composition, predominantly CO_2, can be monitored by using thermal conductivity devices or long-path infra-red analyzers. Paramagnetic analyzers have been utilized for oxygen analysis. Knowing the carbon dioxide evolution rate and the oxygen uptake rate, the respiratory quotients can be determined from their ratio. This quotient is an indicator of unit cellular activity in the fermenter.

Aeration and agitation

Among the important environmental factors aeration and agitation are most essential. This firstly to supply microorganisms with oxygen and secondly, to mix fermentation broth to achieve uniform suspension of microbes and to accelerate the mass transfer rate in the medium. Depending on the mechanism of how the fermentation broth is agitated, fermenters are broadly classified as agitated fermenters (like CSTF) and non-mechanically agitated fermenter (bubble column and air-lift, etc.). Oxygen transfer rate in fermenter is controlled by the demand of oxygen by the microbes and the supply of oxygen from air bubbles to the broth. Basic gas-liquid mass transfer correlations are already discussed in chapter 7.

For stirred tank fermenter, oxygen transfer depends on impeller design, power input and air sparging rate. For mass transfer in bioreactors where three phases exist, dimensionless numbers such as Sherwood number (ratio of total mass transfer to diffusional mass transfer), Schmidt number (ratio of momentum diffusivity to mass diffusivity) and Grashof number (ratio of gravitational forces to viscous forces) are important to consider. In addition, when air is sparged into a bioreactor with or without mechanical agitation, mass transfer rate is influenced by bubble dynamics and dimensions. When liquid is in turbulent motion, coalescence and breakdown determine the bubble size. Bubble break up is caused by shear forces due to turbulent motion of the liquid.

Importance of agitation

- To increase the oxygen transfer rate from gas bubbles to the broth.
- To increase the rate of oxygen and nutrient transfer from medium to cells.
- To prevent aggregation of mycelia or formation of lumps of cells.
- To improve rate of heat transfer between medium and the cooling surfaces of the fermenter.

In addition, efficient agitation causes gas bubbles to follow a more tortuous path and retard their escape from the culture. It also prevents coalescence of bubbles.

Types of agitators

The shape, number and arrangement of impeller are important for proper agitation. In case of laboratory fermenters, operating speed of agitator is kept 50–300 rpm or more for 2–6 LPM inlet gas flow rate. However, it varies from one system to another. Generally following types of agitators are used:

(i) Disc turbine
(ii) Vaned disc turbine
(iii) Open turbine of variable pitch
(iv) Propellers

Important types of air spargers are:

(i) Porous sparger
(ii) Nozzle sparger
(iii) Ring sparger
(iv) Combined sparger-agitator

Porous sparger, also known as orifice sparger is made of sintered glass, ceramics or metals. The bubble size produced from sparger is always 10–100 times larger than pore size of aerator block. In small stirred fermenters the perforated pipes are arranged below the impeller in the form of crosses or rings. Porous spargers are used in yeast manufacturing, effluent treatment and in SCP production. In nozzle sparger either a single open or a partially closed pipe provides stream of air bubbles. The single nozzle sparger causes a lower pressure drop than any other sparger and normally does not get blocked. Selection of impeller for any bioreactor depends on several factors such as mixing time, power input, reactor dimension eddy size, etc. Major types of spargers are shown in Figure 9.4.

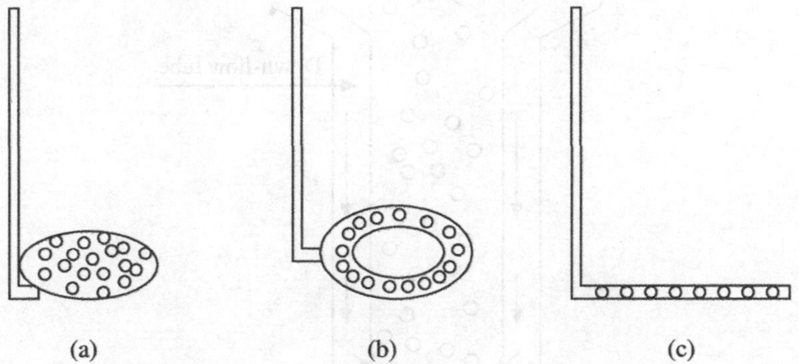

Figure 9.4 Various types of spargers (a) porous, (b) Ring, and (c) nozzle.

Air lift fermenter

In order to meet other specific process requirements, the bubble column has been modified to different form. Air lift bioreactor is one such modification to introduce better liquid circulation for enhanced mass transfer with less agitation. Mainly two types are used: one with internal liquid circulation loop and the other with external liquid circulation. This is shown in Figure 9.5. They utilize the specific gravity difference between the air-rich part of the liquid in the riser and the denser air-depleted part of the liquid in the down corner to obtain continuous liquid movement. As the circulation flow rate is high, both mixing and oxygen transfer occur simultaneously without any mechanical agitation.

The liquid circulation rate depends on the design of the air lift column. The extent to which the gas is separated from the liquid at the top of the column prior to entering the down flow selection can significantly affect the liquid circulation rate. According to the work of Hatch (1973) and Gutierrez and Ericson (1977), the dispersion enters the down-flow section with more gas than can be carried downward by the liquid. Some of the larger gas bubbles rise against the down-flow of dispersion and escape, since the average gas velocity is much smaller than liquid velocity in the down-flow section.

For the disengagement of gas in the down-flow, the actual liquid velocity in the down comer must be larger than the rise velocity of small bubbles and smaller than the rise velocity of larger bubbles. Hatch (1973) reported superficial liquid velocities ranging from 10 to 30 cm/s. Orazem et al.(1979) reported values ranging from 17 to 21 cm/s. Liquid circulation velocity increases with gas superficial velocity in the riser until the liquid circulation velocity is limited by the gas bubble disengagement process at the top of the down comer.

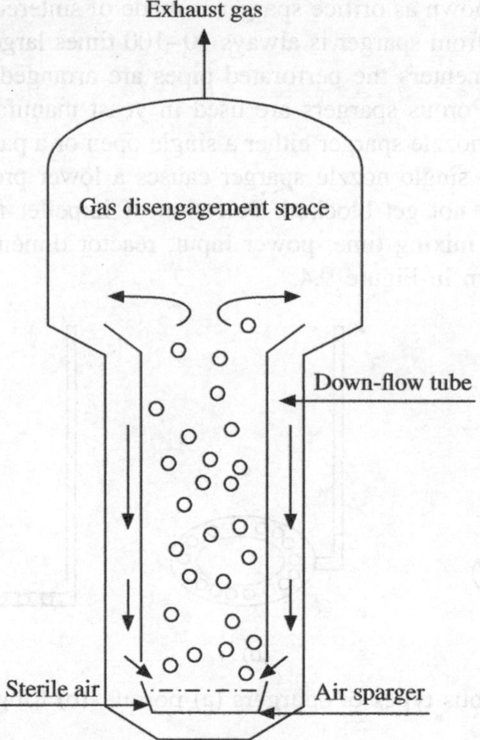

Figure 9.5 Schematic diagram of an air-lift fermenter.

Fluidized-bed fermenter

Fluidized-bed bioreactor (Figure 9.6) is fundamentally a reactor where granular or powdered solids behave like a fluid under gas or liquid flow conditions. Here, a gas or liquid circulated through the bottom of the reactor, to make the solid particles to become buoyant and rise. It consists of a bed of solid particles, usually granular or powdered material. A fluid either gases like air, nitrogen or a liquid is introduced into the reactor from the bottom. At the beginning the fluid velocity is kept low but it is increased gradually to attain minimum fluidization velocity at which upward force exerted by the fluid overcomes the gravity force on the particles.

Consequently, the particles start to expand and transform to a loosely packed state. This type of reactor enjoys several advantages such as improved heat and mass transfer, maintenance of uniform temperature and composition and improved contact and mixing. Pressure drop in fluidized-bed reactors is less than packed-bed. However, there is chance of carry-over of microorganisms in such reactors. Some applications of such type of reactors include denitrification of wastewater, and anaerobic wastewater treatment.

Packed-bed fermenter

This is also called packed-bed column fermenter where enzymes or whole cells of microorganism are immobilized onto a solid matrix. The bed provides conducive growth environment for the microorganism. The matrix in the packed-bed provides a large surface area for the cells or microorganisms to attach and grow, facilitating more mass transfer and exchange of nutrients and

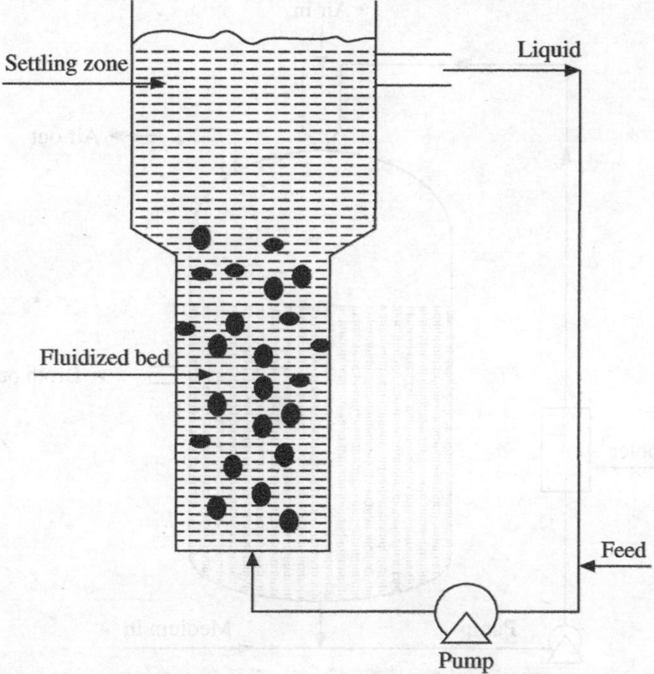

Figure 9.6 Fluidized-bed bioreactor.

oxygen between the cells and the medium. The liquid nutrient is allowed to flow continuously through the immobilized bed. Products are released in the fluid and are removed in the outflow stream. It has several advantages compared to other bioreactors.

These include less shear stress, low risk of contamination, high cell densities, easy scale-up and so on. Packed-bed are commercially used in various fields, including biopharmaceuticals, food and beverage, environmental biotechnology, and bioenergy.

Deep jet fermenter

Deep jet fermenter (Figure 9.7) consists of cylindrical vessel in which air bubbles are dispersed with a nozzle or a frit, together with pump under hydrodynamic effect. The fermentation medium is circulated by pump. The gas mixed with high power jet of liquid is injected into the fermenter. A high power jet is used for entraining the gas into the liquid present in the fermenter. The exhaust gas is removed from the hole present at the top and from the circulation pump which is involved in passing the degassed liquid which then pass to the supplementary cooler. Although gas dissolution rate is high, such design consumes more power.

Solid state fermentation

Solid state fermentation (SSF) is a process where fermentation and bioconversion take place on solid substrate or support systems in presence of minimal moisture condition. Here, growth of microorganisms takes place almost in the absence of free flowing aqueous phase. SSF is nowadays a popular method of fermentative production of protein rich food like miso, tempeh, sake, tofu, cheese, etc., in South East Asian countries

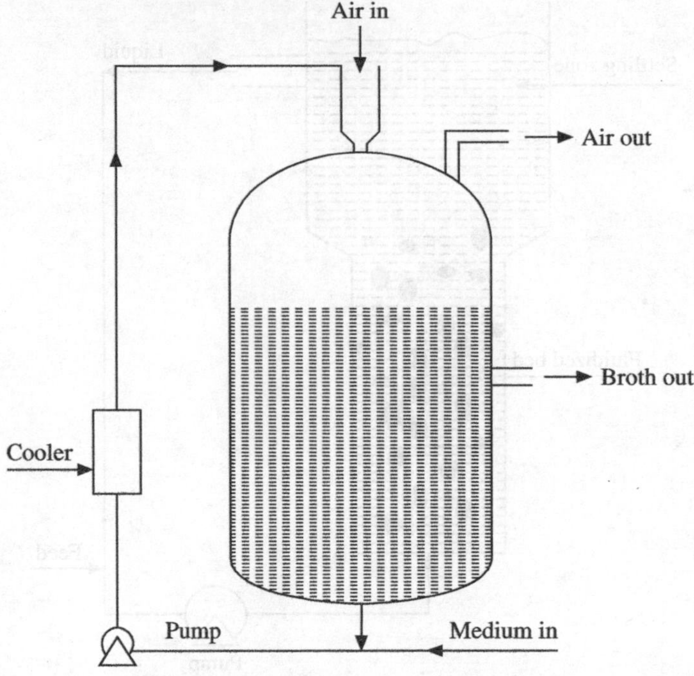

Figure 9.7 Deep jet fermenter.

Unlike traditional submerged fermentation, SSF is carried out in simple and small-scale bioreactor or improvised systems such as wide-mouth flasks, jars, large petri-plates, wooden box and so on. Column type bioreactors are also used for the SSF route of production of glucoamylase. Other reactors include perforated tray-type, paddle mixer, rotating cylindrical drums, etc. A Schematic diagram of tray fermenter for solid state fermentation is given in Figure 9.8.

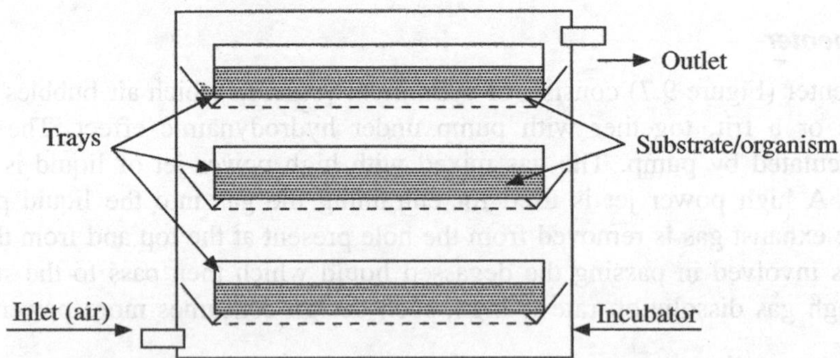

Figure 9.8 Tray fermenter for solid state fermentation.

SSF is broadly used for the synthesis of various organic acids using selected species of fungi. For example, production of citric acid using *Aspergilus niger*, lactic acid and gallic acid by *Rhizopus oryzae*, etc. Certain secondary metabolites like gibberellic acid, aroma, antibiotics, polysaturated fatty acids, polyhydroxyalkanoate, biofuels (ethanol).

Like submerged fermentation process there are certain factors responsible for the product formation in SSF. Among those, selection of suitable microorganism, selection of substrates, physico-chemical parameters and strategies for product isolation and recover are particularly important. Yeast and molds are found to be most suitable microorganism for SSF as they are very much amenable for attached growth techniques as occurred in solid support systems or in immobilized bed. But in some cases by controlling water activity bacterial culture might be made suitable for SSF. Substrates in SSF are generally insoluble solids which can also act as support matrix for growth. Various crop residues, agro waste materials, etc. and even inert materials like polyurethane foams are used as solid substrates. Selection of process parameters, such as particle size, moisture content, relative humidity, pH, temperature, agitation, aeration, etc. are important for efficient operation. If required, nutrient supplementation can also be done.

Advantages of SSF over submerged fermentation

SSF is credited with several advantages as compared to traditional submerged fermentation. SSF eliminates foaming and frothing problems to a great extent. Chances of contamination are reduced as compared to submerged fermentation. It generates almost no wastewater stream. It is also featured with low energy consumption, less cost and simpler techniques. SSF finds its wide applications in the fermentative production of food, agricultural product, synthesis of value added chemicals, bioremediation, biodegrdation of hazardous chemicals and generation of biogas. A photagraph of an indurial scale *in situ* sterilizable glass fermenter is shown in the Figure 9.9.

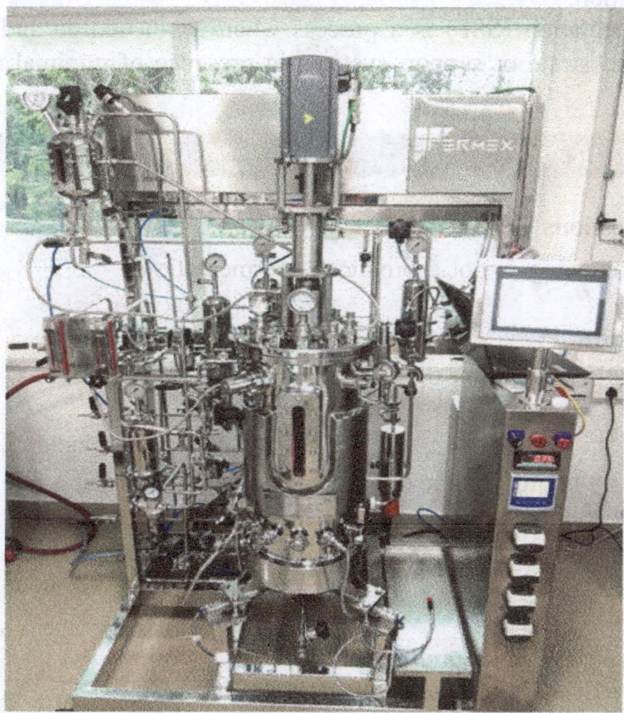

Figure 9.9 A photograph of an *in situ* sterilizable fermenter (Source: with kind permission from M/s Fermex Solutions LLP, Mohali, India).

KEY POINTS TO REMEMBER

- Fermentation is a biochemical process where carbonaceous organic compounds are enzymatically converted into useful products in presence of suitable microorganisms.
- Fermenter provides appropriate, aseptic and pre-defined physico-chemical factors, conducive for the growth of microorganism and synthesis of high-value useful products from low-value substrates.
- Vessel size is dependent on several factors such as required fermentation capacity, cost and ease of fabrication.
- For aerobic fermentation continuous introduction of air (oxygen) into the fermenter is required for growth of organism and also to ensure gas-liquid mass transfer.
- Important types of agitators that are used in fermenter include disc turbine, vaned disc, open turbine and marine type impeller.
- Aseptic operation is of prime importance in any fermentation process. This is achieved by sterilization of the reactor, media to be introduced, air supply or any other accessories.
- Appropriate formulation of medium consisting of carbon, nitrogen and other mineral sources is one of the essential stages of fermentation process.
- Normally inoculums volume is kept 3% to 10% of the medium volume. A relatively large inoculums volume can minimize the length of the lag phase and increase productivity.
- Precise control of various parameters like temperature, pH, foam, DO, etc. are required during fermentation.
- Solid state fermentation (SSF) is a process where fermentation and bioconversion take place on solid substrate or support systems in presence of minimal moisture condition.

QUESTIONS FOR PRACTICE

Multiple Choice Questions

1. Which of the following is not a product of fermentation?
   (a) Lactate (b) Oxygen
   (c) Carbon dioxide (d) Ethanol
2. During alcoholic fermentation the acetaldehyde is reduced to
   (a) Methanol (b) Ethanol
   (c) Lactose (d) Ascorbic acid
3. A chemostat has a liquid volume of 2 litres and is being fed at a rate of 4 litres per hour. Dilution rate for this reactor will be
   (a) 2 litre (b) 2 litres per h
   (c) 2 h^{-1} (d) 4 litres per hour
4. Yield coefficient represents
   (a) total biomass or product produced
   (b) conversion efficiency of a substrate into product
   (c) conversion rate of a substrate into biomass or product
   (d) production time of biomass or product

5. Which of the following is not a sealing used in bioreactor?
(a) O-ring (b) Gasket
(c) Flange (d) Leapseal

6. What is the basic function of a fermenter?
(a) To sterilize the medium
(b) To recover the product
(c) To provide optimum growth conditions to organisms and obtained the desired product
(d) All of the above

7. Which of the following is used as both carbon and nitrogen source?
(a) Corn steep liquor (b) Peanut meal
(c) Soya meal (d) Urea

8. Which among the following is a perforated type of sparger?
(a) Porous sparger (b) Orifice sparger
(c) Nozzle sparger (d) Combined sparger-agitator

9. Baffles are used in fermenter to
(a) Prevent vortex formation (b) Impart turbulence
(c) Maintain uniform mixing (d) Assist in biomass formation

10. Aspect ratio of fermenter vessel is known as the ratio of
(a) Major axis to minor axis (b) Height to diameter
(c) Total height to working height (d) None of the above

11. Agitation is provided in fermenter to
(a) Provide uniform mixing (b) Maintain homogeneity
(c) Facilitate oxygen transfer (d) All of the above

12. Name a suitable pump used in the transportation and metering of liquids in a batch fermenter
(a) Peristaltic pump (b) Centrifugal pump
(c) Gear pump (d) Plunger pump

13. Steam trap is used to
(a) Condense the steam flowing in the pipeline
(b) Remove water resulting from partial condensation of steam
(c) Stop the supply of steam
(d) None of these

14. Which of the following is not a type of steam trap?
(a) Expansion type (b) Compression type
(c) Bucket type (d) Float type

15. While constructing the fermenter, which of the following need not be considered?
(a) High speed agitation and aeration system (b) Temperature control system
(c) pH control system (d) Sampling facilities

16. Which of the following is preferred to be the best material for construction of a small-scale fermenter?
(a) Quartz (b) Glass
(c) Iron Steel (d) Wood

17. The largest diameter for glass fermenter is
(a) 50 cm (b) 70 cm
(c) 60 cm (d) 80 cm

18. Which of the following is responsible for imparting corrosion resistance property of stainless steel?
(a) Cr (b) Fe
(c) Ni (d) Co

19. The AISI grade SS 316 contains ____________ amount of chromium.
(a) 15% (b) 20%
(c) 18% (d) 18.5%

20. Which of the following is not a component of aeration and agitation system?
(a) Impellers (b) Baffles
(c) Stirrer gland and bearing (d) Steam trap

21. Which of the following agitators consist of vanes arranged in a vertical plane or fashion around the circumference?
(a) Vaned discs (b) Disc turbines
(c) Variable pitch turbine (d) Marine propeller

22. Which of the following agitators consist of blades directly attached to a boss on agitator shaft?
(a) Vaned discs (b) Disc turbine
(c) Rectangular turbines (d) Marine propeller

23. Which of the following is not the use of baffles in fermenter?
(a) Increase the effect of agitation
(b) Improve aeration efficiency
(c) Improve cooling capacity
(d) Improve the fermenter capacity

24. Which of the following is not a result of foaming inside the fermenter?
(a) Reduction of effective working volume of fermenter
(b) Hindrance to mass transfer
(c) Product loss due to siphoning of culture broth
(d) All of the above

25. Measuring electrode (glass electrode) of pH probe contains
(a) KCl solution (b) NaCl solution
(c) Glucose solution (d) Distilled water

Answer Key

1. (b) 2. (b) 3. (c) 4. (a) 5. (c) 6. (c) 7. (a) 8. (b) 9. (a) 10. (b)
11. (d) 12. (a) 13. (b) 14. (b) 15. (a) 16. (b) 17. (c) 18. (a) 19. (c) 20. (d)
21. (a) 22. (b) 23. (d) 24. (d) 25. (a)

Descriptive Questions

1. Write the equation for oxygen transfer rate in an aerated fermenter and explain various terms involved therein along with their units.
2. Explain the effects of the following parameters on fermentation and state their measuring and controlling strategies:
   (a) Temperature (b) pH
   (c) DO (d) Foaminess
3. In the methyl red test, does the color of the medium change to red when methyl red is added? Is the methyl red test positive or negative for mixed acid fermentation?
4. Provide three environmental niches where you might find fermentative bacteria. Explain these niches and why it's beneficial for the bacteria to use fermentation.
5. A student designs an experiment to produce an alcohol known as butanol. She mixes a species of *Clostridium* bacteria known to produce this alcohol as a by-product of its metabolism with a broth of water and glucose in excess, and incubates the mixture in constant darkness without any oxygen. For 5 days she notices an increase in bacterial biomass (amount of bacteria) and in the amount of CO_2 and butanol produced. After $1\frac{1}{2}$ weeks when she checks the culture, she observes that there has been a decrease in biomass and in the production of CO_2 and butanol since her first check on the experiment. Plenty of glucose is still present in the culture and no mutations are found. For which of the following conclusions is the evidence the strongest? Explain.
   (a) Since ATP is formed during respiration. ATP has been formed in excess, poisoning the cells.
   (b) The bacterial cells have begun to utilize the CO_2 and water to carry out photosynthesis.
   (c) After $1\frac{1}{2}$ weeks, the cells switched to aerobic respiration.
   (d) The alcohol produced has overpowered and inhibited the growth for the bacterial cells.
   (e) The bacterial cells have totally depleted their energy source and are no longer growing.

   [Ans: (d) *Hints*: Conditions are very much favorable for fermentation by *Clostridium* as it can only live in the absence of oxygen. So, alcohol will be produced in higher amounts and will inhibit the bacteria to do fermentation because it regulates the enzymes of glycolysis indirectly].

   Explanation:

   [The culture medium is well filled up with water and glucose in absence of oxygen. These conditions are very much favorable for fermentation by *Clostridium* because it can only live in the absence of oxygen. So alcohol will be produced in higher amounts and will inhibit the bacteria to do fermentation because it regulates the enzymes of glycolysis indirectly. It makes the plasma membrane leaky so that essential cofactors and coenzymes are moved out from the cells].

   Therefore, the correct option is d.

   Please note that:

   - Option (a), is incorrect as ATP regulates glycolysis when energy is not needed by the cells.

- Option (b), is incorrect *Clostridium* is not a photosynthetic bacterium.
- Option (c), is incorrect as *Clostridium* is an obligate anaerobe.
- Option (e), is incorrect as an energy source means glucose is present in excess in the medium.

6. Within the concept of sugar fermentation in yeast, read the following questions and give explanations to all your answers.
   (i) Does yeast equally utilize all sugars?
   (ii) Hypothesize why some sugars were not metabolized while other sugars were?
   (iii) Why do you need to incubate the yeast before you start monitoring air pressure?
   (iv) How does sugar enter the glycolytic pathway and how does it go through?

10 Recovery and Purification of Products

10.1 INTRODUCTION—INTRA AND EXTRA CELLULAR PRODUCTS

Microbial fermentation leads to the generation of three types of products namely cell mass or biomass, extracellular products and intracellular products. During metabolism and growth of cell in a bioreactor a number of valuable products are generated, in addition to cell mass in various quantities. Some of these are secreted and synthesized by cells and are accumulated in the reactor effluent outside the cell. These are called extracellular products. Examples of extracellular products are ethyl alcohol, lactic acid, penicillin, extracellular enzymes such as peptidase, collagenase, amylase, sucrase, etc. On the other hand, some products are generated within the cellular system and are extracted by applying various cell disruption techniques. These are known as intracellular products. For examples phosphatase, β-galactosidase, immunoglobulin, plasminogen, etc.

The concentration, recovery and purification of products, collectively known as downstream operation, incur the lion's share of the total cost of fermentation. A number of traditional unit operations along with advanced separations are employed for efficient recovery of products, its fragility (whether heat sensitive or thermodegradabale) and quantity. As the chemical composition of fermentation broth is quite complex, a stand-alone separation method is not always helpful in the product recovery. In most of the cases, a combination of two or more separation processes, which are sometimes complementary to each other, are employed. Occasionally the yield of desired bioproducts is so low, that traditional bulk separation methods prove to be unsuitable. In such cases advanced separation methods are used. General methods of recovery of extracellular products are outlined in Figure 10.1

10.2 SEPARATION OF INSOLUBLE PRODUCTS FROM THE BROTH

After fermentation, the broth on average contains from 20 to 25% insoluble or sparingly soluble solids. In some cases a relatively high solids load is also encountered. Insoluble solids are separated from the broth by various methods, such as filtration, centrifugation, sedimentation, etc. These are discussed in the following sections.

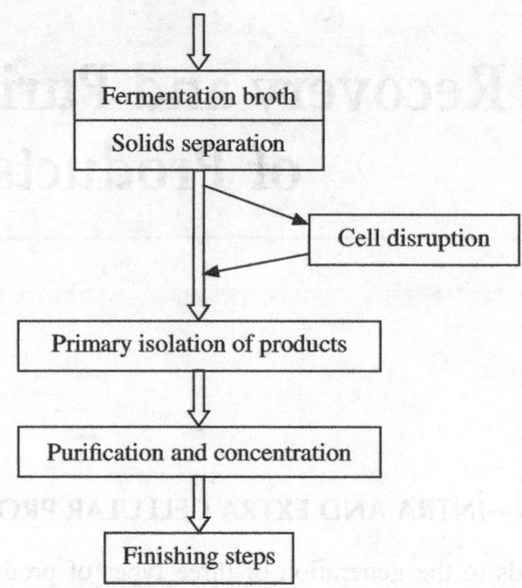

Figure 10.1 General outline of the overall separation steps in fermentation.

10.2.1 Filtration

Filtration is a separation process in which a suspension of small solid particle is forced through a porous media to arrest the solid. The accumulated residual solids also called residue, can form a deposited solid layer (cake). The clear fluid, called filtrate, flows out. The filtering media could be porous membrane, fabric cloth, perforated solids or any other suitable materials. It is also known as septum. The driving force for filtration could be gravity force, centrifugal force or pressure difference. In vacuum filtration sub-atmospheric pressure is maintained at the downstream of the filter septum.

Depending on the volume of fermentation broth and scale of operation, different types of filter are used. For small batches of fermentation, plate and frame filter press is sufficient; while for larger processes continuous rotary vacuum filters are used. Plate and frame filter press is an assembly of alternate solid plates and hollow frames. Both faces of each plate are studded, or perforated to permit drainage. The filter medium is hung over both faces of each plate. As the feed slurry is forced against the filter surfaces, cake is formed inside the hollow frames and grows in size. There are different types of feed and discharge arrangements such as bottom feed and top discharge, top feed and bottom discharge or double feed and double discharge. Plate and frame filter press operates generally up to 700 kN/m^2. It has several advantages to its credit. It is simple in design, inexpensive operation, ease of cleaning and small floor space requirement.

In rotary vacuum filtering operation the drum is rotated slowly (2–5 rpm) while the tank is supplied with steady flow of materials to be filtered (Figure 10.2). A constant level is maintained to ensure a constant depth of submergence of the lower portion of the rotating drum. As the drum rotates, vacuum is applied through the action of the automatic valve. Filtrate passes through the submerged filter disc whereupon cake is deposited on each side of the disc. The cake is then dewatered and discharged.

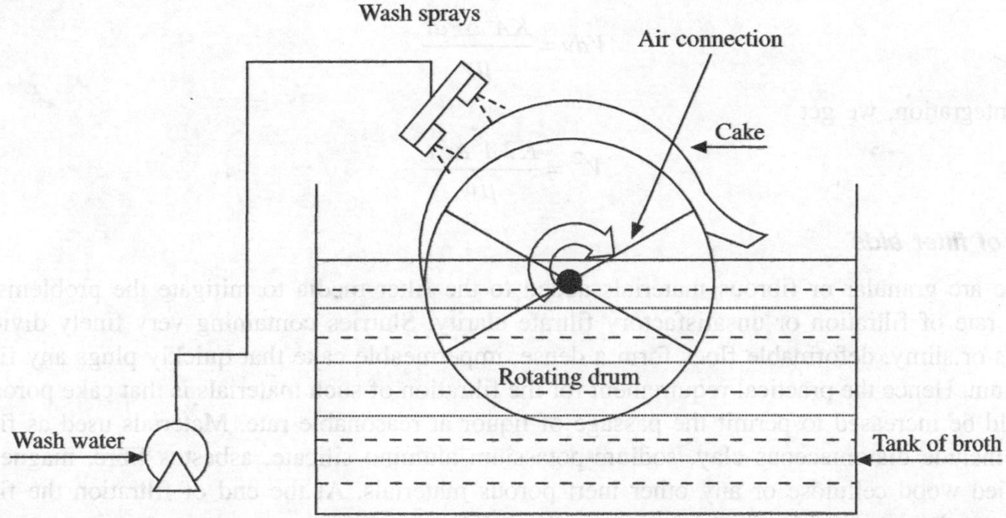

Figure 10.2 Schematic of a continuous rotary vacuum filter (vacuum is applied within the drum resulting in cake formation and dewatering).

A filter cake gradually builds up as filtrate passes through the filter cloth. As the filter cake builds up in thickness the resistance to flow will gradually increase. Thus if pressure applied to the surface of the slurry is kept constant the rate of flow will gradually diminish. Alternatively, if the flow rate is to be kept constant the pressure will gradually have to be increased:

According to Darcy's law

$$\text{Rate of flow} = \frac{dv}{dt} = \frac{KA\Delta P}{\mu L} \tag{10.1}$$

where μ is liquid viscosity; L depth of filter bed; ΔP pressure differential across the filter bed, A area of the filter exposed to the liquid and K is the constant for the system (known as Kozeny's constant).

K itself is a term which depends on the specific surface area's (surface area per unit volume) of the particles making up the filter bed and the voidage ε,

$$K \text{ is defined as } K = \frac{\varepsilon^2}{5(1-\varepsilon^2)s^2} \tag{10.2}$$

Let V be the volume of filtrate passed in time 't', v be the volume of cake deposited by unit volume of filtrate.

$$L = \frac{vV}{A} \tag{10.3}$$

Substituting Eq.(10.3) in Eq.(10.1) we get

$$\frac{dv}{dt} = \frac{KA^2\Delta P}{\mu vV} \tag{10.4}$$

This is the general equation relating rate of filtration to pressure drop. Eq.(10.4) can be integrated for filtration at constant pressure.

$$Vdv = \frac{KA^2 \Delta P dt}{\mu v}$$

On integration, we get

$$V^2 = \frac{K2A^2 \Delta Pt}{\mu v}$$

Use of filter aids

These are granular or fibrous materials added to the filter media to mitigate the problems of slow rate of filtration or unsatisfactory filtrate clarity. Slurries containing very finely divided solids or slimy, deformable flocs form a dense, impermeable cake that quickly plugs any filter medium. Hence the practical requirement for the filtration of such materials is that cake porosity should be increased to permit the passage of liquor at reasonable rate. Materials used as filter aids include diatomaceous clay, sodium-potassium alumino silicate, asbestos fibre, magnesia, purified wood cellulose or any other inert porous materials. At the end of filtration the filter aids may be retrieved by dissolving away the solids.

10.2.2 Centrifugation

The driving force for centrifugation separation is the difference in density of the particles to be separated and that of the medium. Thus, it is a well-suited process of separation of solids of wide range of sizes (0.1 to 100 μm) from broth of fermentation. When solid particles settle in a dense suspension, the settling is called hindered settling. When a particle reaches a terminal settling velocity forces acting on it nullify each other resulting in zero net force. By applying the principle of Stoke's law:

Drag force on a particle = Buoyancy force

$$6\pi\mu_c R u_r = \frac{4\pi R^3}{3} G(\rho_p - \rho_f) \tag{10.5}$$

$$u_r = \frac{dr}{dt}$$

G = acceleration due to centrifugal force = $\omega^2 r$

Stokes law has been used in Eq.(10.5) to express the drag force since particles velocities (and therefore particle Reynold's number) are usually very low. Gravitational force does not appear in Eq.(10.5) because direction of 'r' is normal to the force of gravity.

Integrating Eq.(10.5) with respect to time and distance travelled by the particle gives the time required for movement of particle from r_1 to r_2.

$$6\pi\mu R\left(\frac{dr}{dt}\right) = \frac{4\pi R^3}{3}\omega^2 r(\rho_p - \rho_f)$$

$$\frac{dt}{dr} = 6\pi\mu R \frac{3}{4\pi R^3 \omega^2 r(\rho_p - \rho_f)}$$

Integrating within the position r_1 to r_2:

$$t = \frac{9}{2} \frac{\mu_c}{\omega^2 R^2 (\rho_p - \rho_f)} \ln \frac{r_2}{r_1}$$

Commonly used centrifuges include tubular bowl centrifuge, disc-bowl centrifuge, basket centrifuge, multi-chamber centrifuge, scroll centrifuge and so on. Tubular bowl centrifuge consists of a narrow cylindrical bowl capable of rotating at a high speed of 10,000–15000 rpm with a stationary outer casing. It has provision of cooling to prevent thermal degradation of heat sensitive products. When the liquid feed mixture is subjected to centrifugal force, denser liquid is pushed away from the centre and lighter liquid accumulates near the centre of the rotating bowl. If the solids are present it will also get settled at the bottom. The solid are removed from time to time by water flushing.

Figure 10.3 presents a schematic diagram of a continuous centrifuge. It is also known as disc-stack or disc-bowl centrifuge. Here solid particles are removed in flow between closely stacked cones. Supernatant is withdrawn from the top of the unit. Basket centrifuge is most suitable for slurries having high solid concentration. It contains a perforated basket which rotates at a speed of 4000–5000 rpm. The feed is introduced from the top of the basket into the rotating bowl where it gets centrifuged.

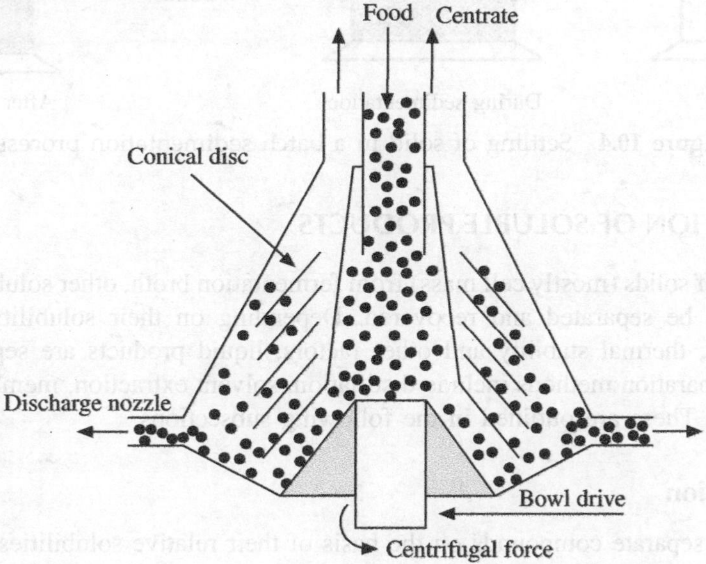

Figure 10.3 Schematic diagram of a disc-stack centrifuge.

10.2.3 Sedimentation

In sedimentation, the cells of microbes including broken cells, non-cellular solids, macroscopic solids and cell debris form a large aggregate or flocs which can be easily settled. Important factors affecting the process of sedimentation include size and nature of individual cells, ionic charges of the constituents of the broth, etc. Addition of various flocculating agents facilitate aggregate formation. Examples of flocculating agents are inorganic polyvalent salts, polyelectrolyte compounds, mineral hydrocolloid, etc. Polyelectrolytes are high molecular weight water soluble organic compounds. Criteria for selection of a good coagulating or flocculating agents are rate of sedimentation, floc size and clarity of the supernatant liquid. After sedimentation, if needed, the settled biomass should be further treated by centrifugation or filtration. Settling of solid in a batch sedimentation process is schematically presented in Figure 10.4. Some strains of yeast are found to be mycelial in nature and thus are highly flocculent type. These are used in the beer brewing industries and single-cell-protein production. These help in effective and inexpensive cell separation.

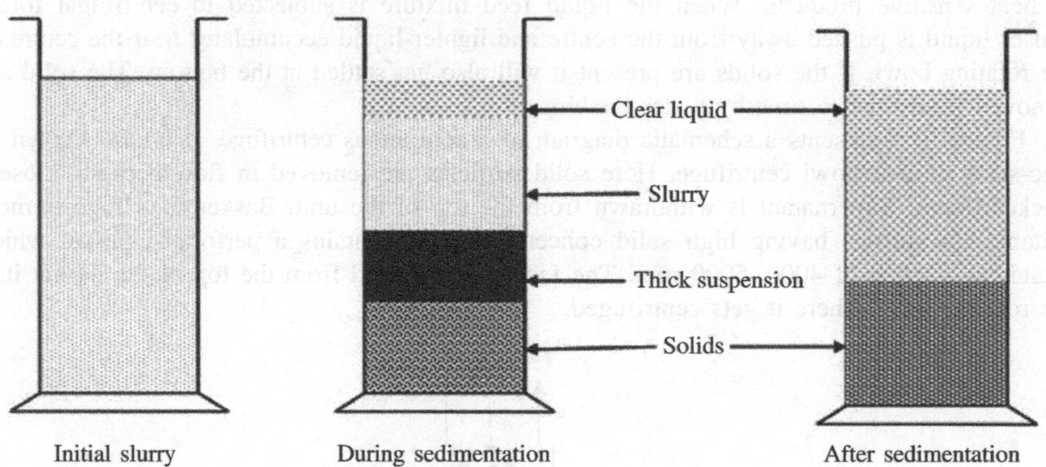

Figure 10.4 Settling of solid in a batch sedimentation process.

10.3 SEPARATION OF SOLUBLE PRODUCTS

After separation of solids (mostly cell mass) from fermentation broth, other soluble or immiscible products need to be separated and recovered. Depending on their solubility, boiling point, molecular weight, thermal stability and other factors, liquid products are separated from the broth. Popular separation methods include distillation, solvent extraction, membrane separation, precipitation, etc. These are outlined in the following subsections.

10.3.1 Extraction

It is a method to separate compounds on the basis of their relative solubilities in two different immiscible liquids. So, an extraction of a substance from one to another liquid phase in which the desired product is preferentially soluble is known as liquid-liquid extraction. It is also known as solvent extraction and partitioning. It is based on the distribution coefficient (or partition coefficient) of any solute or analyte in different solvents.

Partition coefficient or distribution constant is defined as a ratio of the solute (desired product) in the two phases. $K = \frac{C_E}{C_R}$, where:

K = Distribution/Partition Coefficient

C_E = Concentration of Solute in Extract/Organic phase
C_R = Concentration of Solute in Raffinate/Aqueous phase

Liquid-liquid extraction is typically carried out using two different phases, one being aqueous (having water, or an aqueous solution) and the other non-aqueous. The extractable compound of our interest can be available in either of the phases depending on its partition coefficient. In commercial application, the two streams are known as extract and raffinate. The extract is rich in the extraction solvent and the desired components/solute while the raffinate stream has small concentration of the desired solute and the extraction solvent.

Major steps of extraction are:

(i) Contacting solvent, feed (containing desired solute) and extractive solvent by dispersion
(ii) Phase separation by gravity settling
(iii) Solute recovery from extract phase (usually via distillation)

For small volume of broth in batch single stage extractor is used while for better extraction of a large volume of broth multistage operation is most suitable. Continuous multistage operation can be both co-current or countercurrent depending on the flow direction of feed liquid and solvent. In multistage countercurrent operation number of increased stages may increase the efficient of the extraction. Schematic diagrams of mixer-settler device and separating funnel used in liquid-liquid extraction are shown in Figure 10.5.

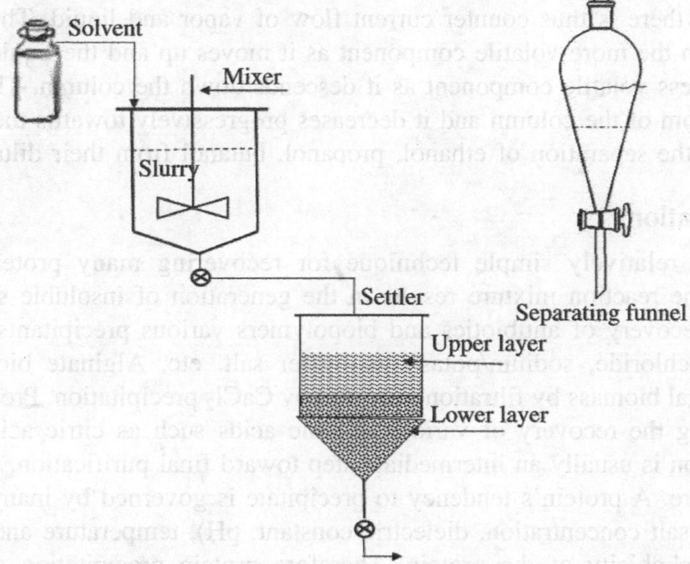

Figure 10.5 Mixer-settler device and separating funnel used in liquid-liquid extraction.

During antibiotic production process liquid-liquid extraction has a major role in the separation of products. In a two stage extraction process, products are first extracted from fermentation broth by using amyl acetate or isoamyl acetate solvent. One of the interesting aspect of penicillin extraction is pH shifting. Penicillin is more soluble in non-aqueous (organic) phase at low pH (2 to 3) while at high pH (8 to 9), it is more soluble in aqueous phase. Thus, it is extracted several times in either of the phases by shifting of pH (also the shifting of partition coefficient).

To improve its yield erythromycin is also extracted using amyl acetate as solvent. Some other antibiotics such as bacitracin are extracted with n-butanol. Commercially penicillin is extracted using centrifugal extractor. Although aqueous-non-aqueous solvent systems are traditionally used in the liquid-liquid extraction, nowadays aqueous-aqueous two phase systems are being used for recovery of certain proteins and other biomolecules. In aqueous two-phase extraction system various combinations of polyethylene glycol-water, dextran-water, potassium phosphate-water and similar others are employed.

10.3.2 Distillation

Distillation is the most prominent unit operations in pharmaceutical, food and beverage industries. The feed to a distillation column may be a binary mixture (containing only two components) or a multicomponent mixture. When a binary liquid mixture is fed to a distillation column (often called a fractionator), the more volatile component (that has lower boiling point) distills off from the top of the column as vapor. These vapors are condensed in the overhead condenser and a part of the condensate is returned to the top of the tower as reflux and the rest is collected as distillate. The liquid that is very rich in the less volatile component is discharged from the bottom of the tower. It is sent to the reboiler where it is partly vaporized, the vapors returned to the bottom of the column and the rest of the liquid is collected as bottom residue. Inside the column there is thus counter current flow of vapor and liquid. The vapor becomes richer and richer in the more volatile component as it moves up and the liquid becomes richer and richer in the less volatile component as it descends down the column. The temperature is highest at the bottom of the column and it decreases progressively towards the top. Distillation is widely used in the separation of ethanol, propanol, butanol from their dilute solutions.

10.3.3 Precipitation

Precipitation is a relatively simple technique for recovering many proteins. Addition of precipitants into the reaction mixture results in the generation of insoluble salt products. For example, during recovery of antibiotics and biopolymers various precipitants are added, such as procaine-hydrochloride, sodium/potassium buffer salt, etc. Alginate biopolymer can be recovered from algal biomass by filtration followed by $CaCl_2$ precipitation. Precipitation method is also used during the recovery of various organic acids such as citric acid, gluconic acid. Protein precipitation is usually an intermediate step toward final purification, since precipitates are typically impure. A protein's tendency to precipitate is governed by many factors, solvent environment (e.g. salt concentration, dielectric constant, pH), temperature and the size, shape, charge and hydrophobicity of the protein. Therefore protein precipitation can depend on a complex combination of environmental parameters. One of the most common strategies to induce precipitation is to alter the solvent's properties and reduce protein's solubility

There are various methods for precipitation:

(a) Salting out
(b) Reduction of solvent dielectric constant
(c) Addition of non-ionic polymers

Salting out

It refers to the protein precipitation at higher salt concentration and the relative effectiveness of salts for salting out follows lyotropics (Hofmeister) series, which, for some common anions, is phosphate > sulphate > acetate > chloride > bromide > nitrate > thiocyanate. The Hofmeister effect appears to be due to the ability of the ions to change the structure of water.

Reduction of solvent dielectric constant

A lower dielectric constant enhances the charge dipole and dipole-dipole attraction between proteins and promotes precipitation. Solvent driven precipitation during various stages of enzyme preparation is done by acetone, alcohol, ammonium sulphate, polyethylenimine, etc. Polysaccharides (like dextran) and biogums (like xanthan gums) are recovered using precipitation with alcohol (methanol or isopropanol).

Addition of non-ionic polymers

Dextran, polyethylene glycols is used and protein-protein interactions due to polymer arise primarily for excluded volume. Although there may be electrostatic effects at low ionic concentrations. The polymer occupies volume and causes the protein to approach each other and ultimately phase separates.

In isoelectric precipitation proteins are precipitated at their isoelectric point. This is defined as the value of pH at which proteins do not carry any net electric charge. Proteins are easily precipitated at the isoelctric point.

10.3.4 Adsorption

Adsorption is a traditional unit operation in which soluble solutes from a dilute aqueous (or liquid) phase can be effectively separated on to the surface of adsorbent materials. It is a physical separation process, where only weak van-der Waals forces plays a major role. On adsorption, transfer of solute takes place from solution to a solid phase. It is an equilibrium separation process, based on distribution of solutes between two phases. Classical examples of adsorbents are activated charcoal, silica gel, Fuller's earth, diatomaceous earth, various polymeric materials, ion-exchange resins and so on depending on the nature of particular feed solution under treatment. The operation is carried out in batch or continuous mode, in column or beds. Packed-beds and moving beds are very popular reactors. Surface area of adsorbents is the most important criteria of a good adsorbent. Ion exchange resins are widely used in protein separation and recovery of streptomycin, cephalosporin, etc. from fermentation broth. Adsorption is also used in butanol and ethanol fermentations and in recovery of lactic acid. Porous adsorbent beads have been traditionally employed in the form of packed beads in process of chromatography.

10.3.5 Membrane Processes

Membrane technologies have emerged as very effective and commercially attractive options for separation and purification of high-value products obtained in various biochemical, and pharmaceutical industries. Membrane process has several advantages to its credit. It is a clean and eco-friendly process which consumes less energy. Due to compact and modular design of membrane systems, it requires lesser space than traditional separation units.

The technically and commercially most relevant processes are pressure-driven processes, such as reverse osmosis, ultra- and microfiltration, pervaporation or gas separation, concentration gradient driven processes such as dialysis, and electrical potential driven process such as electrodialysis. Important features of pressure driven membrane processes are presented in Table 10.1.

Table 10.1 Important features of pressure driven membrane processes

Driving force	Membrane Process	Permeate	Retentate	Type of membrane
Pressure difference	Microfiltration (0.5–2 bar)	Dissolved solutes, water	Suspended particles, water	Symmetric microporous,
	Ultrafiltration (1–10 bar)	Small molecules, water	Polymers, proteins, micelles, colloids particulates	Asymmetric microporous
	Nanofiltration (10–70 bar)	Monovalent ions, water	Small molecules, divalent salts	Thin-film membranes
	Reverse Osmosis (10–100 bar)	Water, small polar solvents, salts.	All solutes, water	Asymmetric skin-type.

Microfiltration

Microfiltration is a low pressure means of separating large molecular weight suspended or colloidal compounds from dissolved solids in the range of 0.05 to 10 microns. It closely resembles conventional coarse filtration or sieving. It is essentially a sterile filtration with pores (0.1 to 10.0 microns) so small that microorganisms cannot pass through them. A microfiltration membrane is generally porous enough to pass molecules of true solutions, even if they are large.

In cross-flow microfiltration (CFMF), the suspension is pumped tangentially over the filtration medium (Figure 10.6). Clear liquid permeates the filtration medium and is recovered as the permeate, while the solids accumulate at the filtration barrier to form a fouling layer, or cake. The cake, constituting an increase in hydraulic resistance, decreases the permeate flux. However, the tangential suspension flow tends to limit the growth of the cake.

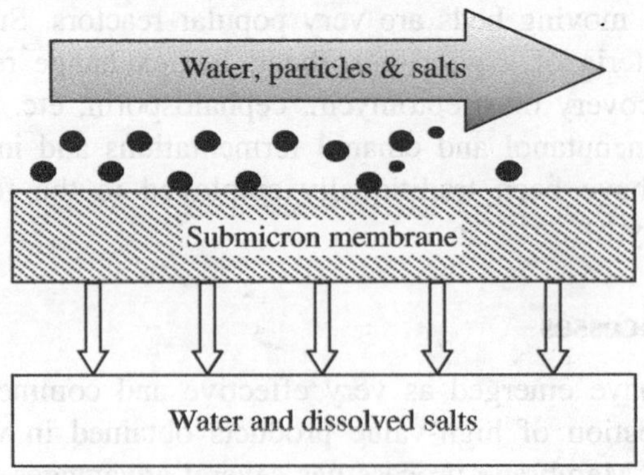

Figure 10.6 Schematic representation of cross flow microfiltration.

Application includes cell harvesting from fermentation broth, fractionation of milk proteins, and corn syrup clarification. Microfiltration with ceramic membranes is widely used for the separation of intact cells from enzymes. Whole cells are retained, whereas the enzyme can pass through the membrane and be recovered from the permeate stream. This is an economically attractive alternative to traditional separation techniques such as centrifugation and rotary vacuum filtration. The traditional recovery methods used for harvesting the cells and purifying the enzymes have been centrifugation or rotary drum filter to remove the bulk of the cells followed by precoat filtration or leaf-press filter. Once separated the enzymes are concentrated and purified using ultrafiltration and finally dried and formulated.

Ultrafiltration

Ultrafiltration (UF), like reverse osmosis, is a cross-flow separation process. Here liquid stream to be treated (feed) flows tangentially along the membrane surface, thereby producing two streams. The stream of liquid that comes through the membrane is called permeate. The type and amount of species left in the permeate depend on the characteristics of the membrane, the operating conditions, and the quality of feed. The other liquid stream is called concentrate and gets progressively concentrated in those components removed by the membrane (Figure 10.7).

The flux equation for diffusion of solvent through the membrane in ultrafiltration is the same as that for reverse osmosis.

$$J_w = A_w(\Delta P - \Delta\Pi) \quad (10.6)$$

In ultrafiltration solutes are generally macromolecular in nature. So, the membrane does not allow the passage of solutes. The concentration in moles/l of the large solutes is usually small. Hence the osmotic pressure is very low and can be assumed negligible. Eq.(10.6) becomes:

$$J_w = A_w(\Delta P) \quad (10.7)$$

Ultrafiltration units operate at about 5 to 100 psi pressure drop compared to 400 to 2000 psi for reverse osmosis.

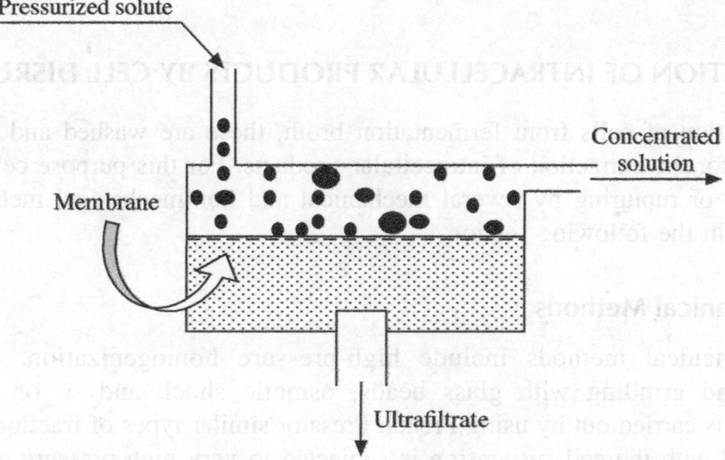

Figure 10.7 Schematic diagram of the basic system of ultrafiltration.

Industries such as chemical and pharmaceutical processing, food and beverage processing, and wastewater treatment, employ UF in order to recycle flow or add value to their products. The main attraction of UF is its ability to purify, separate, and concentrate target macromolecules in continuous systems. UF has successfully been used in protein separation and fractionation, whey processing and clarification of fruit juices.

Nanofiltration

Nanofiltration (NF) membranes have properties between those ultrafiltration and reverse osmosis. The operating pressure for the NF is less than 15 bars. Their separation mechanisms involve both steric (sieving) effects and electrical (Donnan) effects. This combination allows NF membranes to be effective for a range of separation of mixture of organic solutes (either neutral or charged) and salts. Nanofiltration, offers significant advantages such as, lower osmotic pressure difference, higher volumetric flux, higher retention of multivalent salts and molecular weight compounds (>300), relatively low investment, operation, and maintenance costs.

Reverse osmosis (RO) is a high-pressure, energy-efficient technique for dewatering process streams, concentrating low-molecular-weight substances in solution, or purifying wastewater. It has the ability to concentrate dissolved as well as suspended solids. RO is widely used in the desalination of brackish water.

Electrodialysis (ED) is an electromembrane process in which ions are transported through an ion exchange membrane from one solution to another under the influence of an electrical potential gradient. The electrical charges on the ions allow them to be driven through the membranes fabricated from ion exchange polymers. Applying a voltage between two end electrodes generates the potential field required for this. During this transportation anions are able to permeate through anion-selective membranes, but are blocked by the cation-selective membranes. The opposite occurs with cations. As a result, two different solutions are produced inside the stack: one more dilute (the diluate) and the other more concentrated (the concentrate) than the feeding solution. ED has been applied in demineralization of cheese whey, removal of tartaric acid from wine, desalting of fruit juice, dextran solution, purification of fermented solution, serums and vaccines.

10.4 SEPARATION OF INTRACELLULAR PRODUCTS BY CELL DISRUPTION

After the separation of cells from fermentation broth, these are washed and if required, can be further used for the extraction of intracellular products. For this purpose cells are subjected to disintegration or rupturing by several mechanical and non-mechanical methods. These are briefly outlined in the following section.

10.4.1 Mechanical Methods

Important mechanical methods include high-pressure homogenization, ultrasonication, impingement and grinding with glass beads, osmotic shock and so on. High pressure homogenization is carried out by using French Press or similar types of fractionators. A hollow cylinder is filled with the cell suspension is subjected to very high pressure and are extruded through a very narrow orifice to come out to atmospheric pressure. Liquid shear produced as a result of sudden release of high pressure disintegrates the cells. High pressure homogenizers can be used to exert very high pressure in the range of 800–1000 bar and are able to operate continuously with high capacity.

Some mechanical homogenizers use inert beads (20 to 50 mesh; glass, titanium compound, alumina, etc.), whereby the cell suspension is forced through the grinding chamber fitted with beads. Due to high rotational speed, these beads impart shearing and impact forces on the cell walls to break it open. A schematic diagram of a high speed bead mill is presented in Figure 10.8. However, certain physico-chemical factors like size and quantity of the glass beads, residence time within the grinding chamber, cell concentration, temperature and rotational speed are important to get the optimum results. Ball mills, Hughes press, X-press are used for the fractionation of frozen microbial cells. Temperature must be controlled during any of these processes to avoid thermal degradation of cells. There are also chances of oxidative degradation, protein denaturation and these could be avoided by applying certain additives such as, reducing agents or chelating agents.

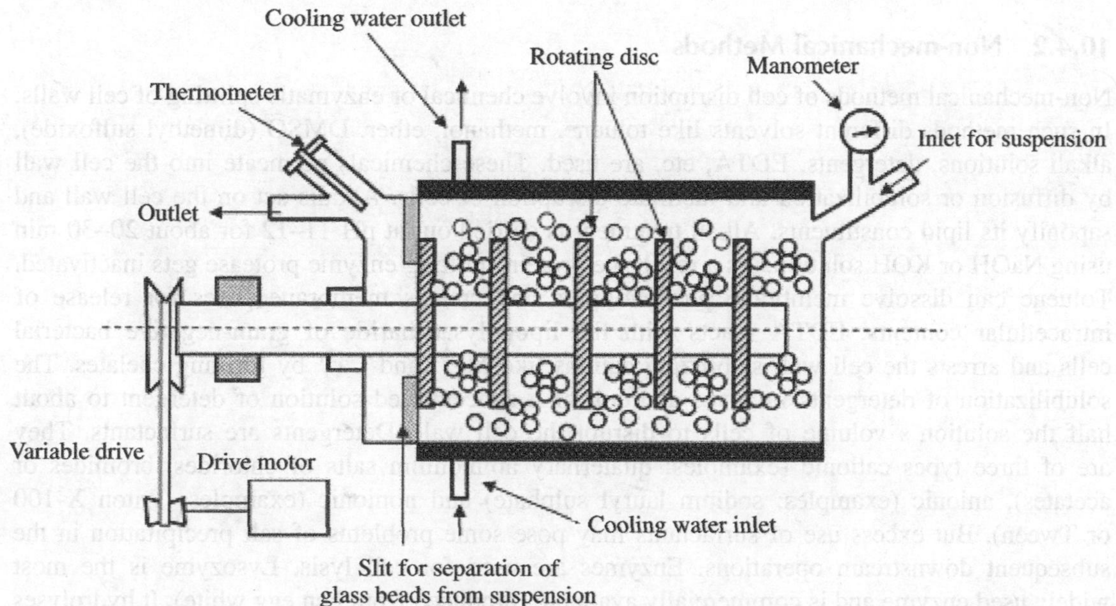

Figure 10.8 High speed bead mill.

In small-scale operation or in the laboratory uses, ultrasonication is an effective method of cell disintegration (Figure 10.9). In ultrasonic bath, frequencies in the range of 50–20 kHz are applied on the cells by an electronic generator. This results in the rupturing of cells by cavitation. Ultrasonic waves create areas of compression and rarefaction alternatively. Due to this compression the bubbles are produced in the cavities and are subjected to several thousand atmospheric pressures. This leads to collapse of bubbles creating shock waves that disrupt the cell wall. In osmotic shock process, cells are dumped in to a large volume of pure water and are allowed stay for some time to render swelling due to osmotic flow of water in to the cells. Consequently the cells burst out. Rupturing the cells with ice crystals, repeated freezing and thawing the cell paste are also practiced for cell disintegration in small-scale laboratory purposes.

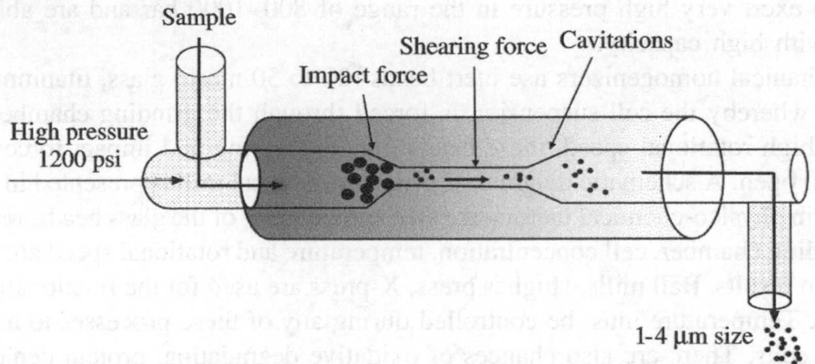

Figure 10.9 High pressure homogenizer for disintegration of cells.

10.4.2 Non-mechanical Methods

Non-mechanical methods of cell disruption involve chemical or enzymatic splitting of cell walls. In such methods different solvents like toluene, methanol, ether, DMSO (dimethyl sulfoxide), alkali solutions, detergents, EDTA, etc. are used. These chemicals permeate into the cell wall by diffusion or solubilization and facilitate disruption of cells. Alkalis act on the cell wall and saponify its lipid constituents. Alkali treatment is carried out at pH 11–12 for about 20–30 min using NaOH or KOH solution. As a result the protein splitting enzyme protease gets inactivated. Toluene can dissolve membrane phospholipids and creates membrane pores for release of intracellular contents. EDTA reacts with the lipopolysaccharide of gram-negative bacterial cells and arrests the cell wall stabilizing cations like Mg^{2+} and Ca^{2+} by forming chelates. The solubilization of detergents involves addition of a concentrated solution of detergent to about half the solution's volume of cells to disrupt the cell wall. Detergents are surfactants. They are of three types cationic (examples: quaternary ammonium salts of chlorides, bromides or acetates), anionic (examples: sodium lauryl sulphate) and nonionic (examples: Triton X-100 or Tween). But excess use of surfactants may pose some problems of salt precipitation in the subsequent downstream operations. Enzymes are used for cell lysis. Lysozyme is the most widely used enzyme and is commercially available (produced from hen egg white). It hydrolyses β-1, 4-glycosidic bonds of the mucopeptide in bacterial cell walls. The Gram-positive bacteria (with high content of cell wall mucopeptides) are more susceptible for the action of lysozyme. In addition there are many other enzymes like cellulase, pectinase, xylanase, chitinase, etc. all of which can effectively disrupt the cell wall of microorganisms.

10.5 FINISHING STEPS FOR PURIFICATION

10.5.1 Crystallization

Crystallization is an important industrial operation in the production of pure products. It offers attractive appearance to the product and many a times is the cheapest technique to produce pure product from an impure solution. There are different methods of crystallization. In all the methods supersaturated solution has to be produced first. If a solution is in equilibrium with the solid phase, it is called as saturated and solutions containing more dissolved solutes than that represented by saturation condition are called as supersaturated solution. The rate of supersaturation is an essential part of all crystallization processes. Liable or unstable and metastable supersaturation are the terms referred to supersaturated solutions. In unstable supersaturation spontaneous deposition of solute occurs in the absence of solid nuclei while in metastable supersaturation the spontaneous deposition of solute in the absence of nuclei does not occur.

Crystallization consists of three steps:

(i) Supersaturation or supercooling
(ii) Crystal nuclei generation
(iii) Crystal growth

Ideally these steps should occur in a controlled manner but in actual practice may occur simultaneously in different parts of crystallization unit. Schematic diagram of a vacuum crystallizer is presented in Figure 10.10.

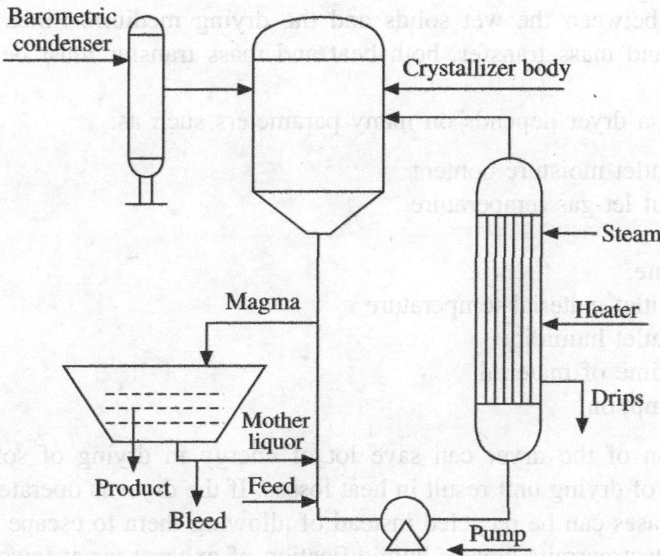

Figure 10.10 Schematic diagram of a vacuum crystallizer.

Supersaturation can be achievd by cooling or evaporation. Supersaturation by cooling is very common. The solubility of solute in a solvent decreases with decrease in temperature. Crystals of uniform size can be produced by a slow and controlled rate of cooling. Solubility data of a solute in solvent is useful in calculation of maximum yield of pure crystals that can be obtained by cooling or evaporation. The yield refers to the quantity of crystals deposited from the solution. In order to form crystals, nuclei generation is necessary. Nucleation may be spontaneous or can be induced artificially. Generally spontaneous nucleation is called as homogeneous nucleation, while artificially induced nucleation is called heterogeneous nucleation.

Crystallization is widely used as intermediate purification step in many pharmaceutical and bioprocess industries. Traditionally citric acid present in citrus fruits was separated in the form of calcium citrate crystals from lemon juice. In the commercial fermentation of citric acid, crystallization is carried out at the last stage to obtain crystals of anhydrous or monohydrated citric acid. Other compounds such as succinic acid, itaconic acid (a precursor of biopolymer), etc. are also separated and purified by crystallization. Crystallization is also applied during the production of penicillin G or V. The crystallization of penicillin is affected by concentration, pH and temperature. Salts of gluconic acid used as additives in polymers and corrosion inhibitor are separated from fermentation broth by crystallization.

10.5.2 Drying

Drying is a diffusional process in which the transfer of moisture to the surrounding medium takes place by the evaporation of water from the surface. Drying does not mean only removal of moisture but during the process, physical structure as well as the appearance has to be preserved. During drying the diffusion of moisture from the interior material to the surface also occurs through capillary action. The driving force for mass transfer results from the humidity difference existing between the wet solids and the drying medium. Since it is a process of simultaneous heat and mass transfer, both heat and mass transfer must be considered in the operation of dryers.

Performance of a dryer depends on many parameters such as:

- Inlet and outlet moisture content
- Inlet and out let gas temperature
- Feed rate
- Gas flow rate
- Inlet and outlet material temperature
- Inlet and outlet humidity
- Residence time of material
- Fuel consumption

Proper operation of the dryer can save lot of energy in drying of solids. Air leakages, improper insulation of drying unit result in heat losses. If the dryer is operated on closed circuit basis, the exhaust gases can be recycled instead of allowing them to escape to the atmosphere. Better solid-air contact permits greater humidification of exhaust air at lower capital costs. An energy balance of any dryer indicates that a lot of energy is lost with the exhaust gases. This loss of energy could be saved by optimizing exhaust air flow rate, humidity and temperature. If the temperature sensitive materials are to be dried in a countercurrent dryer, there is a limit on the inlet temperature. Humidity is also an important factor in the determination of overall energy consumption of a dryer.

Spray drying

Spray-drying is a well-known drying method in food, dairy and pharmaceutical industries. The process is consisted of three major steps such as atomization of feed, droplet-to-particle conversion and product collection. As shown in Figure 10.11 dilute feed solution is pumped to an atomizer, breaking up the liquid feed into a spray of fine droplets. The droplets are ejected into a drying gas chamber where the moisture vaporization occurs, along with formation of dry particles. Finally, using an appropriate device, the dried particles are separated from the drying medium, being then collected in a tank.

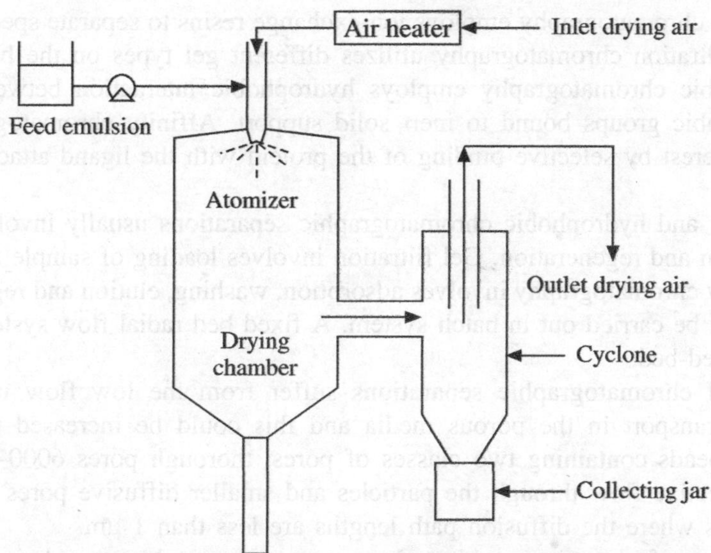

Figure 10.11 A schematic representation of Spray drier.

Freeze drying (lyophilization)

Principle of freeze drying is based on the sublimation process (in which solid directly turns into vapor phase without any intermediate liquid phase). Freeze drying starts with cooling down the product to be dried to the lowest possible temperature. This solidifies the product almost completely. Then the pressure is reduced to below the vapor pressure, near the triple point of water so as to facilitate sublimation of vapors (temp: 0.01°C and pressure 4.6 mm Hg). Unbound moisture is removed during primary drying, while additional time is required to remove bound moisture. Pharmaceutical freeze dryer can dry even liquids. Long shelf-life, preservation of taste and aroma and large retention of texture are the main advantages of freeze drying of biological and food products.

10.6 CHROMATORGRAPHY

Chromatography is used to describe any separation process that depends on partitioning between a flowing fluid and a solid adsorbent. The solid adsorbents are charged resins (e.g. polymers of sulfonated styrene polymer gels containing diethyl-amino-ethyl groups). Dextran gel, cross-linked agarose gels (untreated or derivatized) beaded cellulose, hydroxyapatite, alumina and silica gel. The different adsorbents give rise to different chromatographic techniques, which are generally classified according to the mechanism of separation.

In partition chromatography, the stationary phase is a liquid or a liquid supported on an inert solid. The moving phase can be either a liquid or a gas. The separation of different components is based upon their selective partition between the moving phase and a thin liquid film (stationary phase) held on an inert solid.

Ion-exchange chromatography employs ion-exchange resins to separate species on the basis of charge. Gel filtration chromatography utilizes different gel types on the basis of size and shape. Hydrophobic chromatography employs hydrophobic interaction between the proteins and the hydrophobic groups bound to inert solid support. Affinity chromatography separates the protein of interest by selective binding of the protein with the ligand attached to the solid support.

Ion exchange and hydrophobic chromatographic separations usually involve three stages: adsorption, elution and regeneration. Gel filtration involves loading of sample and elution with buffer and affinity chromatography involves adsorption, washing, elution and regeneration steps. These stages may be carried out in batch system. A fixed bed radial flow system magnetically stabilized Fluidized-bed.

Efficiency of chromatographic separations suffer from the low flow rates due to the restricted mass transport in the porous media and this could be increased to ten times by using perfusive beads containing two classes of pores, thorough pores 6000–8000Å size big enough for convective flow through the particles and smaller diffusive pores line the interior of thorough pores where the diffusion path lengths are less than 1 μm.

For scaling up of chromatographic column, most practical approach is to increase the column diameter while holding the bed height constant; also the elution buffers and the volumetric ratios of the feed are held constant. By utilizing this approach, the scale-up equation for chromatography is given by:

$$\text{SF} = (V_{\text{bed}})_2/(V_{\text{bed}})_1 = V_2/V_1 = A_2/A_1$$

SF is the scale-up factor, the ratio of the column bed volumes (V_{bed}), the volumetric flow rates (V), and the cross-sectional area (A).

On the basis of the type of stationary phase, chromatography is classified as:

(a) *Column chromatography*: In this type of chromatography, the stationary phase is formed into a column in a cylindrical tube

(b) *Paper chromatography*: In this form of partition chromatography, the stationary phase is the absorbed water present in the filter paper and the moving phase is a solvent saturated with water. Filter paper used is made of highly purified cellulose $(C_6H_{10}O_5)_n$.

(c) *Thin layer chromatography*: In this, the stationary phase in thin layers of certain fine solid particles, such as silica gel and alumina, supported on glass plates or plastic plates.

Affinity chromatography

Affinity chromatography is used for the isolation and purification of biological macromolecules, nucleic acid, antibodies, enzymes, etc. It is based on specific biological interaction between an adsorbate molecule and an affinity ligand. Examples of interaction include enzyme and substrate or between antigen and antibody. The ligand must have complementary size and shape of the adsorbate of interest. The interactions are largely reversible. The matrix used in affinity chromatography is in bead form into which a specific ligand can be attached. The matrix must be insoluble in solvent and buffer, chemically inert and mechanically stable. Basic principle of affinity chromatography is explained in Figure 10.12.

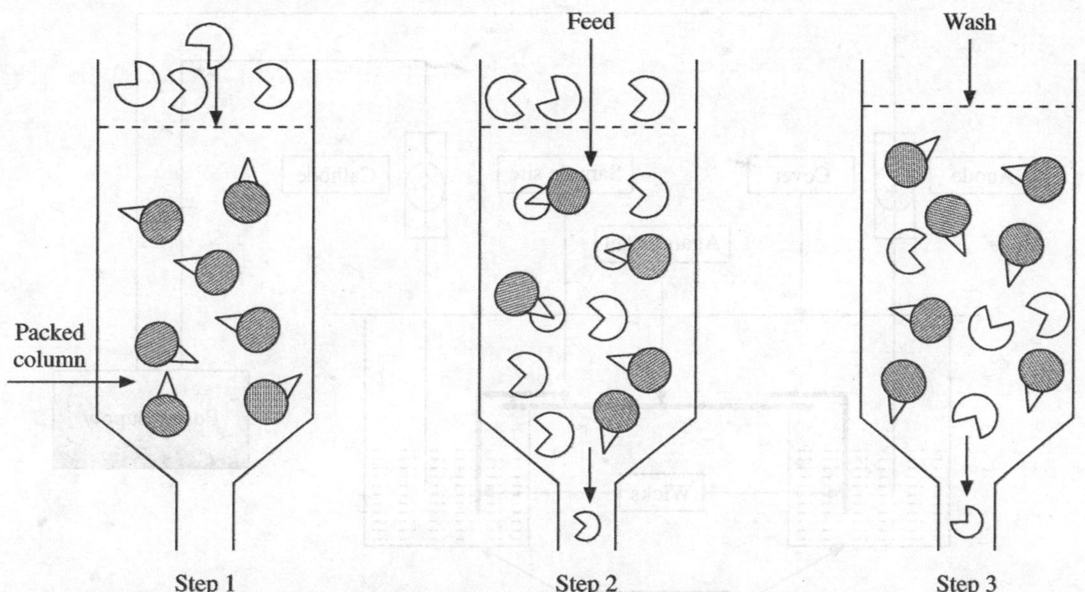

Figure 10.12 Schematic representation of affinity chromatography [Step 1: (Introduce mixture); Step 2: Separation; and Step 3: Elution of bound component].

10.7 ELECTROPHORESIS

Fine grains of solid or liquid or gas held in suspension in a solvent such as colloidal particles of clay, emulsion globules, or foams move towards electrodes when an electric field is applied. This phenomenon of migration of colloidal particles under the influence of an applied e.m.f. is called electrophoresis. Sols of metallic oxides and hydroxides and basic dye stuffs are usually positively charged, while sols of metallic sulphides, acidic hydroxides and acid dye stuffs carry a negative charge. The velocity of migration and the sign of the charge carried by the particles can be determined by macroscopic or ultramicroscopic methods.

Electrophoretic separation is based on different rates of migration of charged molecules or ions under the influence of an electric field (Figure 10.13). Charged molecules have a force exerted on them by the electric field (e.m.f.). If the net charge is 'Q', the force will be $Q.E$, where E is the electric filed strength. Because of this force particles with net positive charge move towards cathode and those with net negative charge move towards anode. Acceleration caused by electrophorectic attraction is opposed by frictional forces, electrophoretic retardation and relaxation effects. However, at steady state the sum of these forces are zero and particles move at a constant migration velocity (m/s) 'u'.

So, at steady state, $u = \mu E$; where E is the electric filed strength (volts/m) and μ is the electrophoretic mobility (m^2/volt.s). Electrophoresis is widely used in the separation of proteins, DNA fragmentation and analysis, detection of proteins, testing of antibodies, etc. Certain gels such as agar or polyacrylamide are used for protein separations by gel electrophoresis. The process is known as polyacrylamide gel electrophoresis (PAGE).

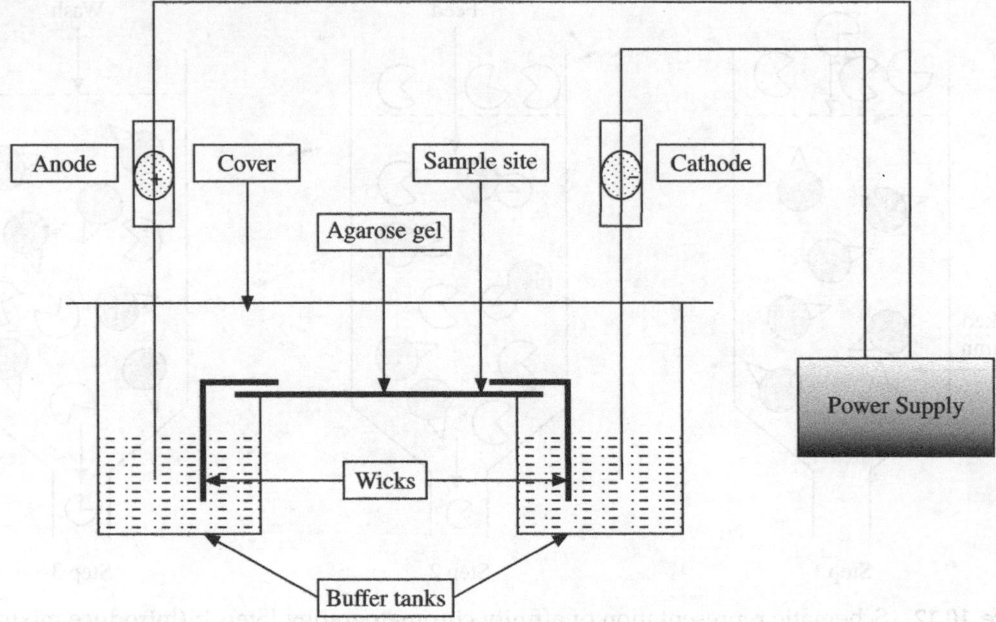

Figure 10.13 A schematic diagram of a laboratory electrophoresis unit.

KEY POINTS TO REMEMBER

- Separation and purification of products are the essential steps in the downstream processing operation.
- Broadly there are two types of products formed during fermentation, extracellular products (example: ethyl alcohol, lactic acid, penicillin, etc.), and intracellular products. (examples: phosphatase, β-galactosidase, immunoglobulin, etc.).
- Insoluble products from the fermentation broth can be separated by filtration, membrane process, centrifugation or coagulation.
- Soluble or miscible products can be separated by using various traditional unit operations such as distillation, solvent extraction.
- Important filtering equipment include plate and frame, rotary vacuum, cartridge, and sand filters.
- Commonly used centrifuges include tubular bowl centrifuge, disc-bowl centrifuge, basket centrifuge, multi chamber centrifuge, scroll centrifuge and so on.
- Addition of various flocculating agents facilitates aggregate formation during sedimentation.
- Extraction of a substance from one to another liquid phase in which the desired product is preferentially soluble is known as liquid-liquid extraction. It is based on the distribution coefficient (or partition coefficient) of any solute or analyte in different solvents.
- Distillation is widely used in the separation of ethanol, propanol, butanol from their dilute solutions.

- There are three important methods of precipitation such as salting out, reduction of solvent dielectric constant and addition of non-ionic polymers.
- Application of membrane filtration includes cell harvesting from fermentation broth, fractionation of milk proteins, corn syrup clarification.
- Cell disruption can be carried out by both, mechanical and non-mechanical means.
- Important mechanical methods include high-pressure homogenization, ultrasonication, impingement and grinding with glass beads, osmotic shock, etc.
- Non-mechanical methods of cell disruption involve chemical or enzymatic splitting of cell walls using toluene, methanol, ether, DMSO (dimethyl sulfoxide), and so on.
- Chromatography is the separation of mixtures by distribution between two or more immiscible phases.
- Electrophoretic separation is based on different rates of migration of charged molecules or ions under the influence of an electric field.

? QUESTIONS FOR PRACTICE

Multiple Choice Questions

1. Which of the following is considered to be an extracellular products?
   (a) Phosphatage (b) Plasminogen
   (c) Penicillin (d) Pepsin
2. Filtration is a unit operation that is commonly used for collecting
   (a) Filtrate (b) Particulate matter
   (c) Precipitate (d) Slurry
3. In the clarification process, which factor is more important
   (a) Depth of the media (b) Pore size of the filter media
   (c) Surface area of the filter (d) Volume of slurry
4. The fluid flows through the filter medium by virtue of
   (a) The potential difference across the filter
   (b) Pressure difference across the filter
   (c) Temperature difference across the filter
   (d) Volume difference across the filter
5. Which one of the following is not a mechanism of filtration?
   (a) Entanglement (b) Impact
   (c) Impingement (d) Straining
6. The cellulose membrane filter is an example of which type of filtration?
   (a) Cake (b) Depth
   (c) Decantation (d) Screen
7. Which among the following filter belong to continuous filter?
   (a) Rotary vacuum (b) Vertical metal leaf
   (c) Plate and frame (d) Stacked disc

8. Centrifugation is based on which of the following law?
(a) Pascal's law (b) Stokes law
(c) Stain law (d) Patrick's law

9. Which of the following force is not used in centrifugation?
(a) Electrostatic force (b) Gravitational force
(c) Centripetal force (d) Centrifugal force

10. Differential centrifugation is based on the differences in ________ of biological particles of different ________.
(a) Size, density (b) Sedimentation rate, sizes and density
(c) Size, structure (d) Mass, size

11. After centrifugation of milk, the supernatant is
(a) Fat (b) Whey
(c) Casein (d) Water

12. The type of sedimentation process of water treatment in which particles settle as individual entity and there is no significant interaction with neighboring particles is referred to as:
(a) Hindered settling (b) Compression settling
(c) Flocculent settling (d) Discrete settling

13. Protein separation techniques are often based on the following properties except
(a) Solubility of the protein (b) Viscosity of protein
(c) Charge of protein (d) Specific binding affinity of protein

14. Primary steps in protein purification include
(a) Homogenization (b) Differential centrifugation
(c) Solubilization (d) All of these

15. Salting out process involves
(a) Precipitation of protein using ammonium sulphate
(b) Precipitation of protein using copper sulphate
(c) Precipitation of protein using sodium chloride
(d) Precipitation of protein using ammonium chloride

16. The stationary phase material in the column chromatography is also called
(a) Adsorbent (b) Adsorb
(c) Elute (d) Absorbent

17. Paper chromatography is an example of ________ chromatography.
(a) Solid-solid (b) Liquid-Liquid
(c) Gas-liquid (d) Solid-liquid

18. If the pressure drop (ΔP) is 1000 units, the flux (J) is 50 units, what is the hydraulic membrane permeability?
(a) 0.02 (b) 0.04
(c) 0.05 (d) 0.06

19. What is the driving force in microfiltration?
(a) Pressure difference (b) Concentration difference
(c) Temperature difference (d) Size difference

20. What is the driving force in dialysis?

(a) Pressure difference (b) Concentration difference
(c) Temperature difference (d) Size difference

21. Read the given statements below and select the correct ones.

(i) A membrane which permits the passage of pure solvent molecules to pass through it and not the solute particles is called semipermeable.

(ii) A membrane which allows some substances to pass through it more readily than others is known as selectively/differentially permeable.

(iii) All living biological membranes are perfectly semipermeable.

(a) (i) and (iii) (b) (ii) and (iii)
(c) (i) and (ii) (d) All are correct

22. Which of the following statements is true about migration of biomolecules during electrophoretic separation?

(a) The rate of migration is directly proportional to the resistance of medium
(b) Rate of migration is directly proportional to current
(c) Low voltage is used for separation of high mass molecules
(d) Rate of migration is inversely proportional to current.

23. The electrophoretic mobility denoted as μ is mathematically expressed as

(a) VE (b) V/E
(c) E/V (d) 1/EV

24. Which of the following factors does not influence electrophoretic mobility?

(a) Molecular weight (b) Shape of molecule
(c) Size of molecules (d) Stereochemistry of molecules

25. In which of the following cases electrophoresis is not used?

(a) Separation of proteins (b) Separation of amino acids
(c) Separation of lipids (d) Separation of nucleic acids

26. Which biological product needs cell disruption after cell culture or fermentation?

(a) Intracellular products
(b) Extracellular products
(c) Both intracellular and extracellular products
(d) Non-cellular products

27. Which cell component acts as a barrier for cell disruption in the cell suspension?

(a) Nucleus (b) Cell wall
(c) Cytoplasm (d) Mitochondria

28. The total volume of material, both solid and liquid, in the column is known as

(a) Void volume (b) Bed volume
(c) Elution volume (d) Retention volume

29. Isolation of eukaryotic m-RNA from a mixture of total cellular RNA is carried out by

(a) Thin layer chromatography (b) Column chromatography
(c) Affinity chromatography (d) Immuno-blotting chromatography

30. A globular protein of molecular weight 50 kDa exists as a mixture of monomers and dimers in solution. The most appropriate techniques for the separation of these two forms of the proteins is

(a) Gel filtration chromatography (b) Ion-exchange chromatography
(c) Thin-layer chromatography (d) Paper chromatography

Answer Key

1. (c)	2. (a)	3. (b)	4. (b)	5. (b)	6. (d)	7. (a)	8. (b)	9. (a)	10. (a)
11. (b)	12. (d)	13. (b)	14. (d)	15. (a)	16. (a)	17. (b)	18. (c)	19. (a)	20. (b)
21. (c)	22. (b)	23. (b)	24. (d)	25(c)	26. (a)	27. (b)	28. (b)	29. (c)	30. (a)

Descriptive Questions

1. How to separate the biological macromolecules based on the biological properties?
2. What are the basic differences between upstream and downstream processing?
3. What are the different techniques one can adopt for separation of insoluble after fermentation?
4. Differentiate between ion exchange and gel filtration chromatography.
5. What is ligand?
6. How to assure that during manufacture of any targeted product no contamination has taken place when the process is a pure culture one.
7. Which unit operations are used in the separation of antibiotics and alcohol from fermentation broth?
8. How does the intended use of a fermentation product influence the choice of recovery process used?
9. What physical property enables one to use distillation for the separation of ethyl alcohol enriched from the water mixture?
10. What is an azeotrope? Explain how can azeotropes be used to purify mixtures?
11. Azeotropes are also defined as liquid mixtures having a vapor pressure at constant temperature as a function of composition. Explain with your own words why vapor pressure depends on the composition of the mixture.
12. If the distillation temperature is different from the theoretical distillation temperature, why do you think this is so?

11 Important Industrial Bioprocesses

11.1 ANAEROBIC BIOPROCESSES

Anaerobic bioproceses are widely applied for commercial production of various industrial solvents, organic acids, etc. These include ethyl alcohol, acetone, butanol, lactic acid, etc. These are discussed in the following sections.

11.1.1 Ethanol

Ethanol is a useful organic chemical which finds its major applications as solvent in pharmaceutical, cosmetics and perfumery industry; food and beverage and as an additive of gasoline in energy sector. Approximately 4 million tons of ethanol is produced globally per annum. 80% of this ethanol is produced by microbial fermentation method.

Raw materials

Any carbohydrate-based materials like sugars, starch or lignocellulosic, which are abundantly and inexpensively available can be used as raw materials. These include grains like maize, corn, barley, sweet sorghum, sugars like grape sugar, cane sugar, sugar cane juice, rice, potato and so on. In India, molasses, a byproduct of sugar industry, is the principal raw material for distillery industries. Molasses contains nearly 40–50% fermentable sugar. Cellulosic materials, woody biomass or even agro-waste materials can be used provided these are effectively hydrolyzed by various enzymes. However, hydrolysis of starch or cellulose to produce fermentable sugars is a cumbersome process.

Microorganisms

Yeasts are the most preferred microorganisms for ethanol fermentation. *Saccharomyces cerevisiae*, *Candida brassicae*, *Candida utilis,* etc. are generally used. These apart, bacteria such as, *Zymomonas* sp. is also used on various sugar substrates. Yeast can transform hexose sugars into ethanol and CO_2 is produced as a byproduct.

$$C_6H_{12}O_6 \rightarrow 2C_2H_5OH + 2CO_2$$

There could be formation of certain byproducts such as acetic acid, glycerol, succinic acid, etc. along with ethyl alcohol during fermentation reaction.

Reaction conditions

Optimum temperature of fermentation is maintained at 35°C and pH is kept at 4–6. Use of thermophilic organisms allows the reaction to take place at high temperature up to 50–60°C for commercial production. Fermentation takes place in absence of oxygen.

Feed solution consists of a carbon source (any sugars, preferably molasses or glucose), nitrogen sources like ammonium salts or urea and minerals like potassium dihydrogen phosphate, magnesium sulphate, calcium chloride and yeast extract. Glucose concentration should be maintained otherwise substrate inhibition of growth might take place. It should be less than 100g/L in continuous fermenter. The operation could be both in batch or continuous modes under aseptic conditions. Steam sterilization is performed prior to fermentation.

Process

Reactor is filled with nutrient media up to 70% of its volume. Freshly prepared sterilized seed culture of yeast is inoculated aseptically into the fermenter vessel. Total volume of batch takes 30–40 h. A part of media is also recycled if required. Continuous operation with cell recycle is credited with high conversion (nearly 95%) and less residence time. Figure 11.1 represents a simplified flow diagram of a continuous ethanol fermentation process. Mechanically agitated, glass lined SS reactor is generally used.

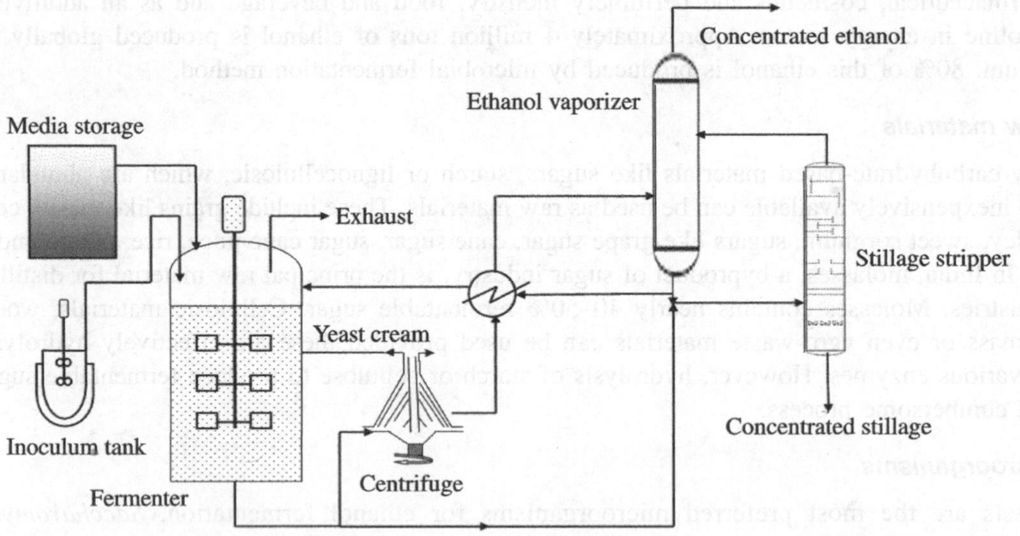

Figure 11.1 Flow diagram for the production of ethanol by fermentation.

After reaction gets over, the contents of the fermenter is processed for separation of cell mass by centrifugation. The solid free dilute liquor is then rectified using a distillation column. Low temperature vacuum distillation is used to produce up to 95% ethanol. Ethanol water mixture produces azeotrope which prevents further separation by normal distillation. An entrainer (such as benzene) is added to break the azeotrope or some other methods like extraction with CO_2, dehydration via adsorption onto molecular sieves, combined distillation–pervaporation are useful to make 100% alcohol. However, selection of a suitable dehydration purification process is finally dictated by the overall energy consumption.

11.1.2 Lactic Acid

Lactic acid is commonly known as 2-hydroxypropionic acid. It is widely used in food and biochemical industries. The principal uses of edible grade lactic acid include flavoring agent, preservatives and acidulant in food and beverage processing, cheese, salad dressing, pickles, etc. The commercial grade of lactic acid are used in deliming for leather and wool processing, solvent in pharmaceutical applications, household cleaning and descaling, and as polymer. Polylactic acid (PLA) is used as biomaterial in various biomedical applications such as suture threads, bone fixation screws and so on. PLA is a biodegradable polymer which has immense application potentials.

The bacterial species, such as, *Lactobacillus, Streptococcus* are common for fermentative production lactic acid. These bacteria follows glucose catabolism route of Embden–Meyerhof–Parnas (EMP) pathway to produce two moles of pyruvate from one mole of glucose according to the following reaction:

$$C_6H_{12}O_6 \rightarrow 2C_3H_6O_3$$

Certain species of Lactobacillus synthesizes lactic acid, from one mole of glucose via controlled fermentation according to the following reaction. Ethanol and carbon dioxides are the byproducts.

$$C_6H_{12}O_6 \rightarrow C_3H_6O_3 + C_2H_5OH + CO_2$$

Commercially lactic acid is produced by controlled fermentation of sugar from various sources. The selection of raw materials depends on certain factors such as availability, cost effectiveness, and conversion or yield. Sucrose from cane sugar or sugar beet juice, lactose from cheese whey, maltose from starch, dextrose sugars from corn, etc. are considered to be popular raw materials for lactic acid production as carbon sources in the fermentation. In addition, various nitrogen sources such as corn steep liquor, malt extracts, barley, yeast extracts are used in the fermentation media.

Traditionally the industrial fermentation of lactic acid is carried out in batch fermenter, made of high-quality stainless steel (sometimes glass lined reactors are also used) equipped with heat transfer coils as the reaction is mildly exothermic in nature. Prior to the fermentation, the reactor is steam sterilized to ensure aseptic conditions. Agitation is provided to ensure homogeneity of the reaction mass. Although fermentation conditions vary to a little extent for various licensed processes, the temperature in general is maintained at 45–50°C and pH in the range of 6–7. The time of fermentation varies between 1 to 5 days depending primarily on the carbon source used. Process flow diagram for lactic acid production is presented in Figure 11.2.

The important parameters for the yield of lactic acid are temperature, pH, initial sugar concentration, etc. The lactic acid produced in the fermentation is neutralized usually by addition of calcium carbonate or calcium hydroxide slurry. CO_2 is continuously released which in turn creates anaerobic conditions in the fermenter. The yield of lactic acid is about 90% of the initial sugar concentration.

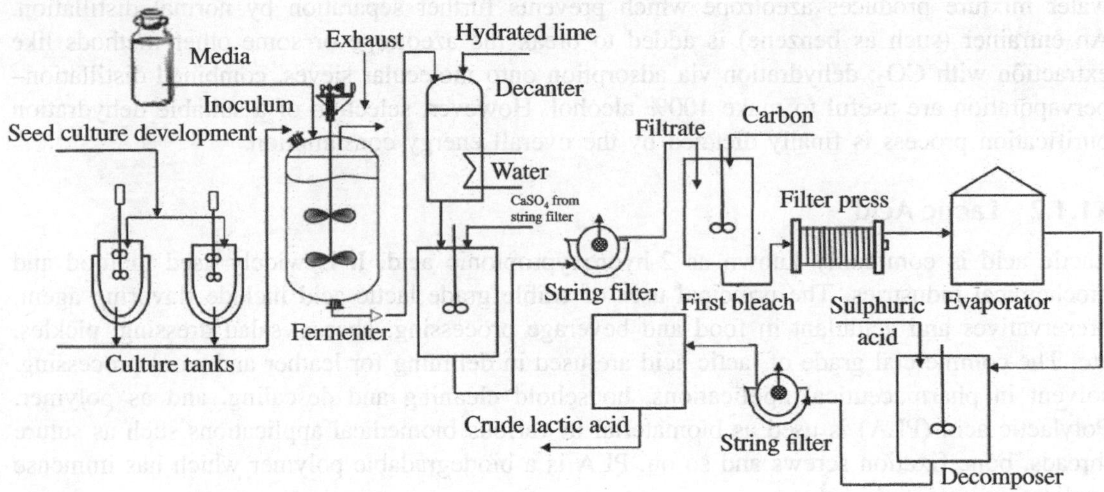

Figure 11.2 Process flow diagram for lactic acid production.

Product recovery and purification

Downstream processing of lactic acid involves its separation and purification from the fermentation broth and this constitutes the major part of production cost. The fermentation with pure sugar substrate and minimum nitrogen sources simplifies the product recovery. At the first step the temperature of the fermenter is increased to 80–100°C and pH up to 10–11. As a result, protein coagulation and solubilization of calcium lactate are facilitated. Some amount of residual sugar is degraded during the process. The liquid is then filtered using a suitable filter to remove biomass. Dilute sulphuric acid is added to obtain lactic acid. Insoluble precipitates of calcium sulphate are filtered off. Alternatively the purification can also be carried out by calcium lactate precipitation. Further purification of lactic acid to obtain various grades is done by various advanced process such as adsorption with suitable adsorbents, ion exchange purification, electrodialysis and so on.

11.1.3 Acetone and Butanol

Acetone and butanol are widely used as industrial solvents in paints, lacquer, enamel, rubber and plastic, detergent, adhesive, printing inks and many other process industries as well as in the manufacturing of antibiotics, hormones and vitamins. Butanol is also used as an additive of automotive fuel.

Although acetone and butanol can be manufactured in petrochemical industries, the processes are more energy intensive. Moreover, finite resources of fossil fuels like crude oil warrant alternative routes of production, which are clean and environmentally benign. Biochemical production of acetone and butanol by anaerobic fermentation thus, assumes great importance. Generally, various species of *Clostridium* such as *Clostridium acetobutylicum* is popularly used. It can grow on carbon substrates like glucose, lactose, fructose, starch, molasses, whey and so on. Presence of different amylolytic and saccharolytic enzymes facilitates the fermentation.

Traditionally, Weizmann process has been used for production of acetone and butanol. In this corn starch is used as the precursor, which undergoes gelatinization at 60°C for about 20 min, followed by steam sterilization. The mash thus obtained is taken to the fermenter. It is inoculated with 5% inoculum of *C. butylicum*. Fermentation continues for 2 to 2.5 days. pH is maintained at 6.5, however it bounds to drop in course of reaction due to the formation of acetic and butyric acid. Gaseous carbon dioxide and hydrogen are also evolved. Gradually the acetic acid and butyric acid are converted into acetone and butanol. A simplified flow diagram of acetone-butanol production process is shown in Figure 11.3. There are various methods to improve solvent acetone/butanol yield in fermenter. Both continuous and fed-batch modes of operation are in practice. Recovery of acetone and butanol is done by fractionation.

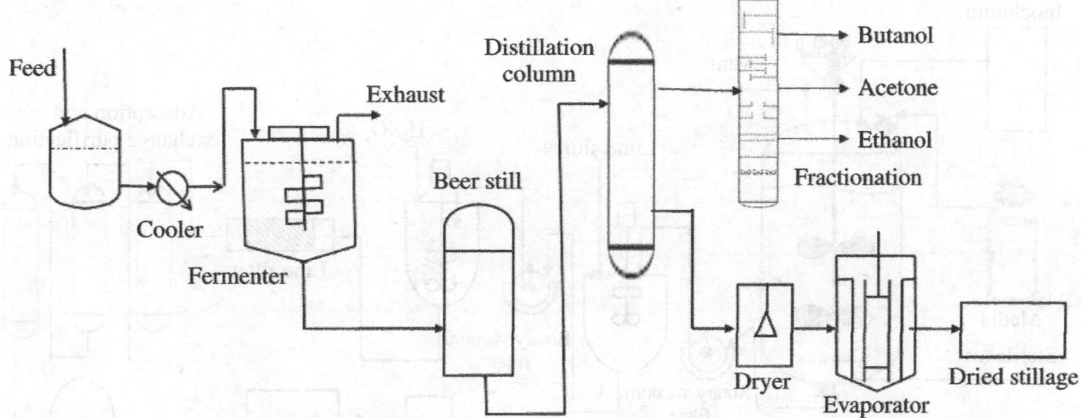

Figure 11.3 Flow diagram of Acetone-butanol production.

11.2 AEROBIC BIOPROCESSES

Aerobic bioprocesses are extensively used for the production of organic acids such as citric, acetic, gluconic etc., vitamins, antibiotics, flavors and fragrances and so on. Some of the important production processes are briefly described below.

11.2.1 Citric Acid

Citric acid is naturally present in all citrus fruits and vegetables. It can be crystallized in the form of calcium citrate from lemon juice. Originally citric acid was produced from the fermentation using *Penicillum*. But since the yield was very low, the microorganism was replaced with *Aspergillus niger*. Commercial production of citric acid is carried out aerobically in a fermenter using *Aspergillus niger* or some species of yeast like *Candida*.

Uses

Food and beverage industries are the major consumers of the bulk of citric acid produced. It can be used as acidulant, flavoring agent and preservatives. It also acts as stabilizer in fruit juices, gelling agents in jams and jellies, turbidity preventer in wine and ciders and as an effervescent in pharmaceutical tablets and powders in combination with calcium bicarbonate.

Production by fermentation

Surface or submerged fermentation and even solid state fermentation are applied in the production of citric acid. However, submerged fermentation is relatively more popular than other techniques. A variety of carbonaceous raw materials like cane molasses, beet molasses, starch, even hydrocarbons are used a carbon source for fermentation media. Several other raw materials that have been used by various researchers include bagasse, rice husk, wheat straw, pineapple waste, sugar beet cossette, fruit peel and so on. Molasses is considered to be the preferred raw material on account of its cost effectiveness and appreciable sugar concentration (as high as around 55%). A generalized flow diagram for the production of citric acid by fermentation is shown in Figure 11.4.

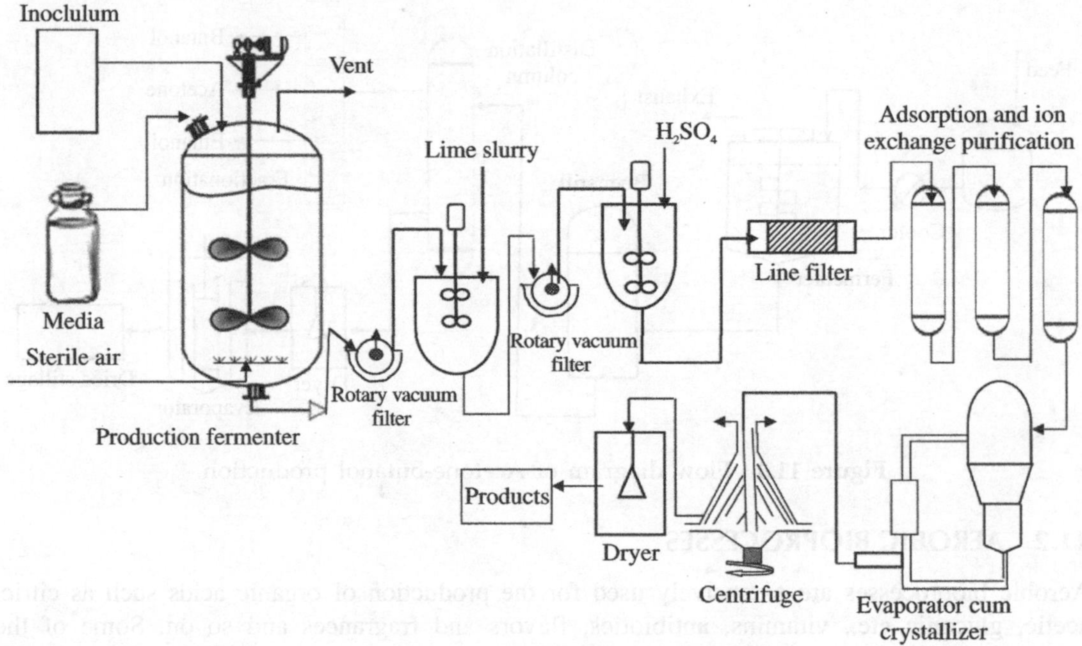

Figure 11.4 Flow diagram for the production of citric acid by fermentation.

About 80% of total world's production of citric acid is carried out by this submerged method. Advantages are higher yields, productivity and low labor cost. Both stirred tank fermenters and tower fermenters are used. The material of construction of the fermenter is high-grade stainless steel. As the process involves aerobic fermentation arrangement for aeration and agitation are provided to maintain a high level of dissolved oxygen. Adequate sterilization is done to maintain aseptic conditions. Cooling is provided externally through cooling jacket. Pretreated raw materials along with other nutrients (like nitrogen sources, phosphorus trace elements) are sterilized prior to their introduction into the vessel. Inoculation is done by pre-cultivated mycelia. *Aspergillus niger* is the most widely used microorganism for citric acid production.

It can be easily cultured; using many different low cost carbon-sources and gives high yield. Nowadays mutant strains are used for industrial production. Inoculum size of 10% of fresh medium is generally used for inoculation. Submerged fermentation takes 5 to 10 days depending on the process conditions. Fermentation can be carried out in batch, continuous or fed batch system.

Process parameters

During the course of fermentation several organic acids are produced apart from citric acid as co-products. These include lactic acid, acetic acid, etc. As a result pH of the broth drops. In some cases pH gets reduced even less than 3.0. pH in the range of 2–3 is desirable for optimum production of citric acid. Low pH prevents the formation oxalic and gluconic acid which are inhibitory to citric acid production. Uniform and high aeration rate is required for effective fermentation.

Product recovery and purification

Downstream processing of citric acid from the fermentation broth is traditionally carried out by precipitation method. In this method lime solution is added to form sparingly soluble precipitate of tricalcium citrate tetrahydrate. The precipitate is separated by filtration and is washed further with distilled water to remove any adherent impurities. It is followed by treatment with dilute sulphuric acid to produce calcium sulphate which is again filtered out using rotary vacuum filter. The brown filtrate, containing citric acid (known as crude acid) is treated with high quality activated carbon for removal of colour. Sometimes it is also passed through the ion exchange resin beds. Several anion-exchange and cation-exchange resins such as sodium polystyrene sulfonate, polymers with quaternary ammonium groups, etc. are commercially used. Lastly, vacuum crystallizer is used to concentrate the liquor. Citric acid monohydrate is formed. This is followed by centrifugation, drying and packaging after size classification. Anhydrous citric acid is obtained by crystallization above the transition point of 36.6°C, while the citric acid crystallizes as the monohydrate under this temperature.

11.2.2 Penicillin

Antibiotic is defined as a substance produced by microorganism and which can suppress the growth of unwanted microorganisms. The antimicrobial potency of antibiotics has made it a widely accepted drug in a large number of clinical applications. Penicillin erythromycin, tetracycline, cephalosporin are some of the popular antibiotics used in many diseases.

Accidental discovery are those in which scientists end up in discovering something else instead of the one he or she is looking for. It is also known as serendipity—the fact of finding interesting or valuable things by chance. There are many examples of such kind of discoveries in the history of science and technology. Penicillin was accidentally discovered by Alexander Fleming in 1928. He observed the mold (a type of fungus), *Penicillium notatum* hindered and stopped the basic growth of bacteria, *Staphylococcus* sp. Fleming received the Nobel Prize in medicine in 1945 for this discovery.

Types of Penicillin

Penicillin can be of different types depending on the composition of the alkyl group. Desired penicillin is synthesized by selecting appropriate precursor in the culture, thereby modifying the alkyl side chain. For example if phenyl acetic acid is used as precursor, benzyl penicillin is produced. Two most important types of penicillin are Penicillin F (Fentanyl penicillin) and Penicillin G (Benzyl penicillin, a narrow-spectrum antibiotic). (Figure 11.5) Penicillin is extracted as potassium or sodium salts.

Chemical nature of penicillin

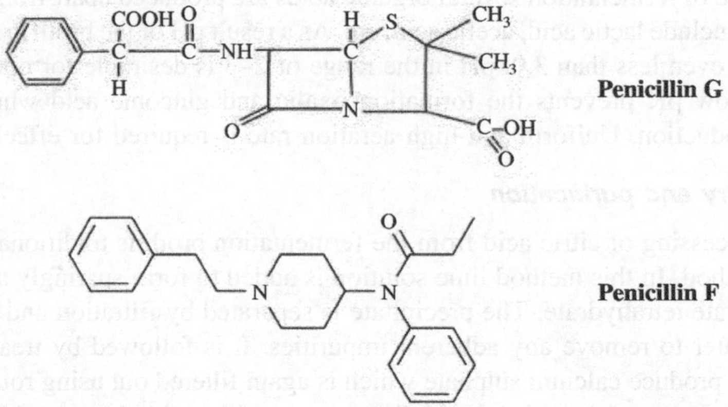

Penicillin G

Penicillin F

Figure 11.5 Chemical structure of Penicillin G and Penicillin F.

Penicillin Production Process

Penicillin is commercially produced by aerobic fermentation process in presence of the pure strain of *Penicillium chrysogenum* as the microorganism. Process flow diagram for the production of penicillin is shown in Figure 11.6. *P. chrysogenum* is a mold (fungus). Since penicillin is not a primary metabolite, it is formed during the secondary phase of growth.

The major steps of the process are as follows:

(i) Inoculum preparation and Culture propagation
(ii) Production by fermentation
(iii) Isolation and purification of products

Inoculum preparation and Culture propagation

The process of fermentative production of penicillin begins with the preparation of inoculum. Starting culture of *P. chrysogenum* from a slant is isolated and further subcultured. Medium for fermentation consisting of carbon, nitrogen and mineral sources is prepared and steam sterilized at 121°C temperature, 15 Psi pressure for a period of 15 min. A common fermentation medium for penicillin is composed of corn steep liquor solids (3.5%), lactose (3.5%), minerals ($CaCO_3$, K_2HPO_4, etc.), edible oil (0.25% as antifoaming agent), and the precursor. In addition, nitrogen source, i.e., yeast extract, lactic acid, inorganic ions, growth factors, etc. are also added depending on the product requirement.

100 ml medium with spores of *P. chrysogenum* strains is inoculated in Erlenmeyer flask. Suspension of highly sporulated working stocks in water is preferred. Theses spore are then added to flask containing wheat bran and nutrient solution for further sporulation. The flasks are incubated for 5–7 days at 25–27°C. Spores are then transferred to seed tank and incubated for 24–48 h at 24°C with aeration and agitation for sufficient mycelial growth. These mycelia can be used for production fermenter and is incubated at BOD incubator by placing them on a rotatory shaker.

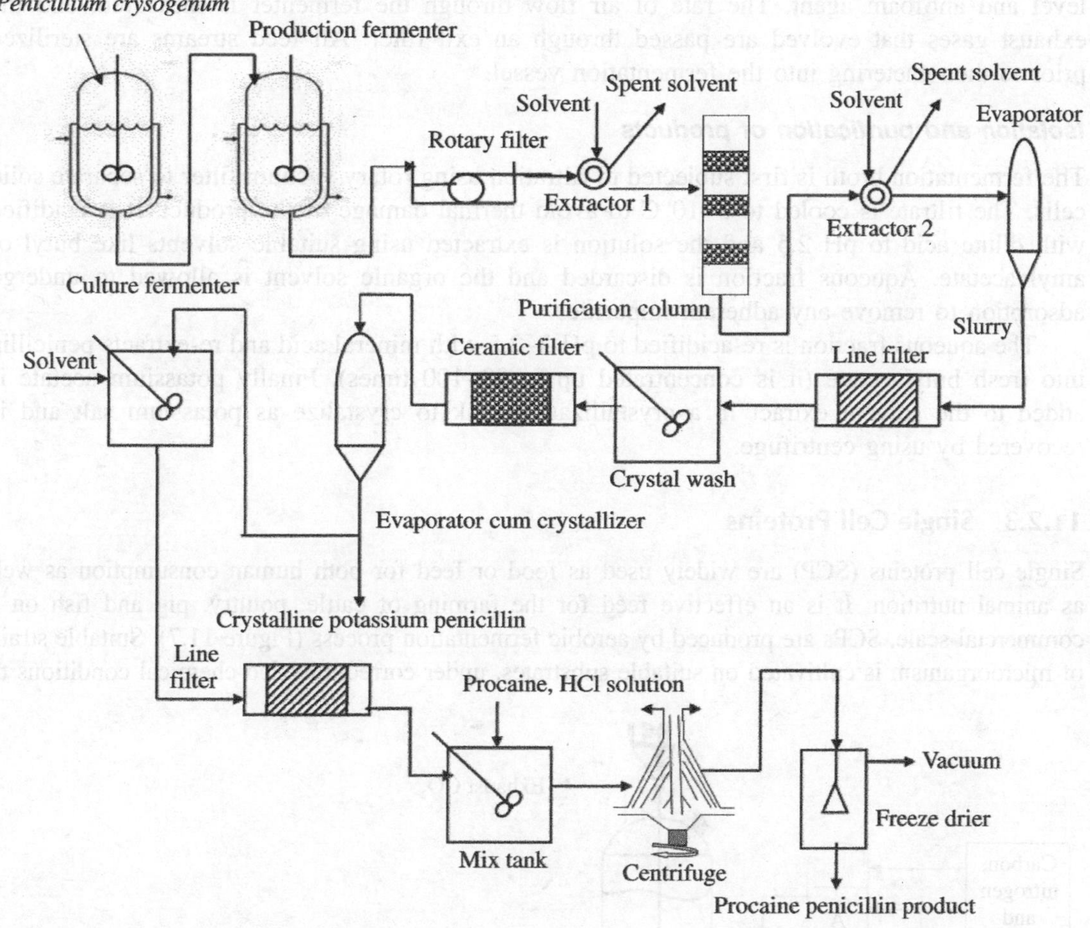

Figure 11.6 Flow diagram for the production of penicillin.

Production by fermentation

The process is aerobic fermentation in fed batch or batch operation. Stirred tank or air lift fermenters are used. pH of the fermentation media is set at 5.5 to 6.0 which may increase up to 7–7.5 (optimum). This might occur due to evolution of NH_3 gas and consumption of lactic acid. If the pH remains still higher buffer solution is added. Temperature of the fermenter is kept at 25–27°C. Since it is aerobic fermentation, sterile air is continuously introduced through an air-sparger. Agitation is provided with the help of an impeller agitator at 120–150 rpm to ensure proper mixing and homogeneity.

Temperature has a pronounced effect on the production of penicillin. Heat generation due to exothermicity and nutrient metabolism should not increase the temperature beyond 1-2°C. Increase in temperature may lead to thermal degradation of enzymes. Thus efficient and well-controlled cooling system is required in the form of either cooling coil or jacket. The fermenter is also equipped with several probes to measure and monitor pH, temperature, dissolved oxygen level and antifoam agent. The rate of air flow through the fermenter is measured, and the exhaust gases that evolved are passed through an exit filter. All feed streams are sterilized prior to their metering into the fermentation vessel.

Isolation and purification of products

The fermentation broth is first subjected to filtration using rotary vacuum filter to separate solid cells. The filtrate is cooled to 5–10°C to avoid thermal damage of the product. It is acidified with dilute acid to pH 2.5 and the solution is extracted using suitable solvents like butyl or amyl acetate. Aqueous fraction is discarded and the organic solvent is allowed to undergo adsorption to remove any adherent impurities.

The aqueous fraction is re-acidified to pH 2–2.5 with mineral acid and re-extracts penicillin into fresh butylacetate (it is concentrated up to 80–100 times). Finally potassium acetate is added to the solvent extract in a crystallization tank to crystalize as potassium salt and is recovered by using centrifuge.

11.2.3 Single Cell Proteins

Single cell proteins (SCP) are widely used as food or feed for both human consumption as well as animal nutrition. It is an effective feed for the farming of cattle, poultry, pig and fish on a commercial scale. SCPs are produced by aerobic fermentation process (Figure 11.7). Suitable strain of microorganism is cultivated on suitable substrates, under correct physico-chemical conditions to

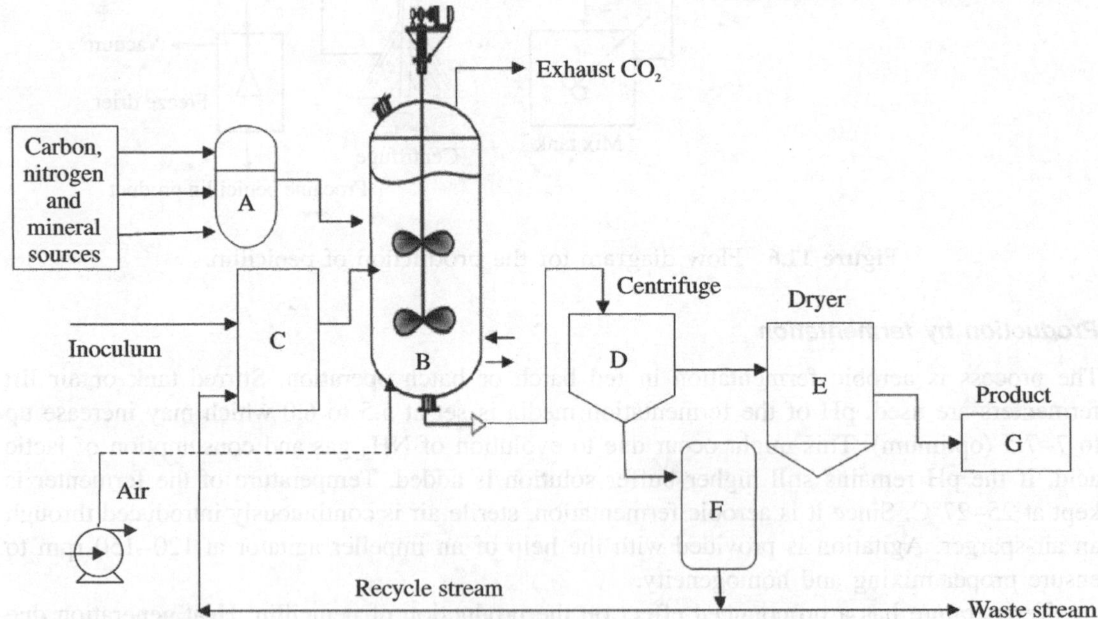

A: Pre-fermentation medium reservoir; B: Fermenter; C: Inoculum store; D: Cells separator; E: Spray drying F: Waste treatment G: grinding (product)

Figure 11. 7 Flow diagram for the production of single-cell-protein.

produce SCP. Traditional raw materials (carbon substrates) for SCP include mono and disaccharide, methane and higher alkanes, methanol, waste biomass, lignin, cellulosic raw materials, sulphite pulping liquor, dairy waste, etc. Important microorganisms that are used include basteria (such as *Lactobacillus* sp., *Bacillus subtilis*), fungi (such as *Aspergillus niger*, *Aspergillus oryzae*), yeast (such as *Candida tropicalis*, *Candida utilis*) and algae (*Spirulina*, *Chlorella*).

SCP has high protein content (crude protein contents of SCP from *Candida utilis* is 52% while that in plant proteins alfa-alfa is only 17%.) and exhibits high growth rate. Its production is ecofriendly, economic and energy efficient.

SCP production process is carried out in the similar route of fermentation followed by product separation. Appropriate inoculation is done with the selected strain of microorganism, which is grown using defined media containing carbon, nitrogen and mineral sources. Air-lift fermenters may be conveniently used. Aerobic fermentation for about 36–48 h produces biomass along with other liquid metabolic products. Broth is separated by centrifugation. Sometimes, it undergoes agglomeration and flocculation. Solid biomass is separated and washed thoroughly with water for several times and is then subjected to drying. Spray drier is used to get powdered form of the dry cell mass. Solid products are then grinded and packaged as per requirement.

11.2.4 Bakers' Yeast

Bakers' yeast is the common name of the microorganism *Saccharomyces cerevisiae* widely used in baking processes. The production of Bakers' yeast is carried out by aerobic cultivation of *S. cerevisiae* using carbohydrates as carbon source. Simple carbohydrates such as glucose, fructose, hydrolyzed starch or even molasses can be used as the carbon source in the growth of Baker's yeast. Aqueous ammonia, ammonium salts and urea are used as nitrogen sources. Apart from 'C' and 'N' sources, different minerals and trace elements are also used in the growth media. These include potassium hydrogen phosphate, soluble salts of Mg, Fe, Ca and some vitamins. Temperature and pH of the fermentation media are kept at 30°C and 6.5 respectively. High concentration of DO is maintained by continuous sparging of air through an orifice from the bottom of the bioreactor. Sometimes, ethanol is produced due to high concentration of carbohydrates however this needs to be controlled.

With the progress of reaction viscosity of the broth increases, thereby requiring efficient stirring. Vigorous aeration and agitation are required to maintain homogeneity of the broth. Multistage fermenter or fed-batch reactor can also be used for Baker's yeast production. Fermentation time is approximately 20–30 h. After fermentation gets over, the contents of the broth are centrifuged for the separation of yeast. Yeasts are separated in the form of light brown coloured cream, which can be stored and a part of it can even be used for inoculation of another batch. The cream is compressed and dried at 45°C. Moisture content of the yeast should be of 5–10%. It can also be extruded in and made as free flowing granules. Fine yeast particles can be obtained using spray or fluidized-bed dryers.

11.3 FLAVOR AND AROMA PRODUCTION

Flavor, fragrance and aroma have always drawn attraction of the consumers while selecting a particular food, consumables or even medicines. Due to growing awareness of using natural products there is a renewed interest in the natural routes of aroma production. It is mainly obtained via biochemical reaction either in solid state or submerged fermentation. Use of microorganisms to produce flavor and aroma during production of fermented food and drinks was a part of traditional knowledge by human beings.

Many of the fermentation processes yield secondary metabolites which are responsible for aroma production. Generally plant components are the major sources of natural flavor. These are naturally available volatile flavor compounds (VFC). So cultivation of plant cells can produce specific flavor and fragrance. Other than this, enzymatic biotransformation process can produce aroma compounds via fermentation. Different microorganisms have different metabolic pathways through which specific flavor components are generated. Sometimes a particular strain itself can be used to produce VFC *in situ* or in presence of suitable substrate. Table 11.1 gives a brief summary of the major types of flavoring compounds. Important types of flavoring compounds and their examples are briefly summarized in Table 11.1.

Table 11.1 Flavoring compounds and their examples

Major class compounds	Examples	Flavor
Lactones	Decalactone, hexalactone, octalactone, jasmine lactone	Flavors of dairy products, buttery, creamy, fruit nut, etc.
Carboxylate esters	Ethyl, butyl, isoamyl acetates, butyrates, hexanoates, isoamyl acetate	Sweet and fruity smell
Alcohols	Nonanol, Propanol, Furfuryl alcohol, 2-octanol	Rose like smell (2-phenyl ethanol)
Ketones	2,3 butadione, 2-butanone, 2-nonanone, 2-octanone	Pungent, cheesy and fruit fragrance
Aldehydes	2-hexanal, 2-methyl butanal, decanal, octanal	Vanilla, cherry and fruity flavor (benzaldehyde)
Pyrazine	3-methoxy-3-isopropyl pyrazine, 2-ethyl 3, 5 dimethyl pyrazine	Chocolate, almond, bell peppers, baked corn, etc.

Tissue culture of vanilla is another biotechnological route of producing vanilla flavor. Role of various phytohormones are important in such production processes. Compound such a phenyl propanoid can produce vanilla flavor metabolites via biotransformation route.

11.4 IMPORTANT ISOLATES FROM PLANTS OR ANIMALS

Many useful and important medicaments and pharmaceutically useful products are traditionally sourced from various plants or even animal organs. Solvent extractions with specific non-aqueous or aqueous solvents, supercritical fluids, ionic liquids, etc. are widely used for the manufacturing of different alkaloids. For example, quinine, an antidote of malaria is extracted from the bark of cinchona. Pain relievers such as morphine, codeine are extracted from opium by non-aqueous solvent. Decaffeination of coffee or waste tea leaves by supercritical extraction or by other methods can produce caffeine—a stimulant used for anti-depression. Reserpine, commonly used as tranquilizer is extracted from roots of *Rauwolfia serpentina*, a climbing shrub (sarpagandha), also known as Indian snake root by extraction with methanol or petroleum naphtha. Solvent extraction is carried out in batch or continuous extractors using concurrent or countercurrent methods. After acidification the solvent is removed by various methods like neutralization followed by evaporation. Generally a combination of evaporation crystallization, centrifugation and drying are employed.

Insulin

Insulin is an important hormone, which helps regulate the level of blood sugar in a human body. It is produced by the β-cells of pancreas. However, diabetic patients need to take regular shots of insulin as a medication. There is a growing demand of insulin as the cases of diabetes are rapidly increasing worldwide. Insulin is a high molecular weight (12000–36000) peptide molecule with 51 units of amino acids. Insulin is traditionally extracted from the pancreas glands of beef or hog. Refrigerated pancreas glands are minced, acidified in slurry, and is treated with alcohol in a grind tank. Acidulated alcohol extracts the gland slurry in a countercurrent extractor. The cake is processed in a fat-fry tank to separate the waste fat. The crude alcohol extract is neutralized with ammonia solution and is filtered in a precoat filter. The filtrate is re-acidified followed by evaporation and fat separation. The extract is then taken to a chill tank, filtered again and is fed into a second evaporator. The concentrated extract from the second evaporator is filtered and salting-out technique is applied. Salting out product is crystallized insulin. Nowadays, human insulin is synthesized by recombinant DNA technology using genetically modified microorganisms such as *E. coli, S. cerevisiae*, etc.

11.5 NUTRITIVE SWEETENERS (GLUCOSE/FRUCTOSE SYRUPS)

Nutritive and low calories sweeteners are produced by enzymatic transformation of starch into dextrose sugars. High fructose corn syrup (HFCS) is a popular sweetening agent used in beverage and confectionaries. HFCS has 1.3 times sweetening power than fructose. It contains nearly 55% fructose. The process starts with gelatinization of corn starch at high temperature (65°C) to convert it into dextrose. Starch undergoes enzymatic hydrolysis in presence of α-amylase—a thermostable enzyme, to produce glucose syrup. Glucose syrup can be further treated with glucose isomerase enzyme and fructose syrup is obtained. In this way, about 42% of glucose is converted to fructose. Immobilized enzyme reactors ensure facile conversion with more yields. Secondary refining of fructose syrup is accomplished by clarification, deionization and adsorption methods. Sometime, saccharification of liquefied starch slurry using glucoamylase can yield more dextrose syrup (glucose). It is filtered in rotary precoat filters to remove oils, proteins or other solid impurities (Figure 11.8).

11.6 BIOSAFETY IN INDUSTRIAL OPERATIONS

Microorganisms have been exploited in the production of industrial chemicals through microbial fermentation for a long time. The versatility of microorganisms is attributed to their ability to synthesize wide spectra of chemical compounds, which range from simple amino acids, fatty acid and monosaccharides to the very complex alkaloids, polysaccharides, nucleic acids and

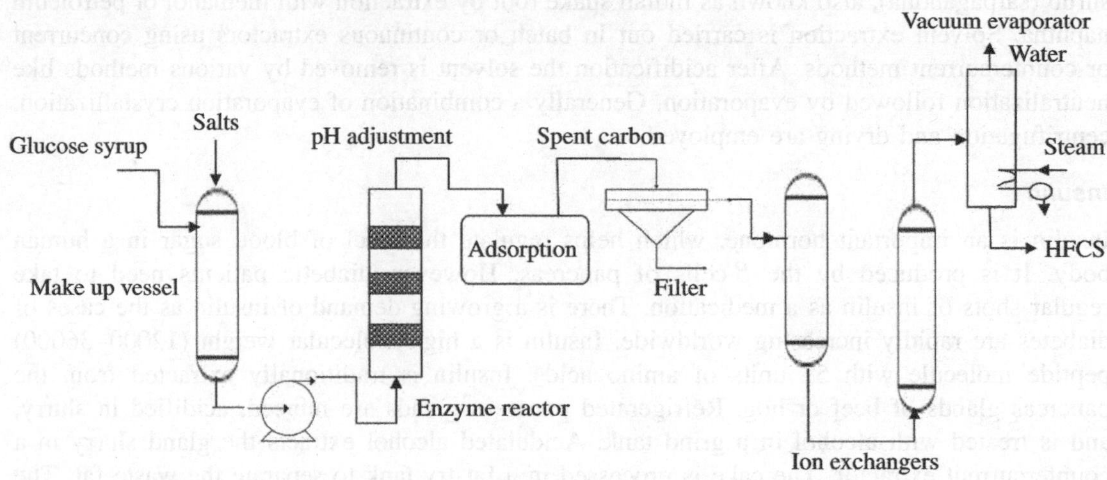

Figure 11.8 Flow diagram for the production of HFC.

proteins. Proteins also include enzymes that catalyze a variety of chemical reactions producing various microbial components and metabolites.

By recombinant DNA technology, the versatility of microorganisms have been further expanded through the creation of genetically modified organisms (GMO), to produce increased amounts of required metabolites, and also to express foreign genes to produce peptides and proteins belonging to higher forms, including the human.

Many countries have devised regulations and guidelines for the safe execution of bioprocesses for products development as well as the use of genetically modified microorganisms. Steps have also been taken towards international harmonization of safety considerations. The regulation of biotechnology and bioprocesses from research through product marketing and disposal, in many countries involves a series of laws. Such regulations govern not only the scientific aspects but also the health and safety of workers, intellectual property rights, the legal issues of clinical trials, the licensing of medicinal products, liability and trademarks.

The commercialization of pharmaceutical products derived from biotechnology is a complex process and is subject to a wide range of regulations. Some are more specific for biotechnology, while some are more general. Governments worldwide have passed legislations to control the production and use of chemicals, pharmaceuticals, medicinal products, etc. The manufacturer has to perform laboratory studies and to submit the results to governmental authorities for assessment of the product quality, as well as any potential hazard to human health.

11.7 GOOD MANUFACTURING PRACTICES (GMP)

All operations requiring the handling of 20 litre or more capacities for research or industrial purposes are deemed to be 'large scale'. For large-scale production of biomolecules from microorganisms by various routes, a set of measures or guidelines prescribe criteria for GMP. These include measures such as proper engineering for containment, quality control, personnel protection, etc.

The workplace or laboratory for bioprocess development should be free from smoke dust, vibration, or any other chemical interference. Temperature and humidity should be precisely controlled. The equipment and the instruments should be well calibrated as per standard protocols. Personal protective equipment such as garments, masks, gloves, goggles, boots, etc. should be provided. Presence of any viable microorganisms should be checked on regular basis. The users and technicians need to be well trained. In addition, the laboratory spaces should be easily cleaned and bench-tops are made to be resistant to chemicals and other disinfectants. The practice of mouth pipetting should be avoided. Production of aerosols should be minimized by proper containment of potential aerosol generating devices such as fermenter, centrifuge, filter, etc.

Physical contamination conditions (to minimize the potential for escape of viable organisms):

- Culture of viable organism shall be handled in a well-protected and sterile environment.
- Culture fluid along with any viable GMO must be deactivated after use by following standard operating procedure.
- Care should be taken while collecting samples for analysis during the course of reaction.
- Exhaust gases from a closed system shall be treated by HEPA filters or incineration.
- Emergency measures should be followed as per standard procedure.

11.8 QUALITY CONTROL OF BIOLOGICAL PRODUCTS

The proposed requirements for controlling the safety, purity and potency must be regarded as flexible and subject to change on a case by case basis, as experience of the manufacturer and use of such products increase. It should take into consideration molecular identity of product, biological potency, purity, toxicity, immunogenicity, and consistency of production. To commercialize a product based on the use of a GMO, a process must be developed for making that product, consistently and reproducibly, batch after batch. To make it profitable, the process has to be economical and has to be scaled up. Further, the final product, particularly when it is a therapeutic, has to meet stringent regulatory requirements, such as, it cannot contain traces of an undesired materials or contaminants. In essence, the objective of good practices in bioprocesses is essentially to ensure quality of the product, safety of the process and protection of consumers, workers and the environment.

KEY POINTS TO REMEMBER

- Yeasts are the most preferred microorganisms for ethanol fermentation Examples are *Sacchararomyces cerevisiae*, *Candida brassicae*, *Candida utilis,* etc.
- Yeast can transform hexose sugars into ethanol and CO_2 is produced as a byproduct.
- Low temperature vacuum distillation is used to produce up to 95% ethanol. Ethanol water mixture produces azeotrope which prevents further separation by normal distillation.
- The bacterial species, such as, *Lactobacillus*, *Sterptococcus* are common for fermentative production of lactic acid using various sugars as substrates.

- The lactic acid produced in the fermentation is neutralized usually by addition of calcium carbonate or calcium hydroxide slurry.
- Acetone and butanol are produced by anaerobic fermentation of glucose or lactose, in presence of *Clostridium acetobutylicum*.
- Commercial production of citric acid is carried out aerobically in a fermenter using *Aspergillus niger*.
- Molasses is a preferred raw material for citric acid production due to its cost effectiveness and appreciable sugar concentration.
- Penicillin is commercially produced by aerobic fermentation process in presence of the pure strain of *Penicillium chrysogenum*.
- SCP has high protein content (crude protein contents of SCP from *Candida utilis* is 52% while that in plant proteins alfa-alfa is only 17%.) and exhibits high growth rate.
- Many of the fermentation processes yield secondary metabolites which are responsible for aroma production. Generally plant components are the major sources of natural flavor.
- Enzymatic biotransformation process can also produce aroma compounds via fermentation. Different microorganisms have different metabolic pathways through which specific flavor components are generated.
- Human insulin is synthesized by recombinant DNA technology using genetically modified microorganisms such as *E. coli, S. cerevisiae*, etc.
- Starch undergoes enzymatic hydrolysis in presence of α-amylase—a thermostable enzyme, to produce glucose syrup.
- The workplace or laboratory for bioprocess development should be free from smoke dust, vibration, or any other chemical interference. Temperature and humidity should be precisely controlled.

? QUESTIONS FOR PRACTICE

Multiple Choice Questions

1. Which of the following is not a primary metabolic product?
   (a) Ethanol (b) Citric acid
   (c) Acetic acid (d) Penicillin
2. Which of the following is a downstream process?
   (a) Screening (b) Product recovery
   (c) Media formulation (d) Sterilization of media
3. Which of the following process is involved in the industrial production of ethanol from molasses?
   (a) Oxidation (b) Hydrolysis
   (c) Esterification (d) Anaerobic fermentation
4. Alcoholic fermentation is carried out by which of the following yeast?
   (a) *Lactobacillus* (b) *Escherichia coli*
   (c) *Bacillus* (d) *Saccharomyces cerevisiae*

5. Which of the following alkanols is commonly used as a fuel additives to reduce emission in gasoline engines?

(a) Ethanol (b) Methanol
(c) Butanol (d) Propanol

6. Which of the following microbial fermentations are anaerobic?

(a) Citric acid and propionic acid (b) Ethanol and acetone-butanol
(c) Penicilin and vitamin B-12 (d) Streptomycin and rifampicin.

7. During the production of alcohol by fermentation budding yeasts, oxygen supply is kept limited, why?

(a) Budding yeasts are obligate anaerobes and cannot tolerate oxygen
(b) Budding yeasts lose mitochondria in the absence of oxygen
(c) Budding yeast are facultative anaerobes.
(d) Alcohol is oxidized further in presence of oxygen

8. Bioethanol, often known as ____________ is a denatured alcohol.

(a) Ethylene (b) Methylated spirit
(c) Ethylene glycol (d) Methylene

9. The pH of molasses solution in the manufacture of ethanol should be

(a) 1–2 (b) 2–3
(c) 7–9 (d) 4–5

10. In lactic acid fermentation, the final electron acceptor is:

(a) Lactic acid (b) Pyruvate
(c) Oxygen (d) NAD

11. Single-cell-protein (SCP) refers to the production of

(a) Extracellular proteins (b) Metabolites
(c) Intracellular proteins extraction (d) Fermentation of waste products

12. The fungus ____________ is most commonly used for industrial production of citric acid.

(a) *Aspergillus niger* (b) *Escherichia coli*
(c) *Gluconobactor suboxidance* (d) *Lactobacillus pentosu*

13. ____________ is the predominant carbon source for citric acid production.

(a) Lactose (b) Yeast
(c) Glucose (d) Saccharin

14. In context of industrial microbiology, what is the purpose of using genetically modified microorganisms?

(a) Enhanced resistance to antibiotics
(b) Increased production of secondary metabolites
(c) Reduced formation efficiency
(d) Inhibition of enzyme activity

15. Which microorganism is commonly used in the industrial production of vinegar?

(a) *Saccharomyces cerevisiae* (b) *Acetobacter aceti*
(c) *Streptococcus thermophiles* (d) *Lactobacillus plantarum*

16. In the production of amino acids through microbial fermentation, what is the role of microorganisms?

(a) Synthesize amino acids from inorganic compounds
(b) Break down amino acids into simpler molecules
(c) Convert sugars into amino acids
(d) Inhibit amino acid production.

17. What is the primary function of enzyme invertase in industrial bioprocesses?

(a) Break down cellulose
(b) Convert sucrose into glucose and fructose
(c) Synthesize antibiotics
(d) Enhance product flavor.

18. Which microorganism is commonly used in the production of antibiotics?

(a) *Streptococcus pyogenes* (b) *Penicillium chrysogenum*
(c) *Escherichia coli* (d) *Candida albicans*

19. HFCS contains nearly ________ % of fructose sugar.

(a) 45 (b) 35
(c) 55 (d) 65

20. Insulin is traditionally extracted from the ____________ of beef or hog

(a) Liver (b) Pancreas
(c) Thyroid gland (d) Pituitary gland

21. Which of the following is the most common source of SCP?

(a) Multi-cellular yeast (b) Single-celled yeast
(c) Unicellular algae (d) Unicellular bacteria

22. Which among the following is not used as raw material for alcohol production?

(a) Corn (b) Molasses
(c) Whey (d) Grapes

23. The leavening or rising of dough is due to which of the following gases?

(a) Oxygen (b) Carbon dioxide
(c) Hydrogen (d) Sulphur dioxide

24. Various measures taken to prevent any risks to normal organisms from transgenic organisms are known as

(a) Biosafety (b) Biopatent
(c) Biodiversity (d) Biopiracy

25. Which of the following metabolite does not impart flavor to beer?

(a) Esters (b) Thiols
(c) Carbonyls (d) Alcohols

26. Which of the following is used to give banana flavor?

(a) Isoamyl acetate (b) Hydrogen sulphide
(c) Dimethyl sulphate (d) Diacetyl

Answer Key

1. (d)	2. (b)	3. (d)	4. (d)	5. (a)	6. (b)	7. (c)	8. (b)	9. (d)	10. (b)
11. (d)	12. (a)	13. (c)	14. (b)	15. (b)	16. (c)	17. (b)	18. (b)	19. (c)	20. (b)
21. (b)	22. (c)	23. (b)	24. (a)	25. (b)	26. (a)				

Descriptive Questions

1. Name a few microorganisms used for ethanol fermentation.
2. Can you use whey directly as substrate for ethanol fermentation? How can it be made usable?
3. Acid hydrolysis of starch is not recommended. Why?
4. Why cellulosic materials are not preferred feedstock for fermentation.
5. 'Glucose concentration in feed for alcohol fermentation should not be more than 100 g/L' Explain.
6. Why is it difficult to increase the concentration of ethanol beyond 95% during distillation?
7. Cite example of a third component that is used to break the alcohol-water azeotrope during azeotropic distillation.
8. Cite examples of some microorganisms used in lactic acid fermentation.
9. State some applications of lactic acid.
10. Which microorganisms are used to produce citric acid from hydrocarbon and carbohydrate substrates?
11. Which strain of mold was originally used by Alexander Fleming for the production of penicillin?
12. 'Continuous addition of ammonium sulphate is required during synthesis of penicillin' –why?
13. State the advantages of corn steep liquor as nitrogen source in penicillin production.
14. Name the type of filter that is most commonly used for penicillin recovery process.
15. Explain the effect of pH during solvent extraction of penicillin.
16. State the types of HFCS based on their fructose contents.
17. Which enzyme is used for gelatinization of corn starch during HFCS production?
18. Which enzyme is used for the conversion of glucose to fructose?
19. Why is it recommended to perform fermentation at low oxygen concentration and what happens to the microorganism after it has secreted our desired product? Use citric acid as an example.
20. How is Cellular Fermentation used at the industrial scale? Why is it important to know how this process functions to optimize its use in careers?
    (a) Why do some microbes specialize to use a different food source than the surrounding microbes in their environment?
    (b) Why do you use a pH indicator to gauge fermentation?
    (c) Why would you need to perform multiple tests in order to identify an unknown bacterium, instead of a single biochemical test?
21. When we talk about bacteria breaking down things (sewage/food) are we referring to respiration and fermentation or not?

22. What are notable examples of industrially implemented continuous fermentation?

23. Fermentation is anaerobic respiration. There are two types of fermentation.
(i) What are the two types of fermentation?
(ii) What are the products of each type of fermentation?
(iii) What kinds of organisms can do each type of fermentation?

24. A 20-year-old bottle of red wine is opened and found to taste like vinegar on the first sip. Is this possible? Give a plausible explanation.

25. Using the balanced equation for fermentation answer the following question.

$$C_6H_{12}O_6(aq) \longrightarrow 2C_2H_6O(aq) + 2CO_2(g)$$

How many grams of CO_2 are formed from 0.10 mol of glucose?

12 Biological Wastewater Treatment

12.1 INTRODUCTION

Water that is generated as the effluent from any chemical, food processing, bio-processing and allied industrial operations is termed as wastewater. There was a time when the effluent water was discharged into the natural environment without proper treatment or even in untreated form. However, with growing public awareness and stringent environmental norms efficient treatment of wastewater has become a mandatory operation for any industry or production house. Wastewater containing toxic and harmful chemicals, pathogenic cells, heavy metals, nutrients, soluble and suspended impurities not only endangers the aquatic flora and fauna, but also adversely affects the human health and environment.

Wastewater generated from bioprocess industry contains different types of contaminants. Important contaminants of wastewater and their examples are presented in Table 12.1.

Table 12.1 Major contaminants of wastewater from bioprocess industries

Major types of contaminants	Examples
Suspended solids	Soil, silt, fecal matter, bacteria, plankton, cell debris.
Biodegradable organics	Remains of dead organisms, biopolymers, proteins, polysaccharides, etc.
Pathogens and parasites	*Giardia* sp., *Entamoeba histolytica*, *Isospora* sp., etc., tapeworm, whipworm, etc.
Refractory organics	Detergents, phenolic compounds and substituted phenols, sequestering agents like EDTA.
Nutrients	*N* and *P* compounds from human waste, food, etc.
Priority pollutants	Toxic pollutants (as defined in EPA), benzene, chloroform, chlorophenol, toluene, etc.
Heavy metals	Pb, Hg, Cu, As, etc.
Dissolved organics	Urea, nitrates, phosphates.

Basic objectives of biological wastewater treatment are as follows:

(i) To reduce the biochemical oxygen demand (BOD) by reducing the organic contents of water
(ii) To remove nutrients mostly N and P, so that pollution of receiving surface water is reduced.
(iii) Destruction of pathogens and other microorganisms

12.2 OVERVIEW OF GENERAL PARAMETERS OF WASTEWATER

We will now define some important parameter of wastewater characterization and briefly explain their effects.

BOD (Biochemical Oxygen Demand)

BOD is the measure of the quantity of oxygen utilized by active microorganisms for the degradation of organic matter present in wastewater at a certain temperature (20°C) over specific time duration. Since the incubation period of the microbial activity is 5 days, it is generally denoted as BOD5. It is expressed in mg/L. BOD gives a direct indication of oxygen requirement and indirect indication of biodegradable organics. If a water body contains a large quantity of decayable organic matter its DO level could drop more (as high as 90%) and this represents high BOD. On the other hand, if a water body contains a small amount of decaying organics, there would be less reduction of DO level (say 10%) and this represents low BOD. BOD is measured by Winkler method involving titrimetric analysis. BOD values of common water samples are given in Table 12.2.

Table 12.2 Typical values of BOD for various water samples

Range of BOD (mg/L)	Type of water
1–2	Clean water
3–5	Moderately clean
6–9	Polluted water
>10	Very poor quality

COD (Chemical Oxygen Demand)

COD is a measure of oxygen required for the chemical oxidation of both organic and inorganic constituents present in a sample of wastewater. A strong oxidizing agent such as potassium dichromate is used. It is expressed in mg/L. The values of COD are generally higher than BOD for any wastewater sample.

TDS (Total Dissolved Solids)

Total dissolved solids indicate the aggregate of organic and inorganic materials dissolved in a sample of wastewater or in drinking water. These solids may include various soluble salts, minerals, metals, silt, etc. having sizes ranging from molecular to micro-granular. It is expressed in mg/L (ppm).

TSS (Total suspended solids)

Total suspended solids (TSS) are the total dry weight of suspended particulate components present in a sample of wastewater. These particles are trapped in a filter of mesh 0.45–0.5 micron. It is also known as non-filterable residue.

Total N

It is the aggregate of amounts of different forms of nitrogen present in the wastewater sample. Various forms of nitrogen include ammonia, nitrite, nitrate and organic nitrogen. This is also called Total Kjeldahl Nitrogen (TKN).

Total P

It is the summation of amounts of all forms of phosphorus present in wastewater. It may be in soluble or in particulate form.

TOC

It is a direct measure of total amount of carbonaceous organic matter. It is determined by the chemical, thermal or photocatalytic oxidation of organic carbon into carbon dioxide. TOC analysis can give some more information other than TOC. These are the total carbon (TC), inorganic carbon (IC), purgeable and non-purgeable organic carbon.

pH

It is indicative of acidity or alkalinity of the wastewater. It is defined as the negative logarithm of hydrogen ion concentration of any aqueous solution.

Alkalinity

Alkalinity indicates the buffering capacity of the wastewater sample. It implies the resistance to pH fluctuation. The total alkalinity is expressed in terms of alkalinity due to carbonate-bicarbonate and hydroxide. Predominant form of alkalinity is bicarbonate alkalinity. The unit of measurement for alkalinity is mg/L $CaCO_3$.

12.3 GENERAL TREATMENT STRATEGY OF WASTEWATER

There are three main strategies of wastewater treatment:

(i) *Primary Treatment*: In the primary treatment settleable bulky solids are removed by sedimentation, comminution, and grit removal. Mostly physical separation methods are employed using screens, grit chamber, etc. It is the initial step of treatment. Principal purposes of primary treatment of wastewater are:

- Reduction of strength of sewage from 30% to 50%
- Maximum solid removal
- 30–35% BOD reduction

Primary treatment aims to make the water suitable for subsequent operations.

(ii) *Secondary Treatment*: It removes soluble organic matter. Aerobic bacteria used in this phase tend to break down complex molecules. The waste and microorganisms are both found in the sludge, and bacterial composition removes both the remaining solids and nutrients. Secondary treatment relies on naturally occurring biological processes, which means that the oxygen level in the wastewater varies throughout and does not remain constant. Depending on the stage, bacteria could be either aerobic (need oxygen) or anaerobic (do not require oxygen). Anaerobic treatment is used for the decomposition of un-oxidized compounds from aerobic process.

(iii) *Tertiary Treatment*: In the tertiary treatment of wastewater, different treatment techniques are used depending on the variety of contaminants. This is done to reduce the load of total suspended solids, total dissolved solids and residual organics present after secondary treatment. In addition, some particular organic and inorganic ingredients from industrial effluent (like phosphate, sulphate, refractory organics, etc.), nutrients, pathogens are also removed to make it suitable for reuse. Physical separation processes like adsorption, deep bed filtration or even membrane based separations are also employed either as a standalone system or in combination.

12.4 AEROBIC PROCESSES

12.4.1 Activated Sludge Process

Activated sludge process involves treatment with an aerobic suspended growth of mixed microbial population that oxidizes organic compounds into CO_2, H_2O, and NH_3 with production of biomass. It is fundamentally a well-mixed aerator or a continuous stirred-tank reactor coupled with a settling or sedimentation tank. In aeration tanks, air is introduced to facilitate aerobic oxidation using various types of agitation and air sparging systems. This is followed by a sedimentation tank where microbial flocks (known as sludge) are produced. A portion of the sludge is recycled back to maintain feed to biomass ratio. Maintaining a proper recycle ratio is required to maximize the BOD reduction rate. A schematic of the process is shown in Figure 12.1 In addition, mixed liquor suspended solids (MLSS) concentration and hydraulic retention time (HRT) are two important parameters.

MLSS: It is the total quantity of influent materials in the aeration tank along with mixed microorganisms and cell debris.

HRT: It is the average residence time of mixed liquor in the aeration tank. HRT is the reciprcal of dilution rate.

$\text{HRT} = \dfrac{1}{D} = \dfrac{V}{Q}$, where V is the volume of the aerator, Q flow rate and D dilution rate.

Daily carbonaceous oxygen requirement (kg/day) in an activated sludge process can be given by Eq. 12.1:

$$0.75\ \text{BOD} + 0.048\ \text{MLSS} \times V \quad (12.1)$$

Air requirement for prolonged aeration in activated sludge system can be written as Eq. 12.2.

$$\text{BOD} + 4.34\ N_H - 2.85 N_T + 0.024\ \text{MLSS} \times V \times r \times 1.07^{T-20} \quad (12.2)$$

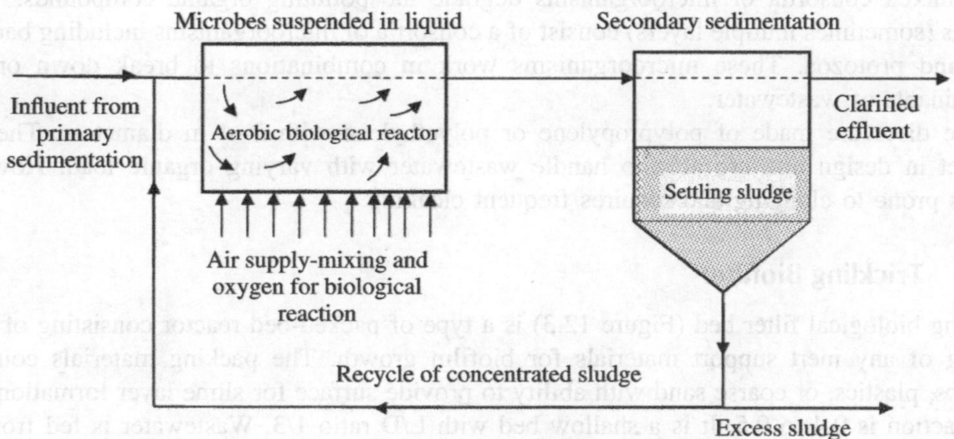

Figure 12.1 Schematic diagram of activated sludge process.

V is the volume (m^3) of aeration tank, N_H: ammoniacal nitrogen removed in kg per day; N_T: total nitrogen removed in kg per day; r: endogenous respiration rate (taken as 3.9 mgO_2/g MLSS/h); T: temperature.

12.4.2 Rotating Biological Contactor

Rotating biological contactors (RBC) use the technique of attached growth of microorganism film on the surface of the rotating disc that comes in contact with wastewater under treatment (Figure 12.2).

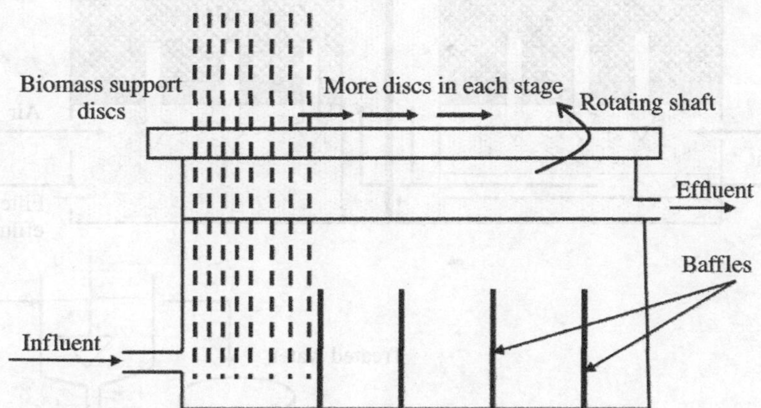

Figure 12.2 Schematic diagram of a rotating biological contactor [Adapted from P.F Stanbury, A Whitaker, S J Hall, Principles of Fermentation Technology (3rd Edition) Elsevier (2017)].

The slow rotational speed ensures efficient contact of the biofilms with the wastewater thereby facilitating its reclamation. Nutrients and dissolved oxygen diffuse through the biofilm and a mixed consortia of microorganisms degrade the polluting organic compounds. These biofilms (sometimes mutiple layers) consist of a consortia of microorganisms including bacterai, fungi and protozoa. These microorganisms work in combinations to break down organic contaminants of wastewater.

The discs are made of polypropylene or polyethylene with 2–4 m diameters. They are compact in design and are able to handle wastewater with varying organic load. However, RBC is prone to clogging and requires frequent cleaning.

12.4.3 Trickling Biofilter

Trickling biological filter bed (Figure 12.3) is a type of packed-bed reactor consisting of loose packing of any inert support materials for biofilm growth. The packing materials could be ceramics, plastics, or coarse sand with ability to provide surface for slime layer formation. Bed void fraction is 0.4 to 0.5. It is a shallow bed with L/D ratio 1/3. Wastewater is fed from the top of the bed while aeration is provided from the bottoms.

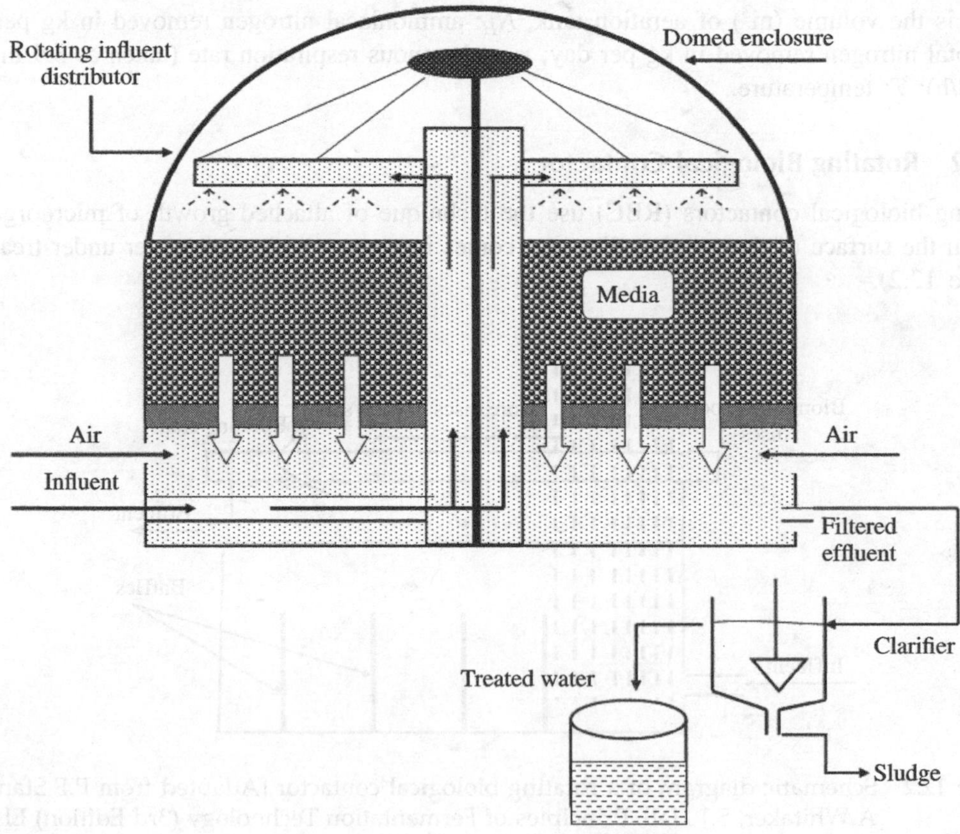

Figure 12.3 A schematic of trickling bed biofilter.

12.4.4 Oxidation Pond

These are small lagoon type reactors, 1–1.5 in depth, having similarity with natural water bodies. Symbiotic growth of bacteria and algae takes place in oxidation ponds. The ponds are provided with inlets and outlets. These also require the presence of oxygen and sunlight. Initially organic waste is oxidized in presence of bacteria. Mainly CO_2, water and ammonia are produced. Algal growth also occurs in presence of sunlight. These algae consume CO_2 and can reduce inorganic waste containing nitrogen or phosphorus compounds. Algal-Bacterial interaction in an aerobic lagoon is schematically presented in Figure 12.4. Since both oxidation and reduction type of reactions takes place it is also known as redox pond. Oxidation ponds are simple, economical and can reduce the BOD up to 90% naturally. However, it requires large land area. Foul smell, sippage and mosquito problems may occur.

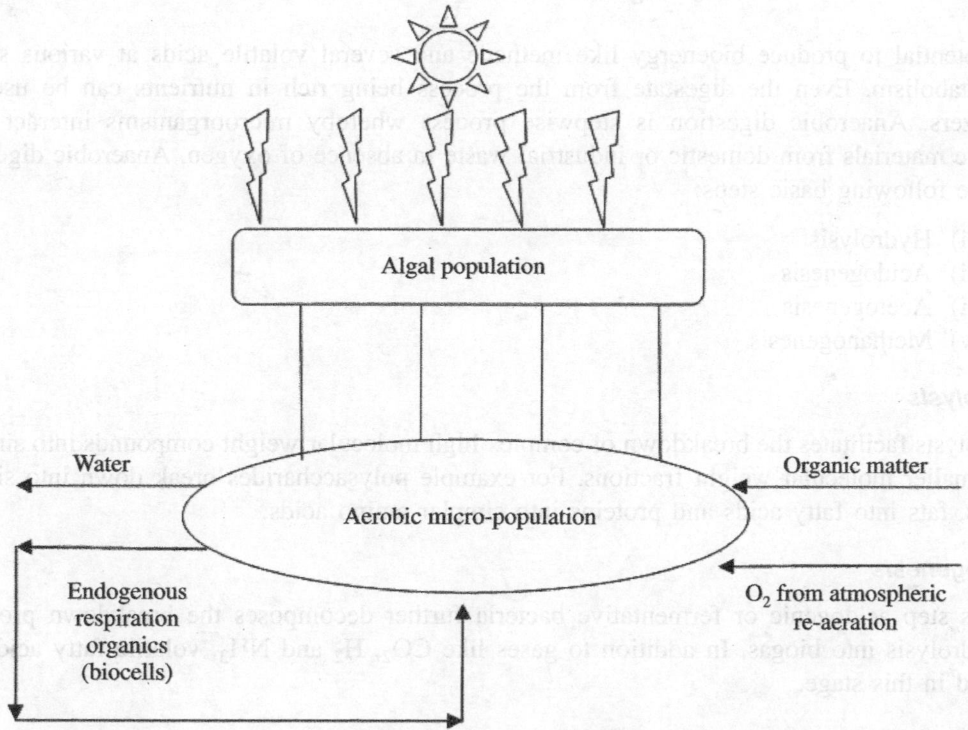

Figure 12.4 Algal-Bacterial interaction in an aerobic lagoon.

12.5 ANAEROBIC PROCESS

Solid waste and sludge generated during different aerobic processes undergo anaerobic digestion for further treatment. Feed for anaerobic digestion includes manure, municipal solid waste, food waste, oils and greases, etc. Anaerobic ponds (Figure 12.5) are deep shallow ponds that exclude oxygen and encourage the growth of bacteria, which break down the effluent. This is a very slow process consuming several days and months. However, anaerobic process has

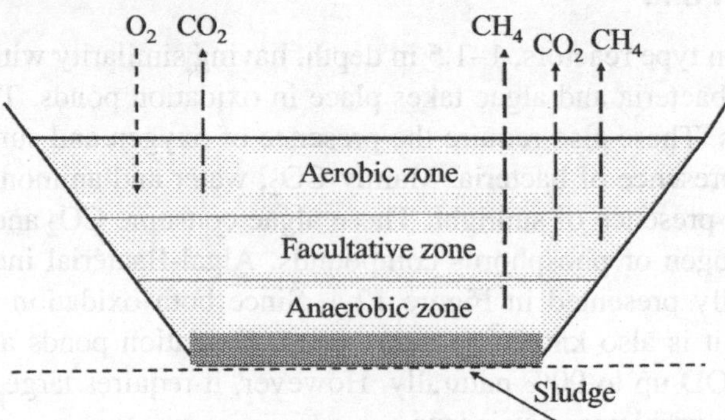

Figure 12.5 Anaerobic pond.

the potential to produce bioenergy like methane and several volatile acids at various stages of metabolism. Even the digestate from the process being rich in nutrients can be used as fertilizers. Anaerobic digestion is stepwise process whereby microorganisms interact with organic materials from domestic or industrial waste in absence of oxygen. Anaerobic digestion has the following basic steps:

(i) Hydrolysis
(ii) Acidogenesis
(iii) Acetogenesis
(iv) Methanogenesis

Hydrolysis

Hydrolysis facilitates the breakdown of complex high molecular weight compounds into simpler and smaller molecular weight fractions. For example polysaccharides break down into simple sugars, fats into fatty acids and proteins into simpler amino acids.

Acidogenesis

In this step acidogenic or fermentative bacteria further decomposes the breakdown products of hydrolysis into biogas. In addition to gases like CO_2, H_2 and NH_3, volatile fatty acids are formed in this stage.

Acetogenesis

The third stage of anaerobic digestion is acetogenesis. Here, acetogenic bacteria transform organic acids into acetic acid along with carbon dioxide and hydrogen.

Methanogenesis

The last stage of anaerobic digestion is the biological process of methanogenesis. Intermediate products from the previous stage are the feed for methanogens which transform them into methane, carbon dioxide and water. The overall equation can be stated as:

$$C_6H_{12}O_6 \rightarrow 3CO_2 + 3CH_4$$

12.6 BIOLOGICAL NITRIFICATION AND DENITRIFICATION

Nitrification and denitrification are two important steps carried out in sequence to remove nitrogenous compounds from wastewater. However, both these reactions are conducted in separate reactors following specific physico-chemical conditions. Nitrogen removal can be accomplished both aerobically and anaerobically. Amino acids and proteins are oxidized biologically to ammonia, which is further converted to nitrite and nitrate by *Nitrosomonas* and *Nitrobacter* respectively.

$$NH_4^+ + 3/2\ O_2 \xrightarrow{\textit{Nitrosomonas}} NO_2^- + H_2O + 2H^+$$

$$NO_2 + ½\ O_2 \xrightarrow{\textit{Nitrobacter}} NO_3^-$$

Overall reaction for aerobic nitrification:

$$NH_4^+ + 2O_2 \rightarrow NO_3^- + 2H^+ + 2H_2O$$

Aerobic nitrification takes place during BOD reduction in presence of these microorganisms. Under anaerobic conditions two types of nitrate reduction take place—assimilatory and dissimilatory reduction. In assimilatory reduction some nitrogen is converted to ammonia to become a part of cell mass. In dissimilatory reduction molecular nitrogen is the final product. Examples of microbes used in denitrification process include the species of *Pseudomonas*, *Acaligens*, *Arthobacter*, etc. At a very low concentration of oxygen, microorganism uses nitrate as the electron acceptor for carrying our denitrification.

Activated sludge, packed-bed, up flow anaerobic sludge blanket (UASB) type of reactors is widely used in anaerobic denitrification process.

Phosphate removal from wastewater is accomplished both under aerobic and anaerobic conditions. Some microorganisms consume phosphate by assimilation along with acetate and fatty acids to produce polyhydroxybutyrate (PHB). *Acinetobacter calcoaceticus* is a widely used microbe for this purpose. This bacterium can store polyphosphate in its cellular mass. PhoStrip is another important process which combines BOD as well as phosphate removal. It consists of both aerobic and anaerobic stages with provision of sludge recycle.

Refractory organic compounds in wastewater include those compounds which are difficult to degrade. Examples of such compounds are aromatic hydrocarbon, alkanes, substituted phenolic compounds, etc. Advanced oxidation processes are generally employed for the reclamation of such compounds.

KEY POINTS TO REMEMBER

- The main objectives of wastewater treatment are to reduce biochemical oxygen demand (BOD) by reducing the organic contents of water.
- If a water body contains a large quantity of degradable organic matter its DO level could drop more (as high as 90%) and this represents high BOD.
- Total dissolved solids indicate the aggregate of organic and inorganic materials dissolved in a sample of wastewater or in drinking water.

- Primary treatment aims to make the water suitable for subsequent operations.
- Secondary treatment removes soluble organic matter. Aerobic bacteria used in this phase tend to break down complex molecules.
- Tertiary treatment is done to reduce the load of total suspended solids, total dissolved solids and residual organics present after secondary treatment.
- Activated sludge process involves treatment with an aerobic suspended growth of mixed microbial population that oxidizes organic compounds into CO_2, H_2O, and NH_3 with production of biomass.
- Anaerobic digestion consists of four major steps namely hydrolysis, acetogenesis and methanogenesis.
- Nitrification and denitrification are two important steps carried out in sequence to remove nitrogenous compounds from wastewater.

QUESTIONS FOR PRACTICE

Multiple Choice Questions

1. BOD measures the
   (a) Amount of bacteria in a given sample of water
   (b) Amount of algae being decomposed in a given sample
   (c) Amount of oxygen bacteria need to decompose organic matter in a given sample
   (d) Amount of oxygen bacteria need to decompose organic matter in a given sample in a fixed period of time
2. Coliforms are used as indicator organisms of sewage pollution because they
   (a) Are pathogens (b) Ferment lactose
   (c) Are abundant in human intestine (d) Grow within 24 hours
3. What is the primary role of microorganisms in the biodegradation of organic pollutants?
   (a) Production of antibiotics
   (b) Synthesis of biofuels
   (c) Break down of complex organic compounds
   (d) Inhibition of microbial growth
4. Effective preliminary treatment can also be used to remove the
   (a) Grit and rag (b) Carbon dioxide
   (c) Nutrients (d) Oxygen
5. Which of the following should be provided in the case where aeration is absent?
   (a) Screening devices (b) Mechanical mixers
   (c) Grit removers (d) Sedimentation tank
6. Which process is employed to gain sufficient head for the wastewater?
   (a) Screening (b) Pumping
   (c) Compressing (d) Aerating
7. What is the most commonly used coagulant?
   (a) Alum (b) Ferric sulphate
   (c) Limestone (d) Coal

8. What is the intermediate zone composed of in aerobic-anaerobic ponds?
(a) Algae (b) Aerobic bacteria
(c) Facultative bacteria (d) Organic solids

9. ______________ is a process which involves further removal of nitrogen.
(a) Nitrification (b) Denitrification
(c) Ammonification (d) Reduction

10. In very first stage of decomposition of organic matter in sewage
(a) Nitrites are formed (b) Nitrates are formed
(c) Carbon dioxide is formed (d) Ammonia is formed

11. Removal of oil and grease from sewage is known
(a) Screening (b) Skimming
(c) Filtration (d) Flocculation

12. Oxidation process results in the formation of
(a) Carbon dioxide (b) Nitrates
(c) Sulphates (d) All of these

13. The standard BOD of water is taken for
(a) 1 day (b) 2 days
(c) 3 days (d) 5 days

14. Trickling filter is the example of
(a) Aerobic process (b) Anaerobic process
(c) Both (a) and (b) (d) Neither (a) nor (b)

15. The sewage treatment in septic tank is due to
(a) Anaerobic decomposition (b) Aerobic decomposition
(c) Parasitic decomposition (d) None of these

16. Why is the sludge aerated?
(a) To avoid bacterial growth (b) To increase bacterial growth
(c) To maintain pH (d) To maintain temperature

17. What are the three ingredients of activated sludge systems?
(a) Cells, sewage and oxygen (b) Cells, sewage and nitrogen
(c) Solids, sewage and oxygen (d) Solids, water and oxygen

18. Trickling filter is an example of
(a) Attached growth process (b) Suspended growth process
(c) Hybrid process (d) All of these

19. Which of the following biological processes can produce alkalinity?
(a) Carbonaceous BOD removal (b) Denitrification
(c) Nitrification (d) Phosphorus removal by chemical addition with ferrous chloride

20. What is the means used for the mixing of the sludge in UASB?
(a) Magnetic stirrer (b) Electrically driven stirrer
(c) Mechanical mixing (d) By means of bubbles

21. What is the rate limiting step in the anaerobic digestion?
(a) Methanogenesis (b) Hydrolysis
(c) Acidification (d) Biogas production

22. What is the temperature which should not be exceeded for the reaction in UASB reactor?
(a) 20°C (b) 32°C
(c) 38°C (d) 45°C

23. The pathogens in digestate are
(a) Highly active (b) Inactive
(c) Partially inactive (d) Neutral

24. In which of the following treatment involve oxidation of the organic constituents of wastewater?
(a) Primary treatment (b) Secondary treatment
(c) Tertiary treatment (d) Advanced treatment

25. Which of the following gases are produced in large amount during sludge digestion?
(a) Methane (b) Carbon dioxide
(c) Hydrogen (d) Nitrogen

Answer Key

1. (d)	2. (c)	3. (c)	4. (a)	5. (b)	6. (b)	7. (a)	8. (c)	9. (b)	10. (d)
11. (b)	12. (d)	13. (d)	14. (a)	15. (a)	16. (b)	17. (a)	18. (a)	19. (a)	20. (d)
21. (a)	22. (c)	23. (c)	24. (b)	25. (a)					

Descriptive Questions

1. Why COD values are greater than BOD values?
2. What is the difference between suspended-growth and attached-growth processes?
3. What does substrate refer to in biological treatment context?
4. How can one determine whether or not membranes are fouled?
5. What does primary treatment refer to?
6. Why are flocs important in biological treatment of wastewater?
7. List and differentiate different types of aerobic attached growth bioreactors and discuss when and under what conditions each of these bioreactors are preferred?
8. State the reasons why the treated effluent is recycled in trickling filters? When there is no recycling of the treated effluent, a trickling filter is reducing BOD of wastewater from 120 mg/L to 50 mg/L. For achieving the treated effluent BOD of 30 mg/L what should be the recycle ratio of the treated effluent?
9. List the characteristics of the wastewater that may be needed in the design, operation and control of a biological treatment unit used for the removal of biodegradable organic matter?
10. Indicate the forms of phosphorus found in water/wastewater samples and indicate which of these forms is considered as available phosphorus.

Suggested Readings

Aiba, S., Humphrey, A.E., and Millis N.F. (1973), Biochemical engineering (2nd edition), Academic Press, New York.

Arnold L. Demain and Julian, Mannual of Industrial Microbiology and Biotechnology, 2nd edition.

Baker F.J., Bailey, J.E., and D.F. Ollis (1986), Biochemical Engineering Fundamentals, 2nd edition, McGraw-Hill, New York.

Baker, F.J., Bacteriological Techniques.

Baldwin W.W. and Wegener W.S. (1986), The kinetics of bacterial cell elongation as measured using the large rod Lineola longa, J. Bacterid.

Benuzzia, D.A. and Segovia, R.F. (1996), Effect of copper concentration on citric acid productivity by an Aspergillus niger strain, Appl. Biochem. Biotechnol, 61, 393–397.

Bonatelli Jr, R. and Azevedo, J.L., (1983), Improved reproducibility of citric acid production in Aspergillus niger Biotechnol.Lett, 4, p. 761–766.

Casida L.E., Industrial Microbiology, New Age International (P) Ltd. publications.

Chaudary, K., Ethiraj, S. Lakshminarayana, and Tauro, P. (1978), Citric acid production from Indian Cane molasses by Aspergillus niger under solid state fermentation conditions,
J. Ferment. Technol, 56, p. 554–557.

Cooper S. (1991), Bacterial Growth and Division, Academic Press, San Diego.

Coulson J.M., and Richardson J.F., (1994), Chemical Engineering, Vol-3, 3rd edition, Elsevier, Great Britain.

Crueger W. and Crueger A., A Text Book of Industrial Microbiology, 2nd edition.

David L. Nelson and Michael Cox, Lehninger Principles of Biochemistry: International edition Paperback–W.H. Freeman; Sixth edition.

E. Davies, Ronald M. Atlas, Gerald Cohen, Charles L. Hershberger, Wei-Shou Hu, David H. Sherman,

Ecker R.E., and Kokaisl G. (1969), Synthesis of protein, ribonucleic acid and ribosomes by individual bacterial cells in balanced growth, J. Bacteriol. 98: 1219.

Garrett, H., Reginald and Charles Grisham, Biochemistry Boston, Twayne Publishers, 2008.

Gunasekaran, Introduction to Microbial Techniques.

Halme A., Kuismin R., and Korteniemi M. (1986), A method to consider delayed laboratory analysis in state and parameter estimation of bioreactor, In: Johnson A (ed) Proceedings of the 1st IFAC Symposium on Modelling and Control of Biotechnological Processes, Pergamon Press, Oxford.

Hambor J. (2012), Bioreactor Design and Bioprocess Controls for Industrialized Cell Processing, BioProcess International, 10(6): p. 22–33.

Ho, C.S. and Oldshue, J.Y. (1987), Biotechnology Processes Scale-up and Mixing, American Institute of Chemical Engineers, New York.

James G. Chappuccino and Natalie Sherman, International student edition: Microbiology—A laboratory Manual, 4th edition.

Jana S., and Deb J.K. (2005), Strategies for efficient production of heterologous proteins in Escherichia coli. Appl Microbiol Biotechnol, 67: p. 289–298.

Joshi V.K., and Ashok Pandey, Biotechnology: Food Fermentation Microbiology, Biochemistry and Technology, Vol. 1 & 2.

Julien C., and Whitford W. (2007), Bioreactor: Monitoring, Modeling and Simulation, BioProcess International, 5(01):S10–S17.

Lee, C.H., Glasgow, L.A., Erickson, L.E., and Patel, S.A. (1987), Liquid circulation in airlift fermenters, in Biotechnology Processes Scale-up and Mixing, American Institute of Chemical Engineers, New York, p. 58.

McDonough, R.J. (1992), Mixing for the Process Industries, Van Nostrand Reinhold, New York, p. 123.

Michael L. Shuler and Kargi F. (2002), Bioprocess engineering: Basic concepts, 2nd edition, Pearson Education, New Delhi.

Modi, H.A., Fermentation Technology: Down Stream Fermentation Technology, Vol-II, Pointer Publications.

Modi, H.A., Fermentation Technology: Up Stream Fermentation Technology, Vol-I, Pointer Publications.

Moran L.A., Scrimgeour K.G., Horton H.R., cOchs R.C., and Rawn J.D., Biochemistry, 2nd edition, Prentice Hall, Upper Saddle River, NJ, 1997.

Mrray Moo Young, Comprehensive Biotechnolog—The Principles, Applications and Rugulations of Biotechnology in Industry, Agriculture and Medicine.

Mulchandani A. and Luong J.H.T. (1989), Microbial inhibition kinetics revisited, EMicrob. Technol., Vol. 11, February p. 66–73.

N.S. Panikov (1995), Microbial Growth Kinetics, Springer Dordrecht, 2nd edition.

Patel A.H., Microbiology, NAROSA publications.

Peter F. Stanbury, Allan Whitaker and Stephen J. Hall, Principles of Fermentation Technology, 2nd edition, Butterworth-Heinemann, An imprint of Elsevier Science.

Prescott and Dunn's, Industrial Microbiology 4th edition, edited by Gerald Reed, CBR publications.

Raven, Peter (2005), Biology, Boston: Twayne Publishers.

Richard C. Willson and David Wu J.H. , Industrial Microbiology—An introduction (By Michael J. Waites, Neil L. Morgan, John S.Rockey and Gary Higton)

Shuler M, Kargi, F. (2002), Bioprocess Engineering Basic Concepts, 2nd edition, Upper Saddle River, NJ: Prentice Hall.

Srivastava M.L., Fermentation Technology, NAROSA publications).

Stanbury, P.F., Whitaker A., and Hall S.J. (1995), Principles of Fermentation Technology (2nd edition), Perganom Press, New York.

V. Lubenova, I. Rocha, and Ferreira E.C. (2003), Estimation of multiple biomass growth rates and biomass concentration in a class of bioprocesses, Bioprocess BiosystEng 25, 395–406.

Wang Z, Ly M, Zhang F., Zhong W, Suen A, Hickey AM, Dordick J.S., and Linhardt R.J. (2010), E. coli K5 fermentation and the preparation of heparosan, a bioengineered heparin precursor, Biotechnol. Bioeng, 107: p. 964–973.

Index

α-helix, 42
β-cells of pancreas, 233
β-1, 4 glycosidic bonds, 38
β-sheet, 42

Absolute specificity, 75
Acetogenesis, 248
Acetone, 224
Acetyl coenzyme, 64
Acidogenesis, 248
Acidophilic, 16, 117
Activated sludge, 22, 249
  process, 7, 244
Activation energy, 141, 142
Active site, 74, 78
Active sites, 75, 79
Adenine, 48
Adsorbent, 205
Adsorbents, 19, 205
Adsorption, 205, 224, 230
  immobilization, 92
Aeration, 186, 231
Aeration tanks, 244
Aerobic, 16
  fermentation, 176, 226, 230, 231
  nitrification, 249
Affinity chromatography, 214
Agar, 106
Agitated batch reactor, 96
Agitation, 176, 186, 223, 231
Air lift bioreactor, 187
Air lift fermenter, 187
Air lift fermenters, 229
Air sparger, 183, 229
Air spargers, 118
Air sterilization, 146
Alanine, 40
Alcohol dehydrogenase, 66, 71, 76
Algae, 12, 19, 20
Alkalinity, 243
Alkaloids, 232
Allosteric enzymes, 76
Amensalism, 32
Amino acid, 39, 41
Amino acids, 40
Ammonification, 28
Amoebae, 22
Amylase, 21
Amylolytic molds, 21
Amylopectin, 38
Amyloplasts, 23
Amylose, 38
Anabolism, 56, 62
Anaerobic, 16
  digestion, 248
Anion-exchange, 227
Anion selective membranes, 208
Antibiotic, 227
  production, 203
Antifoam agent, 230
Antifoaming agents, 183
Apoenzyme, 71
Archaebacteria, 14, 15

Aroma, 231
Aromatic amino acids, 40
Arrhenious equation, 77, 141
Arrhenius constant, 141
Aseptic conditions, 177, 222
Aseptic environment, 143
Aseptic operation, 176, 178
Aspergillus niger, 2, 39, 225, 226
Assimilation, 28
Assimilatory reduction, 249
ATP (Adenosine triphosphate), 23, 62, 71
ATP generation, 65
Autoclave, 178
Autoclaving, 6, 109
Azeotrope, 223
Azotobacter, 28

Bacilli, 15
Bacteria, 12, 15, 247
Bacteriovorous, 22
Baffle, 183
  spacing, 176
Baffles, 176
Baker's yeast, 20, 231
Ball mills, 209
Basic parts of a fermenter, 175
Batch, 227
  culture, 153
  fermenter, 223
  reactor, 153
  sterilization, 143
Beer fermentation, 88
Binary fission, 16, 111
Bingham plastic, 138
Binomial nomenclature, 26
Biocatalysts, 71
Biochemical engineering, 1, 3
Biochemical oxygen demand (BOD), 242
Biochemical reaction, 70
Biochemistry, 36
Bioethanol, 20
Biofilms, 245, 246
Biogeochemical cycles, 27
Biological functions of proteins, 41
Biological potency, 235
Biomass, 56
Biomolecules, 36
Bioprocess, 2
  engineering, 1, 7
Bioprocesses, 2
Bioreactor, 153
Bioremediation, 22
Biosafety, 233
Biotechnology, 1
Biotransformation, 55
Biuret test, 44
Blue-green algae, 18
BOD, 242
BOD5, 242
BOD (Biochemical oxygen demand), 242
BOD incubator, 228
BOD reduction, 249
Bread baking, 88
Bread molds, 21
Broth, 106
Brownian motion, 16
Bubble size, 134
Budding, 20
Butanol, 224

Calcium lactate precipitation, 224
Candida tropicalis, 231
Candida utilis, 231
Cane molasses, 226
Capnophiles, 118
Capsule, 15
Carbohydrates, 36
Carbon cycle, 29
Carbon source, 179
Carotenoids, 23
Casson body fluid, 138
Catabolism, 56, 62
Catalysis, 70
Catalyst, 70
Cation-exchange resins, 227
Cation-selective membranes, 208
Cavitation, 209
Cell aggregation, 159
Cell lysis, 118
Cell maintenance coefficient, 117
Cell organelles, 14
Cellular scavengers, 25

Cellulose, 36, 38
Cell walls, 16
Centrifugation, 200, 207, 222
Centrifuge, 161
Centrioles, 26
Centrosome, 23, 26
Centrosphere, 26
CFU (colony forming unit), 115
Chelators, 181
Chemical, 210
  scavengers, 146
  sterilization, 146
Chemo-autotroph, 108
Chemostat, 153, 156, 159
Chill-proofing, 88
Chitin, 38
Chlorella, 19
Chlorophyll, 18, 19, 29
Chloroplasts, 24
Cholesterol, 46
Chromatography, 213
Chromium, 176
Chromoplast, 23
Chromoplasts, 24
Chromoproteins, 40
Citric acid, 225, 226, 227
  monohydrate, 227
Clarification of fruit juices, 88
Clostridium, 28
Clostridium acetobutylicum, 224
Cocci, 15
COD (Chemical Oxygen Demand), 242
Coenzymes, 71
Cofactor, 71
Colloidal particles, 215
Column chromatography, 214
Commensalism, 32
Competitive inhibition, 85
Competitive inhibitors, 86
Compression, 209
Conjugate, 40
  lipids, 45
  or compound lipids, 46
  proteins, 40
Contamination level, 141
Continuous, 227
  centrifuge, 201
  reactor, 153, 156
  sterilization, 147, 148
  stirred tank fermenters, 159
  stirred tank reactor, 97
Contractile protein, 41
Control of various parameters, 182
Convective mass transfer, 130, 131
Cooling time, 142
Corn steep liquor, 180, 228
Corona discharge, 146
Cottrell precipitator, 146
Countercurrent extractor, 233
Covalent binding, 91, 92
Critical dilution rate, 158
Cross-flow microfiltration, 206
Crude acid, 227
Cryptic growth, 113
Crystallization, 210
CSTF, 156
Culture, 105, 108
Cyanobacteria, 18
Cyclone separator, 179
Cytoplasm, 17, 23
Cytosin, 48
Czapek-dox agar, 109

Damkohler's number, 137
Darcy's law, 199
Deamination, 66
Death, 113
  phase, 110
Decaffeination, 232
Deceleration phase, 110, 112
Decimal reduction time D, 141
Declining phase, 113
Deep jet fermenter, 189
Defense protein, 41
Degree of reductance, 58
Dehydrogenase, 72
Del factor, 142
Delignification, 88
Demerits of immobilization, 91
Denaturation, 43, 77
Denitrification, 27, 29, 249
Deoxyribonucleic acid (DNA), 48
Deoxyribose, 37, 48
Derived lipids, 45, 47
Design equation, 95

Desulphurylation, 31
Dextrin, 180
Dialysis, 44, 206
Diatomaceous earth, 19, 205
Diatoms, 19
Diauxic growth, 111
Dielectric constant, 205
Differential medium, 108
Diffusion, 135, 210
Diffusivity, 131, 135
Dilatant, 138
Dilution rate, 157, 158, 161
Dinitrosalicylic acid, 39
Direct microscopic count (DMC), 113
Disc-bowl centrifuge, 201
Disc-stack centrifuge, 201
Disc-turbine, 176
Disproportionation, 31
Dissimilatory reduction, 31, 249
Dissolved oxygen (DO), 117, 185
Distillate, 204
Distillation, 204
    coefficient, 202
DMC, 113
DO probe, 184
Double helix, 48
Doubling time, 111
Downstream operation, 197
Downstream processes, 3
Downstream processing, 3, 6, 224
Down-time, 158
Dry cell weight, 115
Drying, 212
Dynamic gassing out methods, 134

Eadie-hofstee plot, 83
Ecological interactions, 31
Effectiveness factor, 136
Effector, 76
Electrical (Donnan) effects, 208
Electrodialysis, 206
Electrodialysis (ED), 208
Electronic particle counter, 115
Electrophoresis, 44, 215
Electrophoretic mobility, 215
Electrophoretic retardation, 215
Electrophoretic separation, 215
Elution, 214
Embden–Meyerhof–Parnas (EMP) pathway, 62, 223
Encapsulation, 94
Endogenous respiration coefficient, 122
Endoplasmic reticulum (ER), 23, 25
Enediol, 37
Enolization, 37
Enrichment media, 105
Enrichment medium, 108
Entrainer, 223
Entrapment, 92
Environmental factors, 117
Enzymatic desizing, 88
Enzymatic splitting, 210
Enzyme, 70
    activity, 77, 81
    deactivation, 79
    inhibitors, 85
    stabilization, 77
    -substrate complex, 75, 79, 86
    -substrate complexes, 79
    unit, 81
Enzymes, 41, 70
Equation, 154
ER, 25
Erythromycin, 204
Essential amino acids, 40
Ethanol, 221
Ethyl alcohol, 66
Eubacteria, 14, 15
Eukaryotes, 13
Eukaryotic, 23
    cell, 23
    cells, 13
Exhaust air filter, 184
Exhaust gases, 179
Exit gas composition, 185
Exothermic, 61
Exponential phase, 111
External cross flow filter, 161
Extracellular products, 197
Extract, 203
Extraction, 202
Extremophile archaea, 15

Facultative, 16
FAD (Flavine adenine dinucleotide), 71

Fat, 45
Fatty acid, 45
Fatty acids, 45
Fed batch, 227
 culture, 164
 reactor, 163, 164
Feed pump, 184
Fehling's test, 39
Fermentation, 2, 3, 6, 12, 61, 173, 221
Fermenter, 174, 175, 225
Fibroin, 42
Fibrous, 40
 proteins, 40
Film theory, 131
Filter aids, 19, 200
Filter cake, 199
Filter press, 198
Filtration, 198, 202, 227, 230
 sterilization, 145
Fission, 20
Flagella, 16
Flavor, 231
Flick's law, 131
Flocculating agents, 201
Fluidized-bed bioreactor, 188
Fluidized-bed dryers, 231
Flux equation, 207
Foam control, 184
Foaming, 182
Folin-Ciocalteu reagent, 44
Food processing, 88
Formalin, 146
Functional group specificity, 75
Fractionation, 207, 225
Freeze drying, 213
Freezing and thawing, 209
Fructose, 37
 syrup, 233
Fuller's earth, 205
Fungi, 12, 20

Gaseous sterilants, 146
Gas hold-up, 134
Gas-liquid mass transfer, 130, 176
Gel electrophoresis, 215
Gel filtration, 44, 214
Generation time, 112
Genetically modified organisms, 234
Genus, 26
Geometrical similarity, 166
Geometric similarity, 168
Germicides, 146
Gerontoplast, 24
Gibbs free energy, 138, 139
Glass reference electrode, 185
Globular, 40
 proteins, 43
Gluconic acid, 39
Glucose, 37
 isomerase enzyme, 233
 oxidase, 39
 syrup, 233
Glycerol, 45
Glycerophospholipids, 46
Glycine, 40
Glycogen, 38
Glycolipids, 46
Glycolysis, 62, 65
Glycoprotein, 40
Glycosidic linkage, 37
GMO, 235
Golgi bodies, 23, 25
Good manufacturing practices (GMP), 234
Gram-negative, 16
Gram-positive, 16
Gram reaction, 16
Gram-staining, 16
Grana, 24
Grash of number, 133
Growing of inoculum, 5
Guanine, 48

Half saturation constant, 120
Halophiles, 15, 16
Halophilic, 16
Heating, 142
Heat transfer, 130
Helical structure, 42
Helicobacter pylori, 117
Henry's law, 135
HEPA filters, 235
HEPA (High efficiency particulate air), 146
Heterogeneous nucleation, 211
Hetero polysaccharides, 38

Heterotrophs, 108
High fructose corn syrup (HFCS), 233
High pressure homogenization, 208
High temperature short time (HTST), 144
Hofmeister effect, 205
Holding, 142
time, 144
Homogeneous nucleation, 211
Homogenization, 208
Homo polysaccharide, 38
Hormones, 41
HRT, 244
HTST: High temperature short time, 143
Hughes press, 209
Hydraulic resistance, 206
Hydraulic retention time (HRT), 244
Hydrocarbon metabolism, 66
Hydrogen bonding, 42
Hydrolase, 73
Hydrolysis, 248
Hydrophobic chromatography, 214
Hyperthermophilic, 16
Hyphae, 20

Immobilization, 77, 91, 94
Immunogenicity, 235
Impeller, 183
Importance of agitation, 186
Incineration, 235
Induced fit model, 80
Inducers, 181
Industrial catalyst, 6
Industrial solvents, 224
Industrial uses of enzymes, 87
Inhibitor, 85
Inhibitors, 181
Inoculation, 6, 109, 182
method, 182
Inoculum, 4, 5, 6
volume, 6
Insulin, 233
Integrated bioprocess, 4
International Union of Biochemistry, 72
Inter-particle diffusion, 136
Intracellular products, 197
Intra-particle diffusion, 136
Inulin, 38
In vitro, 71
In vivo, 71
Ion-exchange chromatography, 214
Ionic strength, 118
Ionizing and non-ionizing types of radiations, 144
Irreversible, 85
inhibitors, 85
Isoelectric point, 39, 205
Isoelectric precipitation, 205
Isolates from plants or animals, 232
Isomerase, 73
Isomerization, 73
Isopropyl alcohol, 146

Juice debittering, 88

Katal, 81
Keratin, 42
Krebs, 62
cycle, 64

Lactic acid, 65, 223
Lactobacillus, 223
Lactose, 37
Lag phase, 110, 111
Lanolin, 46
Leucoplasts, 23
Ligases, 74
Limitations of monod equation, 121
Limiting substrate, 119
Lineweaver-burk plot, 83
Lipases, 89
Lipids, 45
Lipoproteins, 40, 46
Lock and key model, 79, 80
Logarithmic or exponential growth phase, 110
Log phase, 111
Lyases, 73
Lyophilization, 5
Lyotropics (Hofmeister) series, 205
Lysosomes, 23, 25
Lysozyme, 210
Lysozymes, 25

Macconky agar, 108
Macroscopic, 215

Maintenance coefficient, 121
Maintenance energy, 121
Maintenance metabolism, 121
Major contaminants, 241
Major products of bioprocessing, 7
Malt extract, 180
Malthus's law, 119, 154
Mannitol, 37
Mass transfer, 130, 133
  coefficient, 132
  resistances, 135
Matrix, 92, 214
Maximum specific growth rate constant, 119
Mechanical homogenizers, 209
Mechanical seal, 177
Media, 179
  preparation, 4, 6
Medium, 105
Membrane filter, 145
Membrane fouling, 145
Membrane technologies, 205
Merits and demerits of continuous culture, 158
Merits of immobilization, 91
Mesophiles, 117
Mesophilic, 16
Messenger RNA (m-RNA), 48
Metabolic heat, 138
Metabolism, 55, 56, 62
  of nitrogen, 66
Metalloproteins, 40
Metastable supersaturation, 211
Methanogenesis, 248
Methanogens, 14
Methods, 215
Michaelis constant, 78
Michaelis-menten equation, 81, 82
Microaerophiles, 118
Microbial diversity, 12
Microbial fermentation, 2
Microdrops, 93
Microencapsulation, 93
Microfiltration, 206, 207
Microorganisms, 2, 12, 56, 105
Million's test, 44
Mitochondria, 23
Mixed liquor suspended solids (MLSS), 244
Modulators, 76
Molasses, 180, 221, 226
Mold, 20
Molds, 21
Molecular sieves, 223
Monod, 154
  equation, 119, 121, 123
  model, 155
Monosaccharides, 36, 37
Motility, 16
Moving phase, 213
Mucilages, 38
Mucopolysaccharides, 38

NAD (Nicotinamide adenine dinucleotide), 71
Nanofiltration, 208
Neutralism, 31
Ninhydrin test, 44
Nitrification, 27, 28, 249
Nitrobacter, 28, 249
Nitrococcus, 28
Nitrogen, 18, 27
  cycle, 18, 27, 28
  fixation, 27
  sources, 180
  uptake, 28
Nitrosococcus, 28
Nitrosomonas, 249
Non-competitive inhibition, 86
Non-newtonian fluids, 137
Non-reducing sugars, 39
Non-synthetic, 107
Nozzle spargers, 176
Nucleation, 211
Nucleic acids, 48
Nucleoproteins, 40
Nutrient agar, 109
Nutrient broth, 109
Nutrient media, 222
Nutrients, 105

Oligosaccharides, 36, 37
Optical density, 116
Optimum bubble size, 134
Optimum growth temperature, 117
Optimum pH, 78, 117
Optimum temperature, 222
Organelles, 23, 25

Orientation effect, 81
Orientation effects, 80
Orifice spargers, 176
Osmoregulation, 25
Osmosis, 118
Osmotic pressure, 118
Osmotic shock, 209
Overall yield coefficient, 154
Oxidase, 72
Oxidation pond, 247
Oxidative phosphorylation, 62, 64
Oxidoreductase, 72
Oxygenase, 73
Oxygenation, 66
Oxygen transfer, 186
    rate, 134
Oxygen uptake rate (OUR), 133, 138
Oxysome, 64

Packed-bed, 249
    column fermenter, 188
    reactor, 98
Paper chromatography, 214
Paramecium, 22
Parasitism or predation, 32
Partial similarity, 166
Particle counter, 115
Particle Reynold's number, 135
Partition chromatography, 213
Partition coefficient, 202
Partitioning, 202
Pasteurization, 143
Pectin, 38
Pectinase, 88
Penetration theory, 132
Penicillin, 2, 203, 227
Penicillin acylase, 87
Penicillin erythromycin, 227
Penicillin F (Fentanyl penicillin), 227
Penicillin G (Benzyl penicillin), 228
Penicillium chrysogenum, 228
Penicillium notatum, 227
Peptide bond, 41
Peptide bonds, 71
Peptidoglycan, 16, 17, 36, 38
Peptone, 181
Permeate flux, 206
Peroxidase, 72
Peroxide method, 134
Personal protective equipment, 235
Pervaporation, 206
Petri dishes, 106
Petroff-hauser, 113
Pharmaceuticals, 8
Phosphate removal, 249
Phospholipids, 46
Phosphorus cycle, 30
Photoautotrophs, 108
Photosynthesis, 19, 24, 29, 36
Photosynthetic, 12
Photosynthetic microorganisms, 12
pH probe, 184
Phrynoderma, 45
Phycobilin, 18
Phycoerythrin, 18
Physical contamination, 235
Phytohormones, 232
Phytoplanktons, 30
Phytosterols, 47
Pili, 16
Pitch control, 88
Plasma membrane, 14, 15, 17
Plastids, 23
Plate and frame filter press, 198
Plate heat exchanger, 147
Plug flow reactor, 153, 164
Polyacrylamide gel electrophoresis, 215
Polyhydroxyalkanoate, 48
Polyhydroxybutyrate (PHB), 48, 249
Polylactic acid (PLA), 223
Polysaccharides, 36, 38
Porous carrier, 94
Precipitation, 204, 205, 227
Precoat filter, 233
Precursors, 181
Predation, 32
Pre-fermentation growth, 6
Preservation of culture, 5
Pressure drop, 207
Primary screening, 4
Primary structure, 41
Primary treatment, 243
Product inhibition, 121, 123
Product recovery, 6

Prokaryotes, 13
Prokaryotic, 14
 cells, 13
Prosphoproteins, 40
Prosthetic groups, 71
Protein precipitation, 204
Protein recovery, 6
Protein synthesis, 25, 26
Proteins, 39
Protist kingdom, 13
Protozoa, 22
Proximity, 80
 effect, 81
Pseudoplastic, 138
Psychrophiles, 117
Pulso-and-shift-technique, 158
Pyruvic acid, 62, 65

QC testing, 7
Quality control, 7, 234, 235
Quaternary structure, 43

Raffinate, 203
Rarefaction, 209
Rate of oxygen transfer, 133
Rauwolfia serpentina, 232
Recombinant DNA technology, 3, 94, 233, 234
Recycle ratio, 153
Recycle reactor, 153
Reducing, 39
 media, 108
 sugar, 37
 sugars, 39
Reserpine, 232
Respiration, 29
Respiratory quotient (RQ), 57
Reverse osmosis (RO), 208
Reversible, 85
Rheology, 137
Ribonucleic acid (RNA), 48
Ribosomal RNA (r-RNA), 48
Ribosome, 23
Ribosomes, 26
Risk factor, 142
Rotameter, 184
Rotary precoat filters, 233
Rotary vacuum filter, 227, 230
Rotary vacuum filtration, 207
Rotating biological contactors (RBC), 245
Rotating biological disc reactor, 22
Rotatory shaker, 228
RuBisCo, 41

Saccharification, 233
Saccharomyces cerevisiae, 20, 221, 231
Safranin, 17
Salting out, 204
Sampling port, 183
Saturated fatty acids, 45
Scale-down, 167, 168
Scale-up, 133, 165
Schmidt number, 133
Screening, 4
Secondary, 18
 metabolite, 55
 metabolites, 18, 112
 screening, 5
 structure, 42
 treatment, 23, 244
Sedimentation, 201
Selection of bioreactor, 165
Selective media, 108
Selectivity, 75, 76
Seliwanoff's reagent, 39
Semi-batch reactor, 163
Semi-batch reactors, 163
Settler-thickener, 162
Sherwood number, 136
Simple lipids, 45
Simple protein, 40
Single-cell-protein, 19
Single-cell-proteins (SCP), 20, 230
Slants, 6, 106
Sludge, 244
Soaking, 89
Solid medium, 105
Solid state fermentation (SSF), 189
Solubilization, 210
Solvent extraction, 202
Solvent extractions, 232
Sorbitol, 37
Soy meal, 181
Spargers, 176

Species, 26
Specificity, 75
and selectivity, 75
Specific maintenance coefficient, 122
Spermaceti oil, 46
Sphingocin, 46
Sphingolipids, 46
Sphingophospholipids, 46
Spirilla, 15
Spirulina, 19
Spray, 231
Spray-drying, 212
Stabs, 106
Stainless steel, 176
Standard Gibbs free energy, 139
Starch, 38
Static method, 134
Stationary phase, 110, 112, 213
Steam injection type sterilizer, 147
Steam sterilization, 6, 178
Steam traps, 178
Stereochemical specificity, 76
Stereo specificity, 76
Steric (sieving) effects, 208
Sterility level, 141
Sterilization, 4, 6, 142, 143, 145, 178
Steroids, 46
Sterols, 47
Stirred tank, 229
Stirrer glands, 177
Stoichiometric coefficients, 57, 139
Stoichiometry, 55, 56
Stoke's law, 200
Storage proteins, 41
Strain, 4, 105
Strict anaerobes, 16
Stroma, 24
Structural polysaccharides, 38
Structural protein, 41
Stuffing box, 177
Sub-culturing, 110
Submerged fermentation, 191, 226, 227, 232
Submerged internal filter, 161
Substrate, 56
Substrate inhibition, 121, 122, 123
Sucrose, 37
Sulphanilamide, 86
Sulphite oxidation method, 134
Sulphur cycle, 30
Superficial gas velocity, 134
Supersaturation, 211
Surface activation, 92
Surface renewal theory, 132
Swelling, 118, 209
Symbiosis, 27, 32
Synthetic, 105

TDS (Total Dissolved Solids), 242
Temperature activation, 77
Temperature probe, 184
Terminal oxidation, 64
Terminal settling velocity, 200
Terpenes, 47
Tertiary structure, 42
Tertiary treatment, 244
Tetracycline, 227
Therapeutic application, 6
Therapeutic applications, 87
Thermal denaturation, 77
Thermoacidophiles, 15
Thermophiles, 117
Thermophilic, 16
Thiele modulus, 136
Thin layer chromatography, 5, 214
Thiomargarita magnifica, 13
Thymine, 48
TOC, 243
Total dissolved solids, 242
Tranquilizer, 232
Transamination, 66
Transferase, 73
Transfer RNA (t-RNA), 48
Transport phenomena, 130
Transport protein, 41
Tricarboxylic acid cycle (TCA), 62, 64
Trickling bed filter, 7
Tricking biological filter, 246
TSS (Total suspended solids), 243
Tubular bowl centrifuge, 201
Tubular reactor, 98
Turbidity, 116, 185
meter, 185
Turbidostat, 156
Turgidity, 26
Turnover number, 75, 91

Turnover numbers, 74, 75
Type of agitators, 135
Types of agitators, 186
Types of enzyme, 72
Types of filter, 198
Types of penicillin, 227

Ultrafiltration, 44, 207
Ultrasonication, 208, 209
Ultraviolet radiation, 145
Uncompetitive inhibition, 87
Unit operation, 3
Unit operations, 3
Unit processes, 3, 4
Unsaturated fatty acid, 45
Unstable supersaturation, 211
Up flow anaerobic sludge blanket (UASB), 249
Upstream processes, 3
Uracil, 48
Urease, 71
Useful bacteria, 17

Vacuoles, 23, 25
Vacuum crystallizer, 211
Vacuum distillation, 222
Valves, 177, 184
Viable cells, 115
Viable organisms, 235
Viable plate count, 113
Volatile flavor compounds (VFC), 232
Volumetric mass transfer coefficient, 133
Volumetric oxygen transfer coefficient, 118, 133

Wall growth, 159
Wash-out, 158
Wastewater, 241
    treatment, 4, 7
Wax-D, 46
Waxes, 45
Weizmann process, 225
Whey, 180

Xanthophyll, 23
X-press, 209

Yeast extract, 181
Yeasts, 20, 221
Yield coefficient, 57, 157

Zoosterol, 47
Zwitter ions, 39